Sustainable Butanol Biofuels

Editors

Anita Singh
Department of Environmental Sciences
Central University of Jammu
Samba, J&K, India

Richa Kothari
Department of Environmental Sciences
Central University of Jammu
Samba, J&K, India

Somvir Bajar
Department of Environmental Science and Engineering
J.C. Bose University of Science and Technology, YMCA
Faridabad, Haryana, India

Vineet Veer Tyagi
School of Energy Management
Shri Mata Vaishno Devi University
Katra, J&K, India

CRC Press
Taylor & Francis Group
Boca Raton London New York

CRC Press is an imprint of the
Taylor & Francis Group, an **informa** business
A SCIENCE PUBLISHERS BOOK

Cover illustrations taken from Creative Commons. The illustrations were prepared by Ms. Kajol Goria, Research Scholar, Department of Environmental Sciences, Central University of Jammu, J and K and Mr. Har Mohan Singh, Research Scholar, School for Energy Management, Shri Mata Vaishno Devi University, Kakryal, Katra, J and K.

First edition published 2023
by CRC Press
6000 Broken Sound Parkway NW, Suite 300, Boca Raton, FL 33487-2742

and by CRC Press
4 Park Square, Milton Park, Abingdon, Oxon, OX14 4RN

CRC Press is an imprint of Taylor & Francis Group, LLC

Library of Congress Cataloging-in-Publication Data (applied for)

ISBN: 978-0-367-76077-9 (hbk)
ISBN: 978-0-367-76081-6 (pbk)
ISBN: 978-1-003-16540-8 (ebk)

DOI: 10.1201/9781003165408

Typeset in Palatino
by Radiant Productions

Foreword

I congratulate the editors on publishing their book, *Sustainable Butanol Biofuel*. Worldwide, fossil fuels are used as the primary energy source, but the environmental consequences of their utilization are increasing daily, like global warming and climate change. Renewable energy options like biofuels attracted attention worldwide due to their cleaner and greener nature. Among biofuels, butanol has gained attention as a renewable transportation biofuel and alternative to gasoline due to its superior fuel properties, higher energy content, and lower Reid vapor pressure.

Butanol can be produced by chemical as well as biological synthesis pathways. Chemical synthesis processes include oxo synthesis, crotonaldehyde hydrogenation, and Reppe synthesis. Biologically biobutanol is produced by acetone–butanol–ethanol (ABE) fermentation using various feedstocks. Butanol, a 4-carbon alcohol, can be produced from the same feedstocks as ethanol like first-generation feedstocks (like cane molasses, cheese whey, and food industry waste), second-generation feedstocks (lignocellulosic biomass), third-generation feedstocks (algal biomass), and fourth-generation feedstock (genetically modified biomass). Innovative biological processes and numerous novel fermentation techniques are used to convert abundantly accessible waste biomasses into biobutanol. The first-generation feedstocks have a food vs fuel problem, which can be overcome by second-generation feedstock. But, the structural complexities of lignocellulosic biomass hinder its conversion into butanol. The Pretreatment step is a prerequisite for the conversion of lignocellulosic biomass into butanol. Pretreatments like physical, chemical, physicochemical, and biological used for pretreatment of biomass have advantages and disadvantages. The butanol toxicity towards fermenting microorganisms is a big challenge in achieving high product titers. There are many challenges with butanol production, like butanol toxicity, feedstock, low product titres, and product recovery, which must be overcome for sustainable butanol production.

The book opens with a chapter on different generations of biofuels, their feedstocks, and global biofuel policies. Sustainable development goals for the bioeconomy are presented in Chapter 2. The authors discuss SDGs-7, which straightforwardly deals with providing affordable energy accessibility to every individual on the planet. The concept of bioeconomy is introduced in the book's third chapter, and Chapters 4 and 5 highlight the current status and challenges associated with butanol as a biofuel. The butanol production from algal and lignocellulosic biomass is discussed in detail in Chapters 6 and 7. The production technologies for butanol are highlighted in Chapter 8. The mechanism of acetone–butanol–ethanol production is presented in Chapter 9, while different pretreatment methods are discussed in-depth in Chapter 10. The genetic engineering aspect for the development of fermentation microbes that can tolerate high concentrations of butanol production is presented in Chapter 11. Chapter 12 presents the nanotechnological aspects of butanol production.

The book presents a comprehensive overview and in-depth technological advancements in biobutanol production. The book *Sustainable Butanol Biofuel* is beneficial for readers working in the butanol production field. The readers can access the information and enrich their knowledge regarding butanol as a biofuel from this collection contributed by experts in this field.

Prof. Sanjeev Jain
Vice-Chancellor
Central University of Jammu

Preface

The policy intervention towards the energy security issue has diverted global attention to exploring alternatives to the traditional fossil-fuel-based non-renewable energy sources. The extensive exploitation of fossil fuels worldwide has resulted in severe environmental concerns–notably environmental pollution and climate change. One of the critical challenges the developing world faces is how to meet its growing energy needs and sustain economic growth without contributing to environmental pollution and climate change. The emission of greenhouse gases (GHGs) on a larger scale from the transportation sector is increasing rapidly, accounting for 15% of the global GHG emissions and 23% of CO_2 emissions generated from energy-related activities. Moreover, the rising emphasis on energy security has also created the need for exploring substitutes for energy sources that could reduce dependency on traditional fossil fuels like petroleum. Among several renewable energy sources, biofuels have emerged as the most potent source, and several countries and international institutions recognized them as "future fuels". Biofuels based on biomass feedstock have a considerable role in offsetting CO_2 production, and that is why several international conventions have excluded CO_2 emission from biofuels combustion from national GHGs inventories. Moreover, the efforts toward producing future fuels like ethanol and butanol are diverting from achieving the targets of sustainable development goals (SDG 7).

Biomass conversion into biofuels seems a novel approach to achieving the carbon-neutral status of the countries, which is the most pressing need of the hour. Butanol, with its extraordinary properties, has emerged as a better fuel for internal combustion engines than gasoline. Moreover, the application of butanol as a chemical substitute in various industrial units and as fuel has raised the demand for it, which attracted the scientific community's attention to explore cost-effective, eco-friendly and sustainable solutions for butanol production. Various substrates are widely utilized for bioethanol production and may also be harnessed for the generation of butanol. Several biomass feedstocks, including lignocellulosic biomass, sugarcane bagasse, and algal biomass, have appeared as a potential cost-effective substrate that has the capability to replace glucose-based substrates. The advantages and limitations of various substrates have been discussed in detail in the book. Integrating biobutanol recovery methods is also an essential parameter for biobutanol production, which has endeavored to increase the viability of commercial production of butanol. However, various possibilities and drawbacks are associated with the integrated biobutanol recovery methods, which are also a key component discussed in the content.

With outstanding characteristics, biobutanol has gained wider recognition globally as an advanced biofuel, which can be used directly as a substitute for gasoline in internal combustion engines. The present book provides readers with in-depth knowledge of the aspects and steps involved in butanol production. Further, the current global status, history, various technologies adopted for butanol production from different feedstocks, and the role of microorganisms in the production process are elucidated in the content for the reader's interest. The whole content is distributed in 12 chapters dedicated to covering every aspect of butanol, from production to applications.

Chapter 1 is focused on different generations of biofuel production. Depending upon the feedstock, biofuels are mainly categorized into four generations. The biofuels policies in various countries are also highlighted in this chapter. Energy security, an important policy issue worldwide, has also been covered to correlate the energy demands with the adequacy of resources for development.

Chapter 2 primarily deals with Sustainable Development Goals (SDG 7) and the role of bioenergy in shifting from a traditional fossil-fuels-based economy to a bio-economy for achieving SDG7. Different green and clean bioenergy options and critical indicators of sustainable bioenergy are also discussed. Challenges and opportunities while implementing the bioenergy-based economy are also highlighted for better knowledge enhancement.

Chapter 3 employs a competitiveness framework on the bioeconomy. Sustainable development, green and circular economy, and societal change are discussed in detail. The second section of the chapter examines the implementation of bioeconomy-based initiatives in different counties around the globe. The content also emphasized synergistic interconnection between a sustainable bioeconomy and the SDGs.

Chapter 4 provides the current status and challenges in butanol production. Various synthesis pathways, including chemical and biological, for butanol production, current advances, and the serviceability of butanol as an alternative fuel are deliberated in detail. The chapter has also emphasized various challenges and future perspectives toward adopting butanol biofuels as a substitute for internal combustion engines.

Chapter 5 assessed the progress in butanol generation along with its production history and synthesis pathways. The chapter also gives insights into the role of microorganisms used, the mechanism followed, and the fermentation systems for the production of butanol and fermentation. The content also elaborates on the progress of several modes of acetone–butanol–ethanol (ABE) fermentation and downstream processing. The limitations of ABE and advances to overcome these issues are also highlighted. The chapter also covers the domain of synthesis pathways, suitable microorganisms, and fermentation processes. The downstream processing, including processes like adsorption, gas stripping, and pervaporation, are depicted.

Chapter 6 deals explicitly with the mechanism and limitations of biobutanol in ABE fermentation and strategies to overcome the limitations and applications of biobutanol. The mechanism of the ABE fermentation process, recent technological advancements, strategies to enhance butanol production using lignocellulosic and algal biomass and their application are detailed in the content. The chapter also provided insights into biobutanol's sustainable and economical process development.

The potential feedstocks and various techniques harnessed for producing butanol are discussed in Chapter 7. Feedstock for butanol production has undergone evolutionary modifications, which has a considerable role in improving biofuel production. Conventional techniques for bio-based butanol production, along with the first, second, third and fourth generation of feedstocks for butanol production, are deliberated upon, considering the exhaustion of conventional fuel resources and rising environmental concerns.

Chapter 8 examines the lignocellulosic biomass as a promising material for large-scale biobutanol production. The content also elaborated on economical and efficient butanol production challenges pertinent to the concentration of stimulators, inhibitors, and sugars. The chapter also provides insights into the technology development for large-scale commercial biofuel production.

Chapter 9 discusses the prospects and the obstacles related to algal butanol. The chapter provides technical guidance for the production of biobutanol from algal biomass with the critical aspects of algal biomass feedstock potential, ABE fermentation, and factors affecting butanol production, fermentation type, and reactor architecture.

The pretreatment and hydrolysis of feedstocks into butanol are deliberated upon in Chapter 10. The content also gives insights into different pretreatment techniques, including mechanical comminution, extrusion, irradiation, pyrolysis, steam explosion, CO_2 explosion, ammonia, liquid hot water, wet oxidation, acid, alkaline, ozonolysis, organic solvent, ionic liquids, biological pretreatments along with hydrolysis processes.

Chapter 11 deals with the genetic engineering aspect of butanol production. The chapter covers the domain of butanol production mechanism through *Clostridium*, genetic and metabolic engineering of the butanol-producing strain, solvent improvement through genetic engineering, and genetic engineering to improve butanol tolerance.

The role of nanotechnology in butanol production is explicitly elaborated upon in Chapter 12. The authors highlighted the application of nanoparticles at different stages of butanol production. The chapter also highlights the impacts of butanol production on the environment, which should be considered for the commercial production of biobutanol.

Technological advances and challenges during butanol production from various feedstocks are presented in this volume contributed to by renowned experts in this field. The editors hope that the readers will find valuable information in this book. The editors are thankful to the team members of CRC press for providing cooperation through this odyssey of *Sustainable Butanol Biofuel*.

Dr. Anita Singh, Department of Environmental Sciences, Central University of Jammu, Samba 181143, J&K, India

Dr. Richa Kothari, Department of Environmental Sciences, Central University of Jammu, Samba 181143, J&K, India

Dr. Somvir Bajar, Department of Environmental Science and Engineering, J.C. Bose University of Science and Technology, YMCA, Faridabad-121006, Haryana, India

Dr. Vineet Veer Tyagi, School of Energy Management, Shri Mata Vaishno Devi University, Katra, Jammu

Contents

1

Current Status and Future Prospective on Different Generations of Biofuel Production

Somvir Bajar,[1,*] *Anita Singh,*[2] *Neha Yadav,*[1] *Kavita Yadav,*[1] *Anjali Prajapati*[1] and *Neeta Rani*[3]

1. Introduction

One of the most pressing concerns confronting the globe today is the energy crises. People are increasingly concerned about global warming as the world's population grows, traditional fossil fuels become depleted, and pollution levels from transportation rise (Kiwjaroun et al. 2009, Narwane et al. 2021, Rather et al. 2021). Energy sources like fossil fuels are left in limited quantities as their exploitation increases at an alarming rate around the world. Such finite energy sources are prone to exhaustion over time (Rather et al. 2021). In the present scenario, there is a clear need to replace energy dependency from fossil fuels with renewable energy sources to address the extraordinary speed of climate change caused by the accumulation of greenhouse gases (GHGs) in the atmosphere (Liu et al. 2021). In order to establish a much more sustainable economy,

[1] Department of Environmental Science and Engineering, J.C. Bose University of Science and Technology, YMCA, Faridabad, Haryana (India).
[2] Department of Environmental Sciences, Central University of Jammu, Jammu & Kashmir (India).
[3] Department of National Security Studies, Central University of Jammu, J&K.
* Corresponding authors: somvirbajar@gmail.com

tremendous efforts have been made in recent decades to improve the availability of non-utilizable biomass as alternative feedstocks and reduce pollution. Several bio-sectors have developed to produce bio-based products to replace fossil-based counterparts, utilizing bioprocesses that use biomass feedstocks (Lee et al. 2021). Several international and national organizations have forecasted worldwide biofuel output in the medium and long term (Jeswani et al. 2020). Currently, biofuels account for around 3.4% of total transportation fuels in the globe (IEA 2019). According to the *Renewables 2019 Worldwide Status Report*, the United States and Brazil lead global biofuel production, accounting for 69% of all biofuels produced in 2018, followed by Europe with 9% (https://www.ren21.net/wp-content/uploads/2019/05/gsr_2019_full_report_en.pdf). According to the International Energy Agency (IEA), biofuels might account for up to one-third of all transportation fuel by 2050 (IEA 2011). India has around 500 million tonnes of biomass accessible every year, with a surplus of 120 to 150 million tonnes. Furthermore, biofuels account for 12.83% of overall renewable energy production (Kumar et al. 2015, Narwane et al. 2021).

Biofuel technology advancements must focus on (i) optimizing advanced technologies for higher efficiency and productivity of conversion of biomass, and (ii) diversifying feedstocks to guarantee the suitability of biofuel production within established ecological and socio-economic constrictions in order to minimize GHGs emissions while meeting worldwide energy requirements and fuel demands, and (iii) broadening the chemical realm to include designer compounds that increase fuel efficiency and reliability while lowering carbon emissions (Liu et al. 2021). Biofuels are the primary energy source for more than 50% of the people worldwide, accounting for up to 90% of energy consumption in developing countries. However, there is no doubt that civilization will require numerous sources of energy in the future, and biofuels will play a significant role in the energy production. Biofuels are divided into four generations: first, second, third, and fourth; these will be discussed in the following paragraphs.

Food-based plants with energy-containing compounds including sugars, oils, and cellulose are used to make first-generation biofuels. The use of 1G is restricted because it creases a debate on food verses fuel and have a detrimental effect on food production. These sources produce the most biofuel and have the highest energy efficiency, but are less effective in reducing GHGs emissions. Second-generation biofuels are made from lignocellulosic feedstocks such as straw, bagasse, forest leftovers (all non-food), and energy crops produced on degraded areas. They're still in the early stages of research and manufacturing. They have the lowest net GHGs emissions, leading to further reduction in the emissions in the atmosphere.

The third-generation biofuels made from algae require more energy to process, which make them less eco-friendly as fossil fuels are consumed to generate electricity to fuel the process. Fourth-generation biofuels, also known as photobiological solar fuels and electrofuels, are derived from genetically engineered crops. They transform solar energy directly into fuel with limitless, inexpensive, and readily available raw ingredients. The third and fourth generation feedstocks have the potential to be a long-term source of biofuel production. The usage of various types of feedstocks is particularly environmentally benign in terms of reducing greenhouse gas emissions and is suitable for large biofuel production.

Incentives, taxes, subsidies and carbon content control are just a few of the policies that have been proposed or implemented to influence the production of biofuels (Rajagopal et al. 2008). Biofuel policies are critical for reducing the world's reliance on fossil fuels and ensuring a better future. In the previous two decades, the Indian government has also implemented a number of schemes and initiatives, including mandating the blend of ethanol with gasoline and diesel with biodiesel for the country's future clean energy goal, as well as rewarding bio-based products and fuels. The National Biofuel Policy is a step ahead, which was designed to take advantage of the many ecological, social, and financial benefits to bring large-scale biofuel production to the country. As per the International Energy Agency (IEA), if appropriate rules and expenditures are in existence, biofuels have the capacity to fulfil more than 25% of global transportation fuel needs by 2050.

2. Generation of biofuels

Biofuel is a type of fuel created by the conversion of biomass. As energy sources, they have the capacity to alleviate a number of issues connected to climatic change and sustainability. It can be classified based on a variety of factors, including the feedstock used, the conversion method, the fuel's technical specifications, and its intended purpose. The details of different feedstocks and processes to produce biofuels and their wider uses are depicted in Figure 1 and Table 1. The first-generation biofuels are made from food crops (such as corn and sugar cane) (Liu et al. 2021). The biofuels are also known as 'conventional biofuels' since they are made using well-established technologies and methods like fermentation, saccharification, evaporation, and enzymatic hydrolysis. Non-food feedstocks, such as energy crops, agricultural residues, forest residues, waste products, and non-edible lignocellulosic leftover of plants, which contain up to 70% polymerized sugars and are the most common type of biomass on the planet (Isikgor and Becer 2015), are used to make second-generation biofuels.

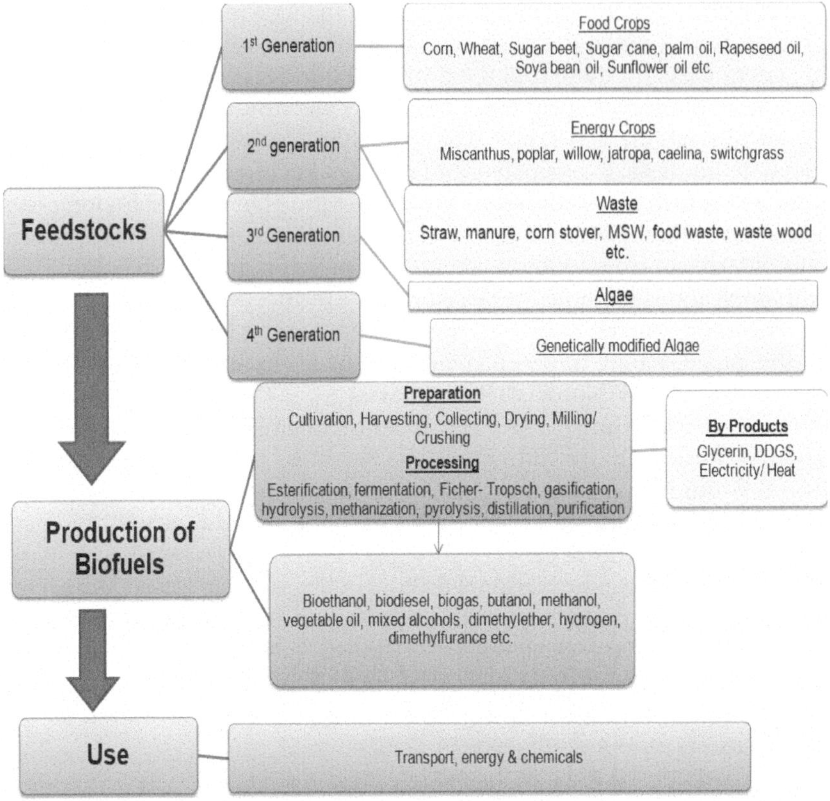

Figure 1. Systematic representation of biofuel production of different generations (Jeswani et al. 2020).

The resultant net carbon footprint (emitted carbon minus absorbed carbon) can often be neutral or even negative, which makes the biofuels much more appealing (Field et al. 2020, Tilman et al. 2006). Third-generation biofuels, often known as 'algae biofuels', are biodiesels made from microalgae using traditional transesterification, gasification, or hydro-treatment of algal oil. Algal biomass, such as macroalgae and microalgae, are used to generate the third-generation biofuel. The manufacturing processes or routes for generation of these biofuels are still in the research and innovation, pilot, or demonstration stages, therefore, second and third-generation biofuels are commonly referred to as 'advanced biofuels' (Jeswani et al. 2020).

Biofuels of the fourth generation are made from genetically modified biomass (Rudra et al. 2021). It is based on microscopic creatures that

Table 1. Overview of biofuels based on their feedstock and production technologies (Magda et al. 2021).

S. No.	Generation of Biofuels	Biofuels	Feedstock	Production Process
1	First	Biodiesel	Oil crops	Transesterification
		Biobutanol	Sugar crops	Fermentation and saccharification
		Bioethanol	Lignocellulosic materials	Advanced enzymatic hydrolysis and fermentation
		Bioethanol	Sugar & crops	Fermentation
2	Second	Biomethanol	Lignocellulosic materials	Gasification, synthesis, & catalytic cracking
		Fischer Tropsch diesel	Lignocellulosic materials	Gasification, synthesis, & catalytic cracking
		Biohydrogen	Lignocellulosic materials	Synthesis, gasification, & fermentation
		Dimethyl ether	Lignocellulosic materials	Gasification, synthesis, & catalytic cracking
		Biogas	Lignocellulosic materials	Synthesis & anaerobic digestion
3	Third	Vegetable oil, biodiesel	Algae	Transesterification
4	Fourth	Biohydrogen, Biomethanol, Synthetic Biofuel	Genetically engineered crops	Pyrolsis, Gasification, & Digestion

have been genetically engineered, such as microorganisms, yeast, fungi, microalgae, and cyanobacteria. Biofuels include biogas, gasoline, bioethanol, biohydrogen, biodiesel, synthesized biofuels, biomethanol, and vegetable oil; these are all made from feedstock or its by-products.

These biofuels, in the form of new cash crops, can aid in rural and agricultural growth, which can ultimately lead to the growth of the nation. Both food and energy can be provided by a properly designed and implemented biofuel solution, which would alsobe a boon for energy production worldwide. Herein, we discussed the generation of biofuel, types, feedstock, and worldwide policies in a comprehensive way.

3. Current status and future prospects on different generations of biofuel production

3.1 First-generation biofuels

The majority of today's biofuels are first-generation biofuels. Most of the first-generation biofuels are obtained from edible materials like corn, sugarcane, grains, wheat, vegetable oils, oilseeds, and condensed animal fats (Zeng et al. 2016, Ambaye et al. 2021). The fermentation of organic sugars, taken from starches from crops such as sugarcane (*Saccharum* sp.), beets, and corn (Zea mays) by yeast (*Saccharomyces cerevisiae*) produces first-generation bioalcohols (bioethanol). Figure 2 represents the various feedstock options that are widely utilized to produce first-generation biofuels. However, the social and environmental unviability of first-generation biofuels limits their potential development. The enhanced usage of fertilizer to raise crops for biofuels has resulted in significant levels of phosphorus and nitrogen reaching waterways. First-generation biofuels have a detrimental impact on biodiversity, GHGs emissions, water usage, land use, and cause water fouling (Singh et al. 2015). As a result, there is a need to shift away from first-generation biofuels. Due to the rising population worldwide, it is more practical to create second-generation biofuels using food source by-products, also known as second-generation feedstock (Suganya et al. 2016, Dahman et al. 2019).

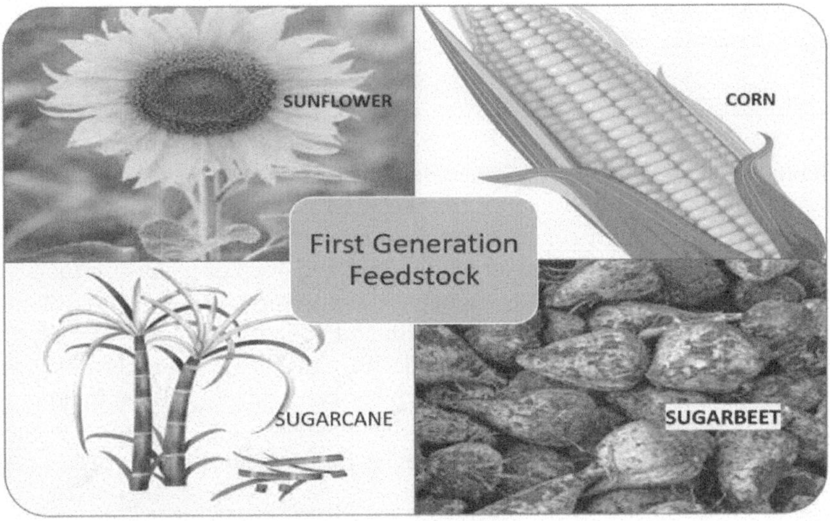

Figure 2. First-generation biomass widely utilized for fuel production.

3.1.1 Types of first-generation biofuels

The different types of first-generation biofuel are discussed below in detail:

Biodiesel: Biodiesel manufacturing began in the early 1990s and has now been widely used as a sustainable energy source for the transport industry, output has continuously expanded since then (Hirani et al. 2018). This is perhaps the most often utilized form of biofuel in European countries. The biodiesel is mostly made through a process known as transesterification (Datta et al. 2019). This fuel is scientifically termed fatty acid methyl and is quite identical to mineral diesel. After combining the feedstock with methanol and sodium hydroxide through the defined chemical process, biodiesel is generated. After being mixed with mineral diesel, biodiesel is widely employed in different diesel engines. Numerous diesel engine producers now assure that their engines function effectively with biodiesel in several areas.

Biogas: Since 1958, biogas has been manufactured and used in both rural and urban locations. Crop residues, livestock manure, sludge, solid waste, dump, and industrial organic material have all been used as feedstock for biogas production. Biogas produced in the United States is expected to be raised around 7.9 million tonnes per year from various feedstocks (Hirani et al. 2018).

Biogas is generated primarily by the anaerobic breakdown of organic matter. Sometimes, the fuel can be created by the biological degradation of waste products that are introduced into anaerobic digesters. Furthermore, the leftover can be used as compost or fertilizer in agricultural applications. The biogas is enriched with methane, which may be reclaimed easily with mechanical biological treatment equipment. Landfill gas, which is generated by using naturally existing anaerobic digesters, is a less pure form of biogas and has the potential to enter the atmosphere with fugitive emission of gases.

Bioalcohols: These are alcohols produced by the fermentation of carbohydrates and sugar with the help of enzymes and microbes. The most prevalent kind of bioalcohol is ethanol, while propanol and butanol are less well-known. Biobutanol is a promising alternative for gasoline as it can be used in variety of engines with less or without modifications. Butanol is made by Acetone-butanol-ethanol, and some trials have shown that butanol is a high-efficiency fuel that may be utilized directly in different gas engines.

Syngas: The first-generation biofuel 'Syngas' is a formed on combining the processes of gasification, combustion, and pyrolysis. The biofuel is transformed first into carbon monoxide, which is then pyrolyzed to provide energy. To maintain combustion within control, relatively less air is provided during the operation. Organic molecules are transformed into gases like carbon dioxide and hydrogen in the final process referred as 'gasification'. Syngas, the resultant gas, could be used for a variety of functions.

3.1.2 Feedstock

The first generation of biofuels is predominantly composed of edible food stocks such as sugarcane, sugar beet, rice, wheat, barley, potato, corn, sugarcane, and vegetable oil, for example, soybean oil, sunflower oil, olive oil, canola oil, mustard oil, etc., and as a result, these feedstocks are likely to be outlawed in the European Union in order to protect the food supply.

3.2 Second-generation biofuels

Second-generation biofuels are already commercially available with recent improvements in research and innovation. Bioethanol and biodiesel generated from non-food crops such as leftovers from agricultural crops, crop residues, wood, and waste cooking oil are referred to as second-generation biofuels (Dahman et al. 2019). Crops, agricultural residues, and other non-crop species of plants will have no or little rivalry with these biofuels in the end (Hirani et al. 2018). The use of an inevitable by-product of the agriculture business for second-generation biofuel production has advantages such as it requires no additional fertilizer, water, or acreage to grow. Although part of this non-edible residue is utilized to make animal feed, there is still a significant quantity that might be used to make biofuel (Begum and Dahman 2015). This biofuel pathway is plagued by arguments about the expensive procedures required for biofuel synthesis from second-generation feedstock. In any case, second-generation biofuel development and policies have the potential to turn this biofuel pathway into a profitable biofuel source (Dahman et al. 2019).

3.2.1 Types of second-generation biofuels

Second-generation biofuels are made from a variety of sources, including trees, grasses, shrubs, agriculture waste, and so on. Based on the methods used to produce them, advanced biofuels come in a variety of forms.

Cellulosic ethanol: Fermented sugar generated from cellulose and polyose components of lignocellulose are used to harvest this biofuel (Deora et al. 2022).

Algae-based biofuels: Algae is the fastest-growing material for biofuel manufacture, and it's a promising alternative to traditional biofuel extraction (Thanigaivel et al. 2022). Algae may grow in both open and closed systems (like lakes, ponds, etc.) (Rebello et al. 2020) and has the advantage of being able to be turned into a variety of biofuels, including biodiesel, biogas, and hydrogen (Makut et al. 2021). The biofuel is produced from a variety of biomass separation and concentration procedures include aggregation, centrifugation, purification, floatation, and flocculation (Branyikova et al. 2018, Singh and Patidar 2018).

Alcohol: Catalytic formulation, which is drawn up from syngas, is used to recover methanol or mixed alcohols. Syngas can also be made into alcohol by fermenting biomass only with the help of a specialized microbe (Susmozas et al. 2020, Deora et al. 2022).

Dimethylfuran: It is also recognised as the 'sleeping giant' of renewable chemicals (Chandel et al. 2020). Dimethylfuran is oxygenated hydrocarbon with a 17% (Deora et al. 2022) gravimetric oxygen concentration, making it an excellent contender for lowering toxins from engine emissions. It can also be used as a diesel fuel enhancer.

Biosynthetic natural gas (Bio-SNG): Anaerobic digestion and certain bacteria can be used to produce biogas. Mash gas and carbonic acid gas are often used to create the biogas. Bio-SNG is also utilized in automobiles in the form of CNG and LNG, and also used to replenish natural gas cylinders (Deora et al. 2022).

3.2.2 Feedstock

The second-generation biofuels are made from readily available lignocellulosic and organic waste materials (Figure 3, such as wood, straw, and switchgrass, as well as oilseed-bearing trees like jatropha) (Lee et al. 2019, Ambaye et al. 2021).

***Grasses*:** Grass species such as switchgrass, miscanthus, Indiangrass, and others have been alternately highlighted. The grass chosen for biofuel production is usually determined by the region, as different grasses are better suited to certain temperatures. Switchgrass is popular throughout the United States (Alexopoulou et al. 2018). Miscanthus is the preferred plant in Southeast Asia (Hirani et al. 2018).

***Jatropha and other seed crops*:** Biodiesel can be produced with the help of seed crops. A plant known as Jatropha became quite attractive among biodiesel supporters in the early twenty-first century. The plant was praised for its high yield per seed, which may reach 40% in some cases (Eloka-Eboka and Maroa 2021). When contrasted to soybeans, which

Figure 3. Feedstocks utilized for the production of second-generation biofuels.

contain 15% (Updaw and Nichols 2019) oil, Jatropha appears to be a miracle crop. Other seed crops with similar characteristics have suffered the same fate as Jatropha. Camelina, oil palm, and rapeseed are some examples. The initial advantages of the crops were quickly realized to be countered by the necessity to use farmland to obtain appropriate yields in all circumstances.

***Waste Vegetable Oil (WVO)*:** WVO has been used as a source of energy for over a century. Some of the early diesel engines, in fact, ran entirely on vegetable oil. Because its wider utility as a food has been depleted, waste vegetable oil is classified as a second-generation biofuel. Recycling it for fuel, in turn, can assist to reduce its overall carbon footprint (Datta et al. 2019).

***Municipal Solid Waste*:** This type of biomass is now used in the production of biofuels. This includes landfill gas, human waste, and grass and yard clippings, among other things. In many circumstances, all of these forms of energy are simply being squandered. Although not quite as efficient as solar and wind, these fuels have substantially lower emissions than conventional fossil fuels. Municipal solid waste is frequently burned in cogeneration units to generate both heat and power (Datta et al. 2019).

3.3 Third-generation Biofuels

Third-generation biofuels are ones that are very new to the energy sector and mostly consist of algae and quickly-growing trees. Biofuels have

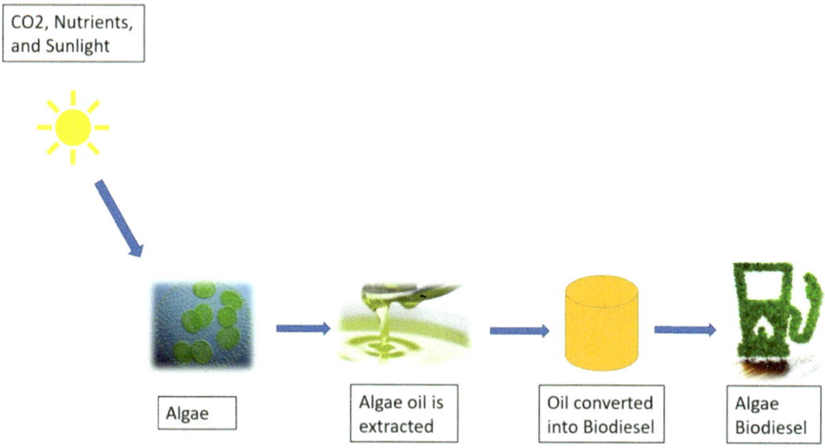

Figure 4. Process to generate third-generation biofuels from microalgae.

been hailed as a viable option since they can cut emissions by up to 90% (Yamakawa et al. 2018), based on the feedstock used. Biofuels have the advantage of being able to be combined with conventional fuels or utilized wholly on their own. The majority of third-generation biofuels are still made on a commercial level using algae as a feedstock. Algae's potential to trap CO_2, generate a lot of lipids, thrive in a variety of environments, and develop magnitudes faster than all land plants (including second-generation feedstock) make it an excellent feedstock source for biofuel production (Ribeiro et al. 2017).

3.3.1 Feedstock

Algae is used as a feedstock in third-generation biofuels, providing a significant amount of lipids for the production of biodiesel as well as other biofuels (Ambaye et al. 2021). The generalised process to utilize algal biomass for production of biodiesel is represented at Figure 4.

Algae have been demonstrated to have consistent properties that make them an appealing option over many other biomass materials. The algal biomass is also rich in starch, protein, and lipids, all of which are used to make biological products, including bioethanol, biodiesel, biobutanol, and biogas (Salama et al. 2017). Algae appears as a viable option with its properties covering modest growth requirements, sequesters carbon, and improves air quality (Medipally et al. 2015). In comparison to other second-generation biofuels, algae offer a number of advantages like high rate of growth and do not require arable land for cultivation. Moreover, algal biomass produces more biomass per hectare as their growth is not limited to seasonal production (Medipally et al. 2015, Dahman et al. 2019). In the face of mounting demand to boost GHGs emission reduction by developing a more efficient biorefinery process, it is also worth noting that algae have a greater carbon sequestration potential. Furthermore, because it uses nitrogen and phosphorus from wastewater sources, algal biomass growth aids in wastewater bioremediation (Dahman et al. 2019).

3.3.2 Algal biofuels

Algae are a broad collection of aquatic photosynthetic organisms that produce a wide range of compounds, including a variety of oils that can be used to make biodiesel, avoiding some of the technological difficulties associated with converting lignocellulose to liquid fuels. They do not require freshwater to grow and may be grown in wastewater or seawater, and are projected to produce great yields under ideal conditions. Algal biodiesel is non-toxic and has lower levels of particles, carbon monoxide, soot, hydrocarbons, and sulphur oxides in comparison to diesel. The literature also support it being utilized for aircrafts (over first-generation biofuels) because of its unique properties of lower freezing point and higher energy density.

3.4 Fourth-generation Biofuels

To maximize biofuel yield, fourth-generation biofuel (FGB) utilizes genetically modified (GM) algae. In fourth-generation biofuels, modified microorganisms convert CO_2 into fuel by the process of photosynthesis. (Vassilev et al. 2016, Alalwan et al. 2019). Microalgae has significant advantages, such as their fast growth and higher oil content, as well as their low structural complexity, which contribute to their wide range of economic applications (Azizi et al. 2018). Apart from genetic manipulation, several fourth-generation technologies include pyrolysis (400 to 600°C) (Azizi et al. 2018, Alalwan et al. 2019), gasification, upgrading, and solar-to-fuel approaches. The goal of these advanced approaches is to increase hydrocarbon yield and build a manmade carbon sink to reduce or eliminate carbon emissions (Sikarwar et al. 2017). These methods are currently in the initial phases of development; however, they have potential to tackle

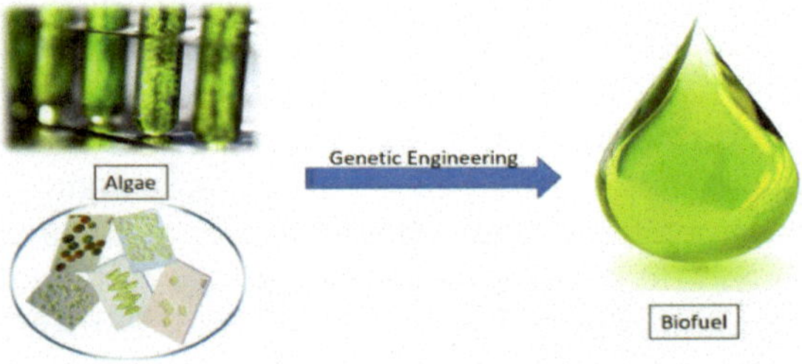

Figure 5. Fourth-generation biofuel production from algae.

issues related to energy and environment (Sikarwar et al. 2017, Alalwan et al. 2019).

3.4.1 Feedstock

Microalgae, macroalgae, and cyanobacteria provide the feedstock for fourth-generation biofuels (Figure 5). Cyno-bacteria contains prokaryotic cell, whereas microalgae and macroalgae contain eukaryotic cells (Ambaye et al. 2021, Deora et al. 2022).

In the bioenergy and biofuel sectors, cyanobacteria have received a lot of interest. Metabolic engineering for numerous photosynthetic species has been developed with advances in Genomic Revolution. The genome of Synechocystis was the very first photosynthetic cell to be completely sequenced (Silva et al. 2019). Synechocystis is a non-filamentous, non-nitrogen-fixing freshwater organism that may grow both autotrophically and heterotrophically. The available genomic, biochemical, and physiological facts are the most useful feature of this variant of cyanobacteria. Due to its tiny genome size relative to larger plant systems, it is well recognized as a physical model used for the exploration of the photosynthesis process in higher plants (Alalwan et al. 2019).

Owing to the accessibility of eukaryotic genetic information, eukaryotic microalgae-based innovation has received much interest recently. Eukaryotic cells are produced when external genes are randomly integrated into the nuclear genome. Genetic transformation into the cellular nuclei, mitochondria, and chloroplasts has effectively produced several types of microalgae (Long et al. 2015). *Chlamydomonas reinhardtii* is the most studied eukaryotic microalgae. It has demonstrated a high level of recombinant protein production, which allows for the commercial manufacturing of a variety of proteins, including complicated monoclonal antibodies and mammal pharmaceutical enzymes, using diverse production platforms (Alalwan et al. 2019).

4. Worldwide policies of biofuels

The production of biofuels at global level has risen significantly to support the pressing needs of government to control carbon emission and reduce the reliance of fuels import. The government policies adopted from time to time are the driving force behind the rising concern over generation of biofuel. Various policies and initiatives established around the world are presented in the section, which provides empirical and theoretical evidence to support the production of advances in biofuels (Table 2, Figure 6). Emerging technologies, social acceptability, infrastructure development, and government support are the key aspects, which aid in the future development of biofuel industries at the local and global level.

Table 2. Biofuel policies in different counties (Das 2020).

Country	Mandate or target
India	By 2030, a blend of 20% ethanol in gasoline and 5% biodiesel in diesel is required.
Brazil	Ethanol blend mandate of 20–25%; biodiesel use mandate of 5% since 2010. (Proposal of use increasing up to 10 % by 2020.)
The United States	By 2022, need 36% gallon of biofuels, with no more than 15 billion gallons coming from cellulosic ethanol.
European Union	By 2020, a mandate requires a 10% biofuel blend in all motor fuels.
China	Biodiesel usage aim of 2.3 billion liters by 2020, non-fossil fuel consumption target of 15% by 2020.

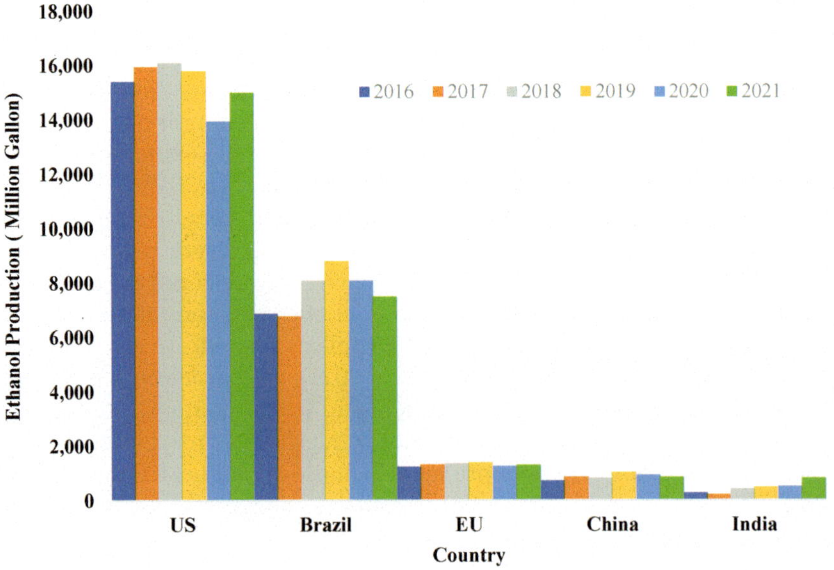

Figure 6. Worldwide annual bioethanol production (RFA).

4.1 United States

The US is the world's biggest producer of biofuel and contributes about 45.5% in the world's biofuel production. The production of biofuel was 1347.3 petajoules in 2020 in US. Austin et al. (2022) has estimated to utilize 36 billion gallons of renewable sources as transportation fuels by 2022 in US. The volume of traditional biofuels (like corn ethanol) can be estimated to be 15 million gallons per year, starting in 2015 (Austin et al. 2022). In United States, corn is used as the primary feedstock for producing biofuels. According to the United States Department of Agriculture (USDA), 5.55 billion bushels of corn were crushed for ethanol production in 2018. Iowa City became the leading state with production of 4.33 billion gallons per year. According to the Energy Information Administration annual fuel ethanol production capacity report, there are 200 fuel ethanol production units with a nameplate capacity of 16,868 MMgal/year with the onset of 2019. Cellulosic biofuels are predicted to grow from 0.1 billion gallons in 2010 to 16 billion gallons in 2022 in US (Tyner 2020).

The lessening of the country's reliance on oil has been the guiding concept of biofuel programs. The Energy Independence and Security Act of 2007 set a goal of reducing gasoline usage by 20% over the next ten years (Das 2020). The 2008 Biomass Program had two key goals: the first was to reduce gasoline use by 30% by 2030, when equated to 2004 levels (Austin et al. 2022) and the second goal was to utilize corn-derived ethanol to generate cellulosic ethanol.

4.2 European Union

By 2020, the Renewable Energy Directive (*RED*) targeted that renewable energy accounts for 20% of all energy used in the EU, along with at least 10% of all energy used in public transportation fuel (Das 2020). In addition to the *RED*, an updated Fuel Quality Directive (FQD) mandated that by 2020, the EU's road transport fuel must be 6% less carbon-intensive than a conventional diesel and gasoline baseline (O'Connell et al. 2019). Despite the fact that renewable energy is used in aviation and shipping, it is not included in the national 10% targets (Das 2020). The production of biofuels in Europe was the highest in Germany (146.3 petajoules in 2020). Germany is third among the world's leading biofuel producers and contributed 2.9% of the worldwide biofuel production capacity by generating 75.8 thousand barrels per day in 2018. According to the German Association VDB of Biodiesel Producers, German firms produced 3.2 million tonnes of biodiesel in 2018 (Singh et al. 2022).

4.3 Brazil

The Federative Republic of Brazil's Nationally Determined Contribution (NDC)[3] towards meeting the goal of the United Nations Framework Convention on Climate Change (UNFCCC) is one of the current driving forces behind the nation's renewable energy policy framework. This treaty, which was first released in December 2015 at the Paris Conference (*COP* 21) and was amended in 2020, established the economy-wide carbon reduction targets (37%, by 2025, 43% by 2030) (Aamodt and Stensdal 2017). The document states that these aims are compatible with the indicated goal of attaining carbon neutrality by 2060 (Gota et al. 2019) under the Paris Agreement.

Since the 1970s, Brazil's major policy enforces ethanol being mixed in gasoline, amount of the ethanol was 27% in 2015 and will reach 31% in 2020 but a slight fall were observed in 2021 ethanol production, i.e., 27% (Teixeira et al. 2020). In 2017, Brazil established the Brazilian Policy for Biofuels (RENOVABIO), a set of regulations to reinvigorate the biofuels market by encouraging energy efficiency improvements in biofuel consumption and production. From 2018 to 2030, the policy intends to minimize the carbon output of the transportation fuel grid by 10% and prevent 620 million tonnes of CO_2eq emissions (Gurgel et al. 2019). Brazil is the second-largest producer of biofuels after United States. In, 2018 the total biofuel production was 693.2 thousand barrels/day in which Brazil's share was 26.5%. Brazil accounts for 14.1% share in biodiesel production, or approximately 99,000 barrels per day (Singh et al. 2022). Brazil is predicted to have produced 30.76 billion liters of ethanol in 2018, up 9% from the corrected total for 2017 (Magarey 2020). Domestic demand for ethanol for fuel and other purposes was estimated to be 28.72 billion liters in 2018 (Singh et al. 2022). Bagasse, a by-product of sugarcane crushing, is widely used as a fuel in sugar mill cogeneration plants to meet onsite energy demand. Bagasse is sometimes utilized to supply extra electricity for export purposes.

4.4 China

China's policies of biofuel emphasized the production of ethanol. China's ethanol manufacturers expanded domestic production capacity by 258 million litres in the last year, bringing total capacity to 5,258 million litres. The National and Development Reform Commission (NDRC) was established with immediate and longer renewable energy development plan in August, 2007. The studies revealed that renewable energy usage as a percentage of primary energy consumption was required to be increased to 10% by 2010 and 15% by 2020 (Zhang et al. 2017). Biofuels are likely to play a significant role in achieving these goals. The output of ethanol in China was expected to increase to 2 million tonnes by 2010 and 10 million tonnes by 2020 (Das 2020). By 2010, biodiesel demand was estimated to be around 200,000 tonnes, which was further expected to raise around 2 million tonnes in 2020 (Chung et al. 2020). China is the world's fifth-largest biofuel producer, with its production capacity of 142 petajoules in 2020 (Fatima et al. 2019). The country's attempts to enhance air quality are primarily driving ethanol production in China. The Ministry of Ecology and Environment (MEE) of China announced strict air pollution reduction measures in Beijing, Tianjin, and Hebei provinces in 2018.

4.5 India

The "National Policy on Biofuels" was firstly published in 2009 with the goal of popularising biofuels by setting a target of blending up to 20% with gasoline and diesel in the transportation sector by 2017. Further, a revised policy, The National Policy on Biofuels 2018, was adopted to achieve the target of 20% biofuels blending with fossil fuels by 2030 (Ministry of Petroleum & Natural Gas 2018). In order to empower the renewal of acceptable fiscal and financial subsidies under each section, the Policy categorizes biofuels as "Basic Biofuels", which includes first-generation (1G) biofuels like bioethanol and biodiesel, and "Advanced Biofuels", which covers second-generation (2G) ethanol, municipal solid waste (MSW), third-generation (3G) biofuels, bio-CNG, and so on. The Policy broadened the range of feedstocks for the production of bioethanol by permitting the use of sugarcane juice, sugar-containing substances such as sweet sorghum, sugar beet, starch-containing substances such as cassava, corn, damaged food grains such as broken rice, wheat, and rotten potatoes, which are unfit for human consumption (Das 2020). During the surplus manufacturing stage of the crops, farmers run the danger of not earning a good rate for their crops. In light of this, the Policy permits the use of excess food commodities for the manufacturing of ethanol for use in gasoline blends with the consent of the National Biofuel Coordination Committee. With a focus on enhanced biofuels, the Policy proposed Rs. 5000 crore viability gap finance packages for 2G ethanol bio-distilleries over six years, as well as additional tax subsidies and a higher purchasing price than 1G biofuels (Prasad et al. 2020). The existing and proposed advanced biofuels plants in India are depicted in Table 3. The policy fosters the establishment of supply chain

Table 3. Existing and proposed advanced biofuels plants in India (Source: Biofuture Platform 2018).

Company	Year	Status	Annual production capacity
Indian Glycols Kashipur	2016	Operational	0.75
Praj Biofuels	2017	Operational	1
Shell Bengaluru	2018	Operational	0.6
Numaligarh Refinery Limited	2018	Planned	60
IOCL Panipat	2019	Planned	30
BPCL Bargarh	2018	Planned	30
HPCL Bhatinda	2017	Planned	30
IOCL Panipat	2019	Planned	0.75
IIOP	2018	Operational	0.01
IOCL	2019	Planned	33

systems for the manufacturing of biodiesel from non-edible oilseeds, used cooking oil, and short rotation crops. The Policy paper captures the roles and duties of all involved departments with regard to biofuels in order to coordinate activities. The policy unveils phase-wise goals with the strategy to meet 10% and 20% ethanol blending by 2022 and 2030, respectively; while the biodiesel blending target was set at 5% for 2030 (Venkatramanan et al. 2020).

4.5.1 Biofuel policy and second-generation biofuels

In India, biodiesel production is mostly based on non-edible oilseed crops such as Jatropha and Pongamia, as well as edible oil waste and animal fats. By 2012, the government planned to plant Jatropha on 11.2 million ha of wasteland in order to meet a 10% blending target (Kataki et al. 2017). Moreover, the national authorities have offered economic rewards for peasants to promote the planting of oilseeds.

The National Policy, 2018 provides a secure channel to private players and promotes 2G ethanol projects. Oil manufacturing companies have decided to sign ethanol purchase agreements with 2G ethanol producers for a duration of 15 years. The program offered financial incentives, increased support for the construction of 2G bio-distilleries, and a price increase for 2G ethanol production relative to 1G ethanol. The financial incentives were set forth in The National Policy, 2018 for the manufacturing of 2G ethanol and elimination of price disparity between 2G and 1G ethanol (Das 2020). Financial rewards for new and second-generation biofuels, as well as a National Biofuel Fund, were part of the national biofuel policy (Saravanan et al. 2018). Subsidies and grants for new and second-generation feedstocks, sophisticated technologies and conversion processes, and production models based on new and second-generation feedstocks are considered on merit, according to the revised biofuel policy.

4.5.2 Third-generation biofuels and National Biofuel Policy, 2018

Algal biofuel is perfectly positioned to fulfil rising transportation fuel demand, while also boosting energy security, creating local jobs, and being carbon neutral. However, there is still a lot of work that needs to be done, both technically and in terms of legislation, before this potential can be fulfilled. If algae biofuels are becoming a feasible alternative source of energy, high production costs and market barriers must be overcome. Policy changes will be critical in accelerating the expansion of this technology so that the tremendous benefits of switching to algal biofuels may be realized (Dahman et al. 2019).

In this regard, policymakers must broaden their definitions of sustainability during the judgment process, taking into account both economic as well as environmental factors. Furthermore, incentive policies are required to enhance 3G production of biofuel since algae, particularly microalgae, have greater promise for biofuels due to fewer negative side effects and possible social benefits (Nazari et al. 2021).

4.6 Opportunities for developing countries

Because of superior climate conditions and lower labor costs, developing countries have a greater potential to produce biomass and biofuel than developed countries. As a result, worldwide commerce in biofuels and feedstocks from developing to developed countries is likely to grow, resulting in major development benefits. These benefits put developing counties in a better position to take advantage of a new and exciting area of the global economy. Agriculture is an ecosystem-based activity, and natural endowment is the most important component affecting agriculture's productive capability. Moreover, developing counties receive additional benefits in term of advanced biofuels which could help to reduce emission in the production and utilization of biofuels.

4.7 Innovative biofuels

Biofuels are new and inventive fuels, which are intended to help in the reduction of net GHGs emissions, improvement in energy security, and promote development. Some features of innovative biofuels like closed CO_2 cycle, no negative impact on food supply and its ability to be generated without affecting the environment or local residents are intended its use more frequently.

4.8 Future of biofuels

Currently, development and usage of 2G biofuels and biobased e-fuels in maritime and aviation are given key priority around the world. National policies and European Energy Directives require decarbonization and the achievement of sustainability targets in the transportation sector. Furthermore, maritime and aviation sectors are becoming greener at a faster rate as expected. The policies for biofuels bring together industry leaders, innovative material, interactive seminars, and networking opportunities. Europe, Brazil, China, and India have each set a goal of replacing 5% to 20% of total conventional diesel with biodiesel. The prospects for biodiesel will be realized faster than expected, if governments continue to aggressively pursue targets for 2G biofuels, and continue to promote research and development for alternative substrates, non-food feedstocks such as grease tallow, Jatropha, Castor, and algae for biofuel production. According to estimates, the United States produces 44%, Brazil 41%, the European Union 13%, and Southeast Asia 2% of the world's total supply of biofuel, which is around 16 (million tonnes oil equivalent, including 80% ethanol and 20% biodiesel). In the previous two decades, the global biofuels market has grown at a rapid rate, and this trend is expected to continue.

5. Energy Security in India

India is a developing country which comprises approximately 18% of the world population. The demand for energy is increasing day by day among the people in India, making it the fastest growing energy consumer in the world. Therefore, a huge amount of energy is required to fulfil that demand and to keep the economy stable.

Currently, India attaches more importance to energy and even the broader concept of national security also imparts a special place to energy security. Bernard D. Cole in his book, *Sea Lanes and Pipelines: Energy Security in Asia*, mentioned that, "The twentieth century was one of the almost continuous wars, both hot and cold, both local and global, energy security was a continuing element in the causation, conduct, and settlement of that long period of warfare (Cole 2008, p. 6)."

The International Energy Agency (IEA) defines energy security as stable supplies of oil and natural gas. At the global level, energy security is meant to ensure adequacy of resources, while at regional level it is seen as ensuring trade can happen without any trouble. At the country level, it can be recognized as the national protection of supply chains, whereas at the consumer level, it assures the satisfaction of consumer demand. Within this holistic overview, energy security appears as a state where a country, including its inhabitants and industries, have access to sufficient energy resources without any disruptions at rational prices.

Energy security has many characteristics and it may be long term and short-term based on different focal points. Long-term energy security is limited to economic advancement and ecological needs and ensuring timely investments of supply energy without interruptions, while short-term energy security is focussed on maintaining the demand and supply balance in case of sudden changes or emergencies.

5.1 Factors influencing energy security

Depending on the circumstances, any change in factors can facilitate or hurt the energy security of a country. There are several factors which can hamper the supply of energy in a country. The first factor is the challenge of physical factors. The geology has a pronounced effect on the availability of energy reserves, particularly fossil fuels like coal, oil, and natural gas.

The second factor involves climate, which determines the use of renewable energy resources at a place suited to it; for example, islands can potentially generate more wind power, solar power can be generated within sunny countries, while wet places may have an abundance of hydroelectric power. The climate helps in deciding the type of renewable energy supplies. Further, environmental conditions can also pose a challenge to energy security, very cold climate or ahigh-wave environment, etc., can make it very difficult to access energy resources, for example.

The third factor is turbulent states where conflicts are happening, leading to disruptions in the supply of energy and posing an extreme challenge to energy security. Energy security can be at a risk if a country had an overreliance on any one particular energy supplier, especially when the supply of energy resources gets disrupted from that country or region.

The fourth factor is dependence on imports; the countries which lack energy sources cannot fulfil their domestic energy demands, and therefore, have to rely on imports for a continuous supply of energy resources. Such countries may face problems when the costs/prices of energy fuels are hiked based on the hegemony of exporting countries.

The fifth factor is access to energy-related technologies. This is important as many countries face a shortage of new technologies in this respect. Therefore, such counties are dependent on technologically-advanced countries every time to fulfil their energy needs.

5.2 Energy consumption at global level

According to the data available in the literature, it is estimated that the level of energy consumption will rise in 2040, and will be 56% higher than in 2010. Most importantly, the consumption level will be higher in developing countries as compared to developed countries. New emerging economies like China and India, high population growth, development of more advanced technologies, and infrastructure development are some of the reasons for the rising energy consumption level.

5.3 India's energy security

India, currently, occupies the third place in the world in terms of energy consumption. Its energy utilization has doubled in the past two decades. India's main sources of energy are coal, natural gas, oil, hydroelectric, nuclear, and several other renewable energy sources. Out of the world's total primary energy, at present, India consumes nearly 6%, due to its fast economic development and demographic explosion, India's utilization level of energy is expected to rise in the coming decades.

Presently, India is looking to various countries to fulfil its energy needs and ensure the smooth functioning of its economy. Nearly 82% of India's oil requirements are accomplished with imports, which by 2030 will increase to more than 90%.

In the time yet to come, Indian policymakers may perceive energy security as a prime security challenge. However, at present, India's shortage of energy is distinguished by four major factors, which includes:

- Fast economic growth, where a reliable supply of petroleum, gas and electricity is required;
- Increase in the level of income, in which affordable and adequate electricity and clean cooking fuels need to be provided;
- The domestic reserves of fossil fuels are limited due to which there is a need of import of gas, oil, and petroleum products and also coal as well;
- And environmental and climatic changes demand the adoption of cleaner fuels and cleaner technologies.

Moreover, India is facing several other challenges such as shortage of energy resources, a rapid increase in population growth, threats from non-state actors, and flaws in energy policy. The lack of a coherent strategy in context to energy security has made India be cooperative in its relations with other countries.

India's economic development has arisen in the context of globalization and liberalization pressures generated by the post-Cold-War world order. India's growth rate could be higher than 5% over the next 30 years and close to 5% as late as 2050, if development proceeds successfully. During the 1990s, India registered a rapid increase in its energy consumption because the country witnessed an extraordinary economic growth. Thus, India needs to develop its own hydrocarbon sector in order to balance its economic goals.

India's known oil and natural gas resources are very less as compared to its demand and use. That is, the major issue is in term of the gap between the energy produced within the country and the energy which is actually required. Based on the extent of requirement and utilization, the energy security of India is practiced under the theoretical structure of 4 As which are: availability, accessibility, affordability, and acceptability. This framework supports India's economic growth and development goals in the midst of environmental concerns.

India faces hard challenges to meet its energy hunger due to low production within the country. The disparity between demand and supply is huge for India, therefore, it can hardly afford to select one preference over the other. There is a chance that in the next few decades, India will

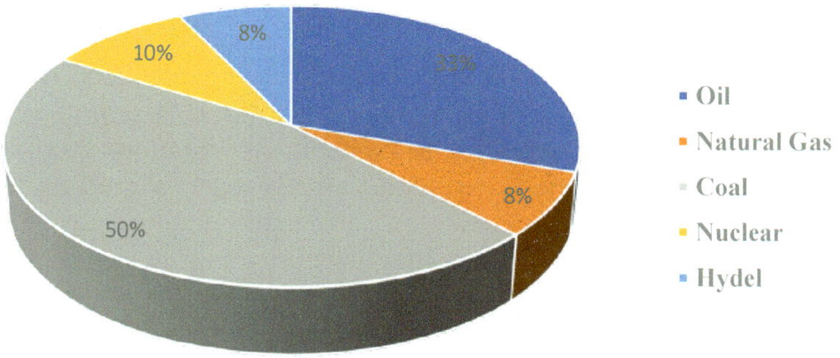

Figure 7. Energy consumption pattern of different energy sources in India.
(Source: International Energy Agency (IEA)]

need to utilize all the available options in order to build up its rational massive capacity base of energy so that the energy requirements can be fulfilled without any shortage. The sector wise share of India's energy consumption is shown in the Figure 7.

There is no doubt that India has already been tapping its own reserves. During the first four years of 10th Five Year Plan, ONGC, OIL, and private and joint ventured companies made 82 oil and gas discoveries. The Krishna-Godavari Basin was the one among them, where most of the gas discoveries were made. The basin is located off the east coast of India. Apart from this, nearly 30 additional discoveries have been made by both private and joint venture companies like Reliance Industries Ltd. (RIL), Cairn, and Essar Oil Ltd (EOL).

Keeping in view the demand and shortage of supply of hydrocarbons, the government of India has encouraged private companies to invest in the foreign oil and gas sectors. Several government and private companies, such as Oil India Limited (OIL), ONGC Videsh Limited (OVL), Indian Oil Corporation Limited (IOC), Gas Authority of India Limited (GAIL), Reliance Industries Limited (RIL), Essar Oil Limited, etc., have been successful in acquiring the overseas oil and gas exploration assets for investment. ONGC Videsh has embarked in significant overseas investments. Currently, it holds interests in thirty-five oil and natural gas projects in fifteen countries (https://ongcindia.com/web/eng/about-ongc/ongc-at-a-glance accessed on 25 September 2022). However, in spite of all this, India needs a huge amount of oil and natural gas to fulfil its energy requirements. For this purpose, India is heavily dependent on imports. In order to meet the demand, most of the imports will have to come from Middle East as well as Africa. At present, India's energy imports mainly come from countries such as Middle East, Africa and Iran. Lone dependence on these regions could cause a threat to India's energy security; India may suffer more as compared to other countries in case of any disruption in energy supplies in these regions (Source - India Energy outlook 2021, World Energy outlook Special Report, International Energy Agency, p. 210, https://iea.blob.core.windows.net/assets/1de6d91e-e23f-4e02-b1fb-51fdd6283b22/India_Energy_Outlook_2021.pdf accessed on 23 September 2022). According to World Energy Outlook 2006, India's dependence on oil imports is expected to increase to 364 million tons by 202425.

Overall, India needs to ensure that its energy supply stays stable and balanced in order to sustain economic growth. Moreover, its rise in status as a global influencer and strong emerging economy will continue with the help of energy security.

6. Conclusion

The modern world faces various issues, including energy security, oil prices, resource depletion, and climate change, all of which directly or indirectly impact the environment. All of these issues have resulted in significant improvements in biomass-derived energy and fuels research and development. As a result, biofuels are likely to be extremely beneficial in addressing such issues in a long-term manner. Biofuels have long been recognized as the most viable solution for lowering carbon dioxide emissions in the transportation industry. Furthermore, biofuels can be simply made from locally available indigenous resources. Algal biofuels have recently received a lot of attention and have been hailed as the most promising method to address the global energy dilemma.

Biofuels are becoming increasingly important as the globe grapples with shifting patterns of energy supply and demand. These fuels have the potential to address local and global concerns such as global warming, carbon sequestration, rising global energy demand, long-term supply insecurity, and fossil fuel repercussions. Biofuel production will aid developing countries by reducing imports and strengthening agricultural economies, resulting in poverty reduction. Many governments are pushing the production and use of biofuels, which converts energy taken from plants to gas, liquid, or oil. Moreover, the reliance on biofuel will certainly help in achieving a greener economy with reduced carbon emission, resource efficiency, and will become bedrock to attain Sustainable Development Goals (SDG 7 and SDG 13).

References

Aamodt, S., and Stensdal, I. (2017). Seizing policy windows: policy influence of climate advocacy coalitions in Brazil, China, and India, 2000–2015. Global Environmental Change, 46: 114–125.

Abdullah, B., Muhammad, S.A.F.A.S., Shokravi, Z., Ismail, S., Kassim, K.A., Mahmood, A.N., and Aziz, M.M.A. (2019). Fourth generation biofuel: A review on risks and mitigation strategies. Renewable and Sustainable Energy Reviews, 107: 37–50.

Alalwan, H.A., Alminshid, A.H., and Aljaafari, H.A. (2019). Promising evolution of biofuel generations. Subject review. Renewable Energy Focus, 28: 127–139.

Alexopoulou, E., Monti, A., Elbersen, H.W., Zegada-Lizarazu, W., Millioni, D., Scordia, D., Zanetti, F., Papazoglou, E.G. and Christou, M. (2018). Switchgrass: from production to end use. pp. 61–105. *In*: Perennial Grasses for Bioenergy and Bioproducts. Academic Press.

Ambaye, T.G., Vaccari, M., Bonilla-Petriciolet, A., Prasad, S., van Hullebusch, E.D., and Rtimi, S. (2021). Emerging technologies for biofuel production: a critical review on recent progress, challenges and perspectives. Journal of Environmental Management, 290: 112627.

Austin, K.G., Jones, J.P.H., and Clark, C.M. (2022). A review of domestic land use change attributable to US biofuel policy. Renewable and Sustainable Energy Reviews, 159: 112181.

Azizi, K., Moraveji, M.K., and Najafabadi, H.A. (2018). A review on bio-fuel production from microalgal biomass by using pyrolysis method. Renewable and Sustainable Energy Reviews, 82: 3046–3059.

Begum, S., and Dahman, Y. (2015). Enhanced biobutanol production using novel clostridial fusants in simultaneous saccharification and fermentation of green renewable agriculture residues. Biofuels, Bioproducts and Biorefining, 9(5): 529–544.

Branyikova, I., Prochazkova, G., Potocar, T., Jezkova, Z., and Branyik, T. (2018). Harvesting of microalgae by flocculation. Fermentation, 4(4): 93.

Chandel, A.K., Garlapati, V.K., Jeevan Kumar, S.P., Hans, M., Singh, A.K., and Kumar, S. (2020). The role of renewable chemicals and biofuels in building a bioeconomy. Biofuels, Bioproducts and Biorefining, 14(4): 830–844.

Chung, C.C., Zhang, Y., Liu, L., Wang, Y., and Wei, Z. (2020). The Evolution of Biodiesel Policies in China over the Period 2000–2019. Processes, 8(8): 948.

Dahman, Y., Dignan, C., Fiayaz, A., and Chaudhry, A. (2019). An introduction to biofuels, foods, livestock, and the environment. pp. 241–276. *In*: Biomass, Biopolymer-based Materials, and Bioenergy. Woodhead Publishing.

Das, S. (2020). The National Policy of biofuels of India–A perspective. Energy Policy, 143: 111595.

Datta, A., Hossain, A., and Roy, S. (2019). An overview on biofuels and their advantages and disadvantages.

de Farias Silva, C.E., Barbera, E., and Bertucco, A. (2019). Biorefinery as a promising approach to promote ethanol industry from microalgae and cyanobacteria. pp. 343–359. *In*: Bioethanol production from food crops. Academic Press.

Deora, P.S., Verma, Y., Muhal, R.A., Goswami, C., and Singh, T. (2022). Biofuels: An alternative to conventional fuel and energy source. Materials Today: Proceedings, 48: 1178–1184.

Eloka-Eboka, A.C., and Maroa, S. (2021). Advanced and sustainable biodiesel fuels: technologies and applications. Sustainable Biofuels, 131–161.

Fatima, T., Xia, E., and Ahad, M. (2019). Oil demand forecasting for China: a fresh evidencefresh evidence from structural time series analysis. Environment, Development and Sustainability, 21(3): 1205–1224.

Field, J.L., Richard, T.L., Smithwick, E.A., Cai, H., Laser, M.S., LeBauer, D.S., Long, S.P., Paustian, K., Qin, Z., Sheehan, J.J., Smith, P. Wang, M.Q. and Lynd, L.R. (2020). Robust paths to net greenhouse gas mitigation and negative emissions via advanced biofuels. Proceedings of the National Academy of Sciences, 117(36): 21968–21977.

Gota, S., Huizenga, C., Peet, K., Medimorec, N., and Bakker, S. (2019). Decarbonising transport to achieve Paris Agreement targets. Energy Efficiency, 12(2): 363–386.

Gurgel, A.C., Paltsev, S., and Breviglieri, G.V. (2019). The impacts of the Brazilian NDC and their contribution to the Paris agreement on climate change. Environment and Development Economics, 24(4): 395–412.

Hirani, A.H., Javed, N., Asif, M., Basu, S.K., and Kumar, A. (2018). A review on first-and second-generation biofuel productions. pp. 141–154. *In*: Biofuels: greenhouse gas mitigation and global warming. Springer, New Delhi.

IEA. 2011. Technology roadmap: biofuels for transport. Paris, France: International Energy Agency.

IEA. 2019. Renewables 2019. Paris. See https://www.iea.org/reports/renewables-2019.

Isikgor, F.H., and Becer, C.R. (2015). Lignocellulosic biomass: a sustainable platform for the production of bio-based chemicals and polymers. Polymer Chemistry, 6(25): 4497–4559.

Javed, M.R., Bilal, M.J., Ashraf, M.U.F., Waqar, A., Mehmood, M.A., Saeed, M., and Nashat, N. (2019). Microalgae as a feedstock for biofuel production: current status and future prospects. Top, 5: 1–39.

Jeswani, H.K., Chilvers, A., and Azapagic, A. (2020). Environmental sustainability of biofuels: a review. Proceedings of the Royal Society A, 476(2243): 20200351.

Kataki, R., Bordoloi, N., Saikia, R., Sut, D., Narzari, R., Gogoi, L., and Chutia, R.S. (2017). An assessment on Indian Government initiatives and policies for the promotion of biofuels implementation, and commercialization through private investments. pp. 489–515. *In:* Sustainable Biofuels Development in India. Springer, Cham.

Kiwjaroun, C., Tubtimdee, C., and Piumsomboon, P. (2009). LCA studies comparing biodiesel synthesized by conventional and supercritical methanol methods. Journal of Cleaner Production, 17(2): 143–153.

Kour, D., Rana, K.L., Yadav, N., Yadav, A.N., Rastegari, A.A., Singh, C., Negi, P., Singh, K., and Saxena, A.K. (2019). Technologies for biofuel production: current development, challenges, and prospects. pp. 1–50. *In*: Prospects of Renewable Bioprocessing in Future Energy Systems. Springer, Cham.

Kumar, A., Kumar, N., Baredar, P., and Shukla, A. (2015). A review on biomass energy resources, potential, conversion and policy in India. Renewable and Sustainable Energy Reviews, 45: 530–539.

Lee, J.Y., Lee, S.E., and Lee, D.W. (2021). Current status and future prospects of biological routes to bio-based products using raw materials, wastes, and residues as renewable resources. Critical Reviews in Environmental Science and Technology, 1–57.

Lee, S.Y., Sankaran, R., Chew, K.W., Tan, C.H., Krishnamoorthy, R., Chu, D.T., and Show, P.L. (2019). Waste to bioenergy: a review on the recent conversion technologies. Bmc Energy, 1(1): 1–22.

Lin, C.Y. (2022). The influences of promising feedstock variability on advanced biofuel production: a review. Journal of Marine Science and Technology, 29(6): 714–730.

Liu, Y., Cruz-Morales, P., Zargar, A., Belcher, M.S., Pang, B., Englund, E., Dan, Q., Yin, K. and Keasling, J.D. (2021). Biofuels for a sustainable future. Cell, 184(6): 1636–1647.

Long, S.P., Marshall-Colon, A., and Zhu, X.G. (2015). Meeting the global food demand of the future by engineering crop photosynthesis and yield potential. Cell, 161(1): 56–66.

Magarey, R.C. (2020). Sugarcane-an old plantation crop that offers new environmentally friendly possibilities. *In:* IOP Conference Series: Earth and Environmental Science (Vol. 418, No. 1, p. 012004). IOP Publishing.

Magda, R., Szlovák, S., and Tóth, J. (2021). The role of using bioalcohol fuels in sustainable development. pp. 133–146. *In*: Bio-Economy and Agri-production. Elsevier. https://doi.org/10.1016/b978-0-12-819774-5.00007-2.

Makut, B.B. (2021). Algal Biofuel: Emergent applications in next-generation biofuel technology. Liquid Biofuels: Fundamentals, Characterization, and Applications, 119–144.

Medipally, S.R., Yusoff, F.M., Banerjee, S., and Shariff, M. (2015). Microalgae as sustainable renewable energy feedstock for biofuel production. BioMed Research International.

Ministry of Petroleum & Natural Gas. (2018). National Biofuel Policy. Controll. Publ. 2018, 1–23.

Narwane, V.S., Yadav, V.S., Raut, R.D., Narkhede, B.E., and Gardas, B.B. (2021). Sustainable development challenges of the biofuel industry in India based on integrated MCDM approach. Renewable Energy, 164: 298–309.

Nazari, M.T., Mazutti, J., Basso, L.G., Colla, L.M., and Brandli, L. (2021). Biofuels and their connections with the sustainable development goals: a bibliometric and systematic review. Environment, Development and Sustainability, 23(8): 11139–11156.

O'Connell, A., Kousoulidou, M., Lonza, L., and Weindorf, W. (2019). Considerations on GHG emissions and energy balances of promising aviation biofuel pathways. Renewable and Sustainable Energy Reviews, 101: 504–515.

Prasad, S., Kumar, S., Sheetal, K.R., and Venkatramanan, V. (2020). Global climate change and biofuels policy: Indian perspectives. pp. 207–226. *In*: Global Climate Change and Environmental Policy. Springer, Singapore.

Rajagopal, D., and Zilberman, D. (2008). Environmental, economic and policy aspects of biofuels. Foundations and Trends® in Microeconomics, 4(5): 353–468

Rather, R.A., and Bhagat, M. (2021). Utilization of aqueous weeds for biofuel production: current status and future prospects. pp. 37–57. *In*: Bioremediation using weeds. Springer, Singapore.

Rebello, S., Anoopkumar, A.N., Aneesh, E.M., Sindhu, R., Binod, P., and Pandey, A. (2020). Sustainability and life cycle assessments of lignocellulosic and algal pretreatments. Bioresource Technology, 301: 122678.

Renewable Energy Policy Network for the 21st Century (REN21). 2019Renewables 2019—Global status report. See https://www.ren21.net/wp-content/uploads/2019/05/gsr_2019_full_report_en.pdf.

Ribeiro, R.L.L., Vargas, J.V.C., Mariano, A.B., and Ordonez, J.C. (2017). The experimental validation of a large-scale compact tubular microalgae photobioreactor model. International Journal of Energy Research, 41(14): 2221–2235.

Rudra, S., and Jayathilake, M. (2021). Hydrothermal Liquefaction of Biomass for Biofuel Production.

Saha, S., Sharma, A., Purkayastha, S., Pandey, K., and Dhingra, S. (2019). 14-Bio-plastics and biofuel: is it the way in future development for end users? Plastics to energy, plastics design library.

Salama, E.S., Kurade, M.B., Abou-Shanab, R.A., El-Dalatony, M.M., Yang, I.S., Min, B., and Jeon, B.H. (2017). Recent progress in microalgal biomass production coupled with wastewater treatment for biofuel generation. Renewable and Sustainable Energy Reviews, 79: 1189–1211.

Saravanan, A.P., Mathimani, T., Deviram, G., Rajendran, K., and Pugazhendhi, A. (2018). Biofuel policy in India: a review of policy barriers in sustainable marketing of biofuel. Journal of Cleaner Production, 193: 734–747.

Sikarwar, V.S., Zhao, M., Fennell, P.S., Shah, N., and Anthony, E.J. (2017). Progress in biofuel production from gasification. Progress in Energy and Combustion Science, 61: 189–248.

Singh, A.R., Singh, S.K., and Jain, S. (2022). A review on bioenergy and biofuel production. Materials Today: Proceedings, 49: 510–516.

Singh, G., and Patidar, S.K. (2018). Microalgae harvesting techniques: A review. Journal of Environmental Management, 217: 499–508.

Singh, R., Prakash, A., Balagurumurthy, B., and Bhaskar, T. (2015). Hydrothermal liquefaction of biomass. pp. 269–291. *In*: Recent Advances in Thermo-chemical Conversion of Biomass. Elsevier.

Suganya, T., Varman, M., Masjuki, H.H., and Renganathan, S. (2016). Macroalgae and microalgae as a potential source for commercial applications along with biofuels production: a biorefinery approach. Renewable and Sustainable Energy Reviews, 55: 909–941.

Susmozas, A., Martín-Sampedro, R., Ibarra, D., Eugenio, M.E., Iglesias, R., Manzanares, P., and Moreno, A.D. (2020). Process strategies for the transition of 1G to advanced bioethanol production. Processes, 8(10): 1310.

Teixeira, A.C.R., Machado, P.G., Borges, R.R., and Mouette, D. (2020). Public policies to implement alternative fuels in the road transport sector. Transport Policy, 99: 345–361.

Thanigaivel, S., Priya, A.K., Dutta, K., Rajendran, S., and Vasseghian, Y. (2022). Engineering strategies and opportunities of next generation biofuel from microalgae: A perspective review on the potential bioenergy feedstock. Fuel, 312: 122827.

Tilman, D., Hill, J., and Lehman, C. (2006). Carbon-negative biofuels from low-input high-diversity grassland biomass. Science, 314(5805): 1598–1600.

Tyner, W.E. (2020, January). Biofuel economics and policy: The Renewable Fuel Standard in 2018. pp. 695–704. *In*: Bioenergy. Academic Press.

Updaw, N.J., and Nichols, T.E. (2019, December). Pricing soybeans on the basis of chemical constituents. pp. 781–799. *In*: World Soybean Research Conference II: Proceedings. CRC Press.

Vassilev, S.V., and Vassileva, C.G. (2016). Composition, properties and challenges of algae biomass for biofuel application: An overview. Fuel, 181: 1–33.

Venkatramanan, V., Shah, S., and Prasad, R. (2020). Global Climate Change and Environmental Policy. Springer Singapore.

Yamakawa, C.K., Qin, F., and Mussatto, S.I. (2018). Advances and opportunities in biomass conversion technologies and biorefineries for the development of a bio-based economy. Biomass and Bioenergy, 119: 54–60.

Zeng, X., Guo, X., Su, G., Danquah, M.K., Chen, X.D., Lin, L., and Lu, Y. (2016). Harvesting of microalgal biomass. pp. 77–89. *In:* Algae biotechnology. Springer, Cham.

Zhang, D., Wang, J., Lin, Y., Si, Y., Huang, C., Yang, J., Huang, B. and Li, W. (2017). Present situation and future prospect of renewable energy in China. Renewable and Sustainable Energy Reviews, 76: 865–871.

2

Sustainable Development Goals (SDGs-7) for Bioeconomy with Bioenergy Sector

Richa Kothari,[1,*] *Kajol Goria,*[1] *Anu Bharti,*[1] *Har Mohan Singh,*[2] *Vinayak V. Pathak,*[3] *Ashish Pathak*[4,*] and *V.V. Tyagi*[2,*]

1. Introduction

Slow economic growth, social inequality, and environmental degradation are the main issues of the current global context. The current scenario of energy consumption and production has the potential to change the global climate patterns because fossil fuel-based energy production contributes to a large portion of greenhouse gas emissions. Climate change is not a regional problem; it is considered a global problem that has continuously shifting the global climate (Gernaat et al. 2021, Singh et al. 2019). In the current scenario, demand for energy is one of the basic requirements for all individuals living on this planet. Meanwhile, the world is facing enormous challenges in economical as well as environmental aspects due to energy dependency on traditionally available fossil energy options (Lin

[1] Department of Environmental Sciences, Central University of Jammu, Rahya Suchani, (Bagla) Samba, India.
[2] School of Energy Management, Shri Mata Vaishno Devi University, Katra, India.
[3] Department of Chemistry, Manav Rachna University, Faridabad, Haryana, India.
[4] Petroleum Research Centre, Kuwait Institute for Scientific Research, P.O.Box 24885, Safat-13109, Kuwait.
* Corresponding authors: kothariricha21@gmail.com; apathak@kisr.edu.kw; vtyagi16@gmail.com

and Zhu 2019). These energy resources are intermittently deteriorating the quality of the environment by emitting harmful carbon emissions.

Clean and green energy technology could be seen as a possible solution in tackling such challenges as it deals with energy options bearing no or zero-carbon emission (Ahmed et al. 2022). Clean energy often refers to energy derived from renewable and non-polluting resources bearing excellent energy efficiency, affordability, reliability, and easier accessibility. These resources play a crucial role in fulfilling the basic energy requirements throughout the globe. Investments in naturally available resources like solar, wind, and thermal power for clean energy extraction could provide a better energy alternative to fossil fuels (Batini et al. 2022). The energy harnessed from the sun by means of solar power collectors like photovoltaic cells and solar thermal collectors refers to "solar energy". The energy derived from wind by means of windmills and wind turbines is termed as "wind energy". Another green and clean energy option is the thermal energy derived by running turbines with the help of steam engines (Sinha et al. 2019). For the prevention of climate change impacts, the utilization of renewable energy is consistently increasing because renewable energy sources are abundantly available on the earth.

In 2016, countries like China (4%) and the United States (5%) used biomass for energy production, whereas Germany (11%), India (21%), and Brazil (33%). It is estimated that the global biomass energy potential of energy crops will be 96 EJ in 2050 (Destek et al. 2021). Wood, municipal waste, agriculture waste, and forest waste are the main sources of bioenergy. The biopower production potential of sugarcane bagasse in India and Brazil is high; in India, the annual biopower from bagasse is 1.93 GW (Kothari et al. 2020). Bioenergy energy technologies are promoted by various international organizations in developed and developing countries, by which energy security can be ensured to each person.

The article is an attempt to understand the bioenergy economy, sustainable development goals, sustainable development goal 7, green and clean bioenergy options, indicators of sustainable development, and the challenges and opportunities presented by a bioenergy economy.

2. Bioenergy economy

The bioenergy economy generally refers to an economy based on bioenergy and bio-waste products derived from naturally available and the transformation of renewable resources. A bioeconomy displays great potential in meeting the necessities of an increasing population. Technology advancement, economic viability, availability of infrastructure, and superior quality biomass serve as the key indicators for boosting the bioeconomy. In the present-day fossil fuels-based economy, environmental threats such as climate change, global warming, the greenhouse effect, and many other similar issues are at a height (Kojima 2019). A huge quantity of carbon emissions in the atmosphere are considered to be a major cause of such issues. These emissions need to be cut down. A bioenergy- based economy also called green economy has been acknowledged as a potential approach for promoting development deprived of harmful carbon emission (Siwal et al. 2021). Long term goals have been set up to achieve the development of a zero or low carbon economy by the year 2050 (Deutch 2020). Bioenergy in this regard could play a crucial role in shifting towards the bioeconomy and influencing zero or low carbon emissions.

Rapidly growing research interest in strengthening the green energy economy has resulted in its incorporation into policies at the global level. The foremost driving force behind such interest could be the advantages of the green economy in tackling the environmental as well as financial crises the world is facing. Subsequently, several extensive investments are encouraged for developing green technology, clean or green goods and services, promoting renewable energy, materials recycling, efficient waste management, and green infrastructure.

A green energy economy mainly comprises four sectors, namely natural or renewable energy sources, a green building and energy-efficient approach, energy-efficient transport options, and recycling and waste-to-energy transformation (Chapple 2008). The green economy is not just promoting clean energy production, but also encouraging technologies that facilitate clean industrial processes, and the growth of markets for products consuming very little energy. Hence, it could be considered that the bioenergy economy delivers products, processes, and services that have a reduced impact on the environment or are capable of improving natural resource utilization. Currently, bioenergy-leading nations such as member countries of the European Union reported bioeconomy market estimations of around 2.4 billion euros (Scarlat et al. 2015). Estimates made by the International Energy Agency (IEA) showed that around 17% of the world's energy demand would be fulfilled by bioenergy by 2060 (Lago et al. 2018).

In the present scenario, many aspects of the bioenergy economy are attracting controversy and debate. Although, with the rapid development of the bioeconomy, new products and services are continually emerging. For instance, renewable sources of energy contribute around a one-fifth share in generation of global power. The bio-based economy has been recognized for developing clean energy technologies, reducing energy dependency on nonrenewable sources, ensuring climate change mitigation and adaptation, employment creation, and maintenance of natural resources sustainability (Ladu and Quitzow 2017). However, the approach also presents some drawbacks in terms of food security. Energy crops production on land unfit for agricultural activities could be considered as the optimum solution to reduce an adverse impact on food security. Thus, it's quite obvious to infer how the bioenergy economy opens up a number of exciting new doors, including possibilities for technological advancement, economic growth, improved ecological sustainability, and more employment.

3. Sustainable development goals (SDGs)

Sustainable development goals (SDGs) are generally referred to as a set of goals further divided into small targets designed to attain sustainable development. These are also called as "Global Goals" and came into existence in 2015 in an agreement created under the aegis of the United Nations (UN). These are comprised of a non-legally binding set of global goals (seventeen in number) aimed at achieving around 169 targets (Cernev and Fenner 2020). The United Nations launched SDGs along with the 2030 Agenda for Sustainable Development to realize sustainable development in the mainstream by the end of 2030.

SDGs calls for collaborative actions worldwide at all fronts, whether it may be public or private organizations. Every institution and individual on this earth has a key role to play in the success of SDGs. The key motivation behind the formation of SDGs is poverty eradication, endorsing environmental protection, ensuring fair and equitable health benefits to both local communities as well as the planetary biosphere, and certifying global peace and prosperity. Figure 1 illustrates seventeen SDGs. These goals have been linked with five different aspects of extreme importance: People, Prosperity, Planet, Peace, and Partnerships (Morton et al. 2017). According to Morton et al. (2017) Goal 1 to Goal 6 correspond to the first aspect called People, Goal 7 to Goal 12 are concerned with Prosperity, Goal 13 to Goal 15 correspond to Peace, and Goal 16 and Goal 17 deal with Peace and Partnerships, respectively.

3.1 Need of sustainable development goals

As far as sustainable development is concerned, it generally deals with harmonizing societal inclusion and economic progress along with consideration of environmental protection. The current scenario calls for accomplishment of a handful of objectives related to SDGs in order to attain total sustainable development. Many of reasons might lie behind the necessity of fulfilling these goals. But highly crucial causes motivating the need of SDGs being accelerating population inclusion, scarcity and depletion of natural resources attributing to over utilization of natural resources, climate change by the aid of increasing global atmospheric

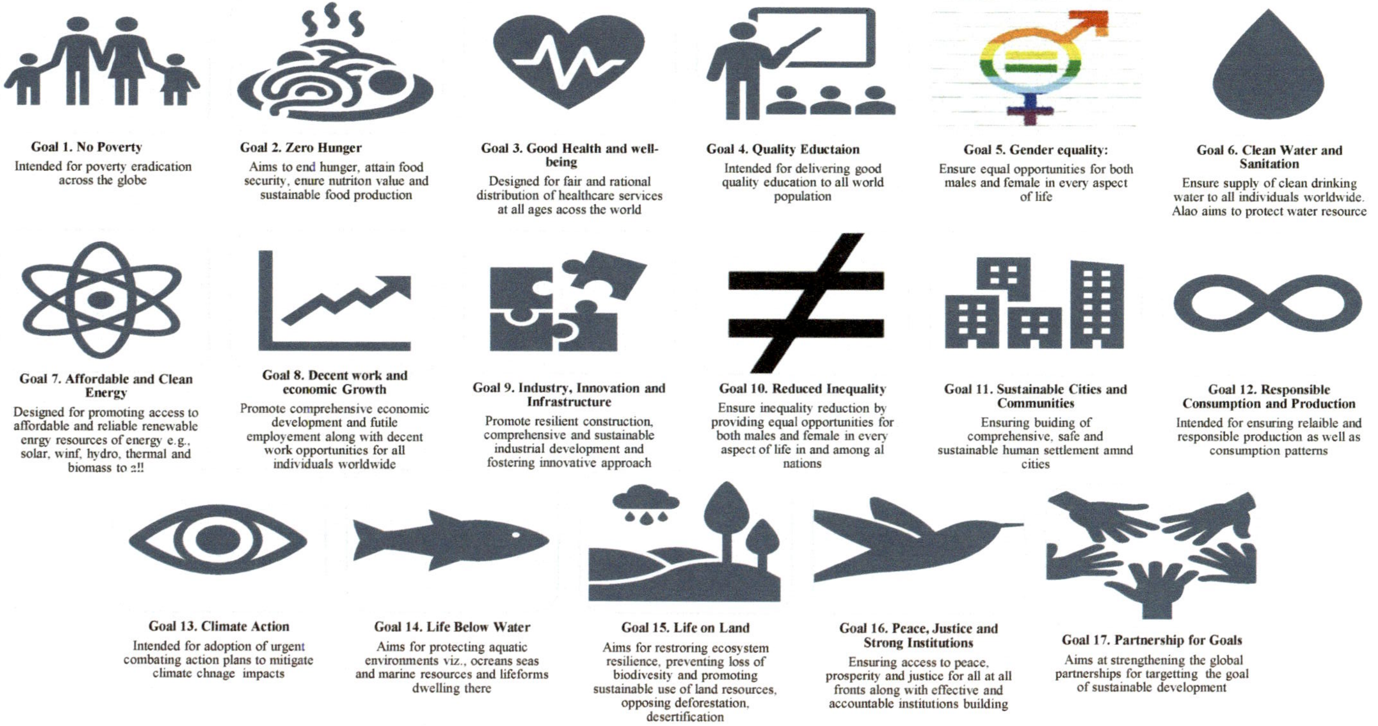
Goal 1. No Poverty
Intended for poverty eradication across the globe
Goal 2. Zero Hunger
Aims to end hunger, attain food security, enure nutriton value and sustainable food production
Goal 3. Good Health and well-being
Designed for fair and rational distribution of healthcare services at all ages acoss the world
Goal 4. Quality Eductaion
Intended for delivering good quality education to all world population
Goal 5. Gender equality:
Ensure equal opportunities for both males and female in every aspect of life
Goal 6. Clean Water and Sanitation
Ensure supply of clean drinking water to all individuals worldwide. Alao aims to protect water resource
Goal 7. Affordable and Clean Energy
Designed for promoting access to affordable and reliable renewable enrgy resources of energy e.g., solar, winf, hydro, thermal and biomass to all
Goal 8. Decent work and economic Growth
Promote comprehensive economic development and futile employement along with decent work opportunities for all individuals worldwide
Goal 9. Industry, Innovation and Infrastructure
Promote resilient construction, comprehensive and sustainable industrial development and fostering innovative approach
Goal 10. Reduced Inequality
Ensure inequality reduction by providing equal opportunities for both males and female in every aspect of life in and among al nations
Goal 11. Sustainable Cities and Communities
Ensuring buiding of comprehensive, safe and sustainable human settlement amnd cities
Goal 12. Responsible Consumption and Production
Intended for ensuring relaible and responsible production as well as consumption patterns
Goal 13. Climate Action
Intended for adoption of urgent combating action plans to mitigate climate chnage impacts
Goal 14. Life Below Water
Aims for protecting aquatic environments viz., ocreans seas and marine resources and lifeforms dwelling there
Goal 15. Life on Land
Aims for restroring ecosystem resilience, preventing loss of biodivesity and promoting sustainable use of land resources, opposing deforestation, desertification
Goal 16. Peace, Justice and Strong Institutions
Ensuring access to peace, prosperity and justice for all at all fronts along with effective and accountable institutions building
Goal 17. Partnership for Goals
Aims at strengthening the global partnerships for targetting the goal of sustainable development

Figure 1. Illustration of Sustainable Development Goals put forward by the United Nations (*Source: https://www.un.org/sustainabledevelopment/*).

temperature leading to extreme weather events and environmental degradation because of environmental pollution (Omisore 2018). Providing an appropriate solution to such issues is the need of the hour. Scarcity of resources might trouble the entire population of the globe in different aspects of food, cloth, shelter and other basic needs of livelihood. In other words, such resource scarcity would induce poverty at greater extent. Studies have reported that by the end of 2025, around two-thirds of the global population would be living in areas exhibiting water-scarce or water-stressed state (Buccola 2020). Also, with the growing population and technological advancement, a huge share of the world's productive land for agriculture would become deteriorated. Food production also needed to be doubled by the year 2050 to meet the food requirements of the growing population (Pastor et al. 2019). Threats like climate change also emphasize the necessity of SDGs to attain sustainable development and reduce these threats' adverse global impact (Zhenmin and Espinosa 2019). Sea-level rise due to climate change is thought to propose an extermination threat to small island nations. Taking into consideration of all economic, societal as well as environmental facets, SDGs are absolutely essential requirement for attaining sustainable development.

3.2 Importance of sustainable development goals

SDGs are acknowledged to be highly important for realizing a sustainable and balanced livelihood, economic growth, and natural harmony. SDGs play a key role in delivering appropriate technological advancement in core areas of sustainable development, i.e., society, economy, and environment (Berawi 2019), that is, SDGs are accredited with endorsing social sustainability, economic sustainability, and environmental sustainability. Social or societal sustainability challenges gender discrimination and encourages gender equality. This might promote the growth of quality livelihood by encouraging the development of local communities (Segerstedt and Abrahamsson 2019). Moreover, fair distribution of quality education and healthcare services might be achieved globally through societal sustainability. Similarly, economic sustainability aims for financial equality without causing any harm to the natural environment. It mainly focuses on equality in economic growth by utilizing natural resources judiciously and providing wealth for all individuals across the globe. Poverty eradication can be achieved by attaining economic sustainability (Roy et al. 2018). Environmental sustainability helps in meeting the current generation needs but ensuring environmental protection also (Chowdhury et al. 2021). This might be achieved in terms of conservation of the natural environment by prohibiting deforestation, countering environmental pollution, promoting ventures engaged with fostering renewable sources of energy, protecting and conserving water resources, encouraging clean and sustainable mobility, and promoting the construction of sustainable structures or sustainable architecture.

4. Sustainable development goals 7

To end poverty and hunger, realize the human rights of all, achieve gender equality and empowerment of women, and ensure the protection of the planet and natural resources, the members of the United Nations pushed the 2030 Agenda in which three core dimensions: human development, sustainable economic growth, and environmental sustainability. Sustainable Development Goal 7 (SDG 7) focuses on ensuring access to affordable, reliable, sustainable, and modern energy services for all by 2030. The provision of access to affordable, sustainable energy to all is not easy because 60% of the global emissions come from energy production, which is the main contributor to climate change (Eras-Almeida et al. 2020, Tucho et al. 2020).

4.1 Target and Indicators of SDG 7

The main target of SDG 7 is to complete by 2030 and the targets are associated with the conditions of countries and contexts, related to policies, funding, location, culture, societal implications, and much more (Eras-Almeida et al. 2020). There are five main targets of SDG 7, which are listed as https://sdgs.un.org/goals/goal7.

- Ensure universal access to affordable, reliable, and modern services.
- Increase the share of renewable energy in the global energy mix.
- Twice the global rate of improvement in energy efficiency.
- Promote international cooperation among counties to access clean energy research and technology in energy efficiency, renewable energy, and cleaner fossil-fuel technology as well as promote investment in the development of clean infrastructure and technology.
- Expansion of infrastructure and development of technology for the supply of modern and sustainable energy services in developing and least developing countries, which are land locked, supporting their respective programs.

SDG 7 is closely associated with Agenda 2030 of sustainable development and with the Paris Agreement on Climate Change. Besides clean energy accessibility, SDG 7 also encourage women empowerment, employment for women and youth, better health and education, many other benefits for equitable and inclusive communities, and protection against climate change. There are some actions are suggested by the High-

Level Political Forum on Sustainable Development (2018), which need to be accomplished on an urgent basis to achieve SDG 7 as:

- Currently, 1 billion people live without electricity: The portion of the global population with access to electricity increased from 78 to 87% from 2000 to 2006. The current growth rate of electricity accessibility means that electricity will not be accessible to 674 million people by 2030 at the global level.
- About 3 billion people do not have access to clean cooking solutions: The improvement of clean cooking solutions has gradually increased but 3 billion people do not have to access clean cooking and the clean fuel access rate for cooking is low. It is expected that around 2.3 billion people will still not be able to access clean fuel by 2030 under current policy and population trends.
- Although the expansion of renewable power production is increasing, more effort will be needed to reach end-use: Support from the cost-competition policies are making it possible to lower the costs of solar and wind energy in comparison to conventional power generation. Thus, the incorporation of renewable power at the global level is increasing. However, the continued consumption of renewable power (which was 17.3% in 2014 and is reached 17.5% in 2015) reflects that the consumption of global renewable power has slightly increased. Thus, there is an urgent need to increase the consumption of renewable power in thermal and transport applications.

The improvement in the efficiency of renewable energy has been accelerating throughout the globe: The global primary energy intensity fell at 2.8% during 2014–2015, which was the fastest decline since 2010. The energy intensity average rate was 2.2% during 2010–2015 but the demand for average energy intensity is currently at 2.7% for the period of 2016-2030. Therefore, an ambitious policy road map is needed at the global level.

4.2 Sustainable development goals and bioenergy

Bioenergy is a key role player of renewable energy, which is in support of the UN-SDGs with regard to climate change and energy security. The IPCC 5th Assessment reports noted that completion of long-term targets of climate change without bioenergy can be high risk. Renewable Energy Policy Network for the 21st Century (REN 21), International Energy Agency (IEA), International Renewable Energy Agency (IRENA) assessed bioenergy at a global level and suggested that bioenergy is three-quarters of all the renewable energy used today and has the potential to count for half of the most cost-effective options for doubling renewable energy use by 2030.

Bioenergy is the main aspect of the bioeconomy and includes agriculture, forestry, and manufacturing. Not only SDGs but bioenergy can also play an important role in implementing the Paris Agreement on Climate Change in which it is mentioned that the "action to conserve and enhance ... sinks and reservoirs of greenhouse gases" and "reducing emission from deforestation and forest degradation" (Bioenergy for Sustainable Development 2017). SDG 2, SDG 13, and SDG 5 are concerned with directly promoting sustainable bioeconomy, that is, combating the climate change implications listed below:

- SDG-2: end hunger, achieve food security and improved nutrition and promote sustainable agriculture.
- SDG-13: take urgent action to combat climate change and its impacts
- SDG-15: protect, restore, and promote the sustainable use of the terrestrial ecosystem, sustainably manage forests, combat desertification, halt and reverse land degradation and halt biodiversity loss.

5. Green and clean bioenergy options

Bioenergy makes up approximately 70% of major renewable energy forms worldwide (International Energy Agency 2016). However, it becomes quite important to consider the usage as well as the production of bioenergy. Presently, its usage is restricted to conventional and domestic ways such as fuelwood, charcoal, or animal waste for cooking as well as heating purposes. This type of usage of biomass has low efficiency and sustainability because the large-scale collection of fuelwood and generation of charcoal can cause significant deforestation. Moreover, utilizing bioenergy in conventional cooking systems is also ineffective, and the incomplete combustion of the fuels can lead to significant emissions (especially particulates) with serious health implications (Sparrevik et al. 2015). One of the most valuable resources obtained through waste valorisation is energy from biowaste (Figure 2). Globally, a population of approximately 2.5 billion, depends upon the conventional use of biomass and out of that approximately 1.3 million people, particularly females and children are subjected to premature deaths annually, due to indoor pollution as well as respiratory problems related to conventional biomass use (Shahsavari et al. 2018).

Modern bioenergy patterns globally, however, show great potential in ensuring sustainability and also being low in carbon energy, paving a way for achieving some of the key SDGs (Hassan et al. 2019). The United

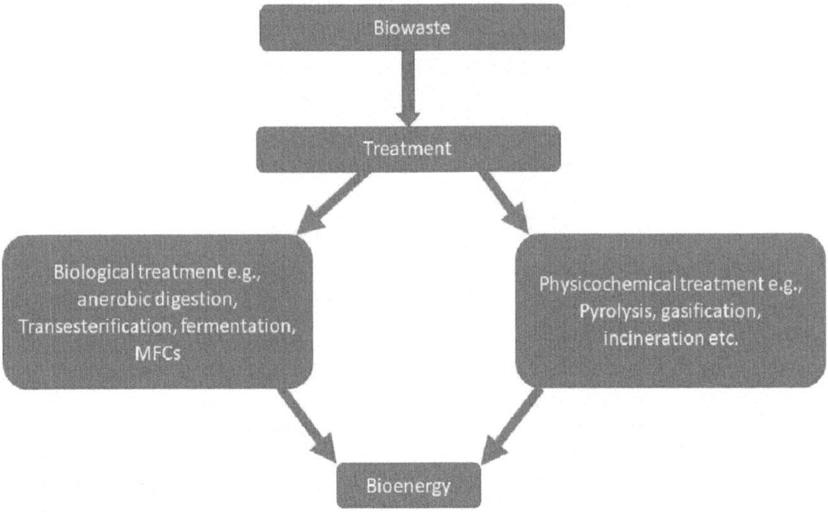

Figure 2. Biowaste to bioenergy.

Nations Environment Programme (UNEP) suggests that anywhere biomass sources exist has the potential for energy production, and policies and strategies need to be framed so that bioenergy pathways can increase development and eliminate problems such as poverty along with safeguarding the ecosystems (Beyene et al. 2015). If implemented in the right way, bioenergy can ensure that urban as well as rural communities have access to renewable sources, which will help in reducing greenhouse gas (GHGs) emissions; provide economic support, ensure infrastructure development, give efficient training, and generate ample employment opportunities and revenue pathways for people who directly or indirectly depend upon the land and agriculture (Owusu et al. 2016).

Presently, several modern bioenergy technologies are capable of generating energy that can replace conventional fossil fuels and decrease carbon emissions, and finally, provide support to national as well as global targets of mitigation. Hence, the stress should be upon the modern utilization as well as application of biomass and bioenergy (Popp et al. 2014).

Bioenergy is an attractive energy option for all developed and developing economies, because of its great flexibility and potential for merging into a wider range of energy systems (Mathiesen et al. 2015). However, compared to other renewable energies like wind, solar, tidal, etc., its flexibility is less, hence making this one of the biggest drawbacks of bioenergy. The versatility of bioenergy can be seen by the fact that it is capable of generating electricity, transporting fuels, and energy carriers from a vast range of biomass feedstock utilizing various bioenergy treatments and conversion technologies. It is also convenient to store bioenergy in the form of pallets, feedstock, or energy carriers which can further be transformed when the need arises. Hence, bioenergy can be regarded as a potent energy option that is renewable and can provide important baseload energy in comparison to other renewable sources. Bioenergy systems can play a pivotal role in complementing the already present energy infrastructures, technologies, as well as applications (Gielen et al. 2019). For instance, if the wood biomass is used in form of wood pellets, it can be more convenient as 5 to 15% of pellets can be co-fired in coal power stations, that too without the need for converting the existing boilers (Nunes et al. 2014). Another example of a bioenergy system utilizing the pre-existing energy infrastructure and technologies is blending bioethanol and/or biodiesel with transport fuels. Biofuels presently is being used in a blended form with those of conventional fossil fuels, leading to reduced carbon intensity (Bergthorson et al. 2015). Hence, bioenergy can be seen as a source of potential alternative for developed as well as developing countries and its utilization can prove to be a striving step towards the achievement of SDGs.

Bioenergy can provide several other important services other than energy. For instance, organic, wood-based biomass waste materials can be used as bioenergy feedstocks and as such, bioenergy can be seen as a potential waste management option. For example, anaerobic digestion can be utilized for the treatment and management of livestock manure, slurry, and other organic waste (Linville et al. 2015). The utilization of residues from agriculture and forests for the generation of bioenergy is again one of the common examples used in the management of waste because usually, the residues are subjected to burning in the fields, which may further cause significant emission of CO_2 and other harmful particles (Bhuvaneshwari et al. 2019). Thus, on utilizing of residues for energy purposes, the direct emissions coming out of burning processes in the field can be reduced to a large extent along with replacing the typical fossil fuels. Therefore, organic wastes, as well as residues from agriculture and forests are very important and potential bioenergy feedstock, that can offer energy and also can allow for waste management simultaneously (Fava et al. 2015). Several researchers have highlighted in their studies that bioenergy can play a crucial role in mitigating climate change, waste management, and facilitating the possibility of turning the climate change goals into a reality (Clarke et al. 2014, Klein et al. 2014). However, more intensive research is still required to make these studies to reality in a more efficient way. Therefore, future research on bioenergy should not only stress upon decreasing the environmental impact and costs, but also the development of policies meant for the goodwill of people.

5.1 Types of bioenergy

Bioenergy has enormous potential in that it can boost the energy supply in developed as well as developing economies such as Brazil, India, China, which have never-ending energy demands (Sharma et al. 2017). Biomass can be directly subjected to burning for the generation of power, or it can also be converted into other substitutes such as oil or gas substitutes. Hence, biomass can be considered to have an enormous amount of energy stored in them which when released can be utilized for several purposes such as the production of renewable electricity or biofuels. Typically, types of bioenergy can be categorized into two main types, i.e., traditional or conventional bioenergy coming majorly from biomass sources that are solid forms and advanced bioenergy that includes conversion of biomass through some conversion technologies into biofuels or bioelectricity. The main categorization of bioenergy has been depicted in the flowchart given in Figure 3.

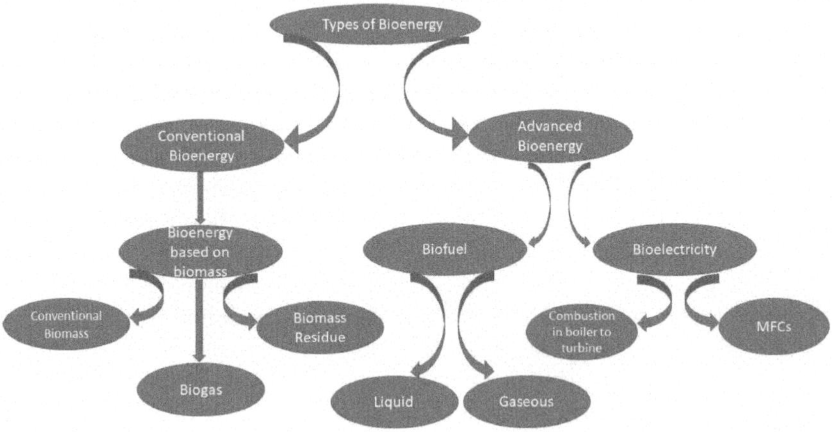

Figure 3. Types of bioenergy.

5.1.1 Conventional bioenergy

Conventional bioenergy includes the domain of bioenergy that is purely based on biomass. Biofuels have entered the market form so long and they are easily available commercially; although several modifications in feedstock generation and in processing is required that can further cut the costs and offer better environmental performance (Sims et al. 2010). Typical conventional bioenergy includes conventional biomass, biogas, and biomass residue. It comprises processes such as incineration of biomass to produce electricity; however, biomass incineration produces electricity by making use of a similar process like that included in coal combustion, and in place of coal, it includes biomass burning of wood. Just like coal combustion, this method is highly polluting (Adami et al. 2020). Hence, advanced technology is highly required.

Another important type of bioenergy includes biogas as biofuel. Biogas is basically the emissions in gaseous form when the organic matter of plant or animal origin is subjected to anaerobic degradation in the presence of a consortium of bacteria. Biogas is generally formed out of wastes but can also be produced using biomass energy feedstock which is a quite efficient process (Bharathiraja et al. 2016). It is a mixture of gases containing methane, carbon dioxide, and other trace gases. Methane gas is the major constituent of natural gas (98%), responsible for making up approximately 55–90% of biogas, according to the origin of organic matter along with conditions for degradation. Biogas can be produced in almost all natural environments possessing the necessary oxygen level and containing a potential degradable organic matter. The natural sources of biogas production involve aquatic sediments, wet soils, buried organic matter, etc. However, human innovation has created some extra sources such as landfills, lagoons, and waste storage infrastructures. However, it is a well-known fact that emissions of biogas from both natural as well as man-made sources have added to the problem of climate change because methane is a major greenhouse gas. Present technology allows the recuperation of biogas by using sealed containers, making biogas easily accessible to be used as a fuel for purposes such as heating, electricity, and other uses. However, the amount of biogas used for transportation is relatively small at present.

Other important types of biomass include biomass residue which can be further categorized into agricultural biomass residue, animal residue, forest residue, and municipal residue. The agricultural biomass residue includes residues from agricultural crops such as stalks, branches, leaves, and also from agricultural byproducts. These can be utilized for energy generation as well (Simonyan and Fasina 2013). Animal waste residue primarily includes waste from poultry and cattle farms and slaughterhouses (Abdeshahian et al. 2016). Forest biomass residues include firewood and forestry residues that can be used for energy purposes (Brašanacet al. 2018). Municipal waste residue includes that fraction of municipal waste that is biodegradable in nature.

5.1.2 Advanced bioenergy

Advanced bioenergy is further categorized as advanced biofuels and bioelectricity. Advanced biofuels basically utilize the pre-commercial methodologies by making use of agricultural as well as forest residues (Ebadian et al. 2020). These residues consist of three main building materials that is, cellulose, hemicellulose, and lignin (Nazari et al. 2021).

Advanced biofuels are sometimes subjected to blending with petro-fuels, which is further made to combust in pre-existing combustion engines, and then either distributed via pre-existing infrastructure or used in slightly modified cars powered by internal combustion engines, such as DME vehicles (Bhattacharya and Datta 2018). Advanced biofuels too can be generated from waste materials such as wheat and corn stalk, wood, and energy crops. Several advanced biofuels are still being researched including ethanol, Fischer-Tropsch diesel, mixed alcohols, and wood diesel (Jenčík et al. 2021). Many of them are in their early developmental stages and can also involve algal and hydrogen-based biofuels out of biomass.

6. Indicators of sustainable bioenergy

It is important to develop indicators for quantification of advantages and expanses involved in bioenergy utilization. A number of researchers have developed indicators for environmental sustainability; however, a few researchers have explored indicators for sustainable bioenergy. In order to develop indicators for sustainable bioenergy, it is necessary to understand the biomass supply chain, transportation stages, processing of by-products, and characteristics of feedstock and its abundance. Biomass supply chain involves procurement, post-production processing, and conversion of biomass to an end product. The bioenergy system is expected to expand in the coming years as it offers various advantages such as it provides localized resources for energy generation, generates rural employment, and is flexible in that it can be used in various energy pathways to generate desired energy products. Although bioenergy offers various advantages over other energy resources but its sustainability is not well established among various stakeholders. Various environmental problems have been found to linked with bioenergy production such as depletion of soil nutrients, loss of biodiversity, emission of harmful gases due to traditional bioenergy applications, and loss of soil productivity (Jordan et al. 2007, Kenney 2008). These disagreements are necessary to develop sustainable bioenergy practices with the support from government policies and regulations (Kline et al. 2009). Various indicators can be observed to evaluate sustainable bioenergy production, which is discussed under following subsections:

6.1 Water consumption

Water use in bioenergy can be explained as a volumetric assessment of water required to produce bioenergy products. Quantification involves the volume of water abstracted, consumed, or altered. Withdrawal of water from various sources to meet the demand of biorefineries, and water consumption in irrigation of energy crops and in the cultivation of aquatic biomass can be used as an indicator. In biorefineries, the large volume of water is consumed in cooling towers and dryers during the distillation process (Wu et al. 2009). Furthermore, water consumption can be categorized in various categories such as blue and green water consumption. Bluewater consumption involves the withdrawal of water from surface water bodies whereas green water consumption involves water evaporated as a part of evapotranspiration during the cultivation of crops.

6.2 Air quality

Conversion of feedstock to bioenergy products involves the direct or indirect generation of gaseous pollutants. Feedstock production and its further processing for the generation of bioenergy products involves the indirect emission of gaseous pollutants. For measurement of air quality parameters, such as oxides of carbon, suspended particulate matter, ground-level ozone, oxides of nitrogen can be considered as the key indicator. Oxides of carbon are released in almost all activities that involve the combustion process. For example, the transportation of biomass feedstock involves the combustion of fossil fuels. On the other hand, tropospheric ozone is formed due to the photochemical reaction of oxides of nitrogen and oxygen. Oxides of nitrogen are also emitted in almost all combustion activities during biofuel production or traditional biomass application (Seinfeld and Pankow 2003).

6.3 Ecosystem productivity

It is essential to assess the primary productivity of the ecosystem with regard to feedstock cultivation. The basis of this indicator is the net flux of carbon from the atmosphere to its fixation in biomass. This indicator can be measured prior and after feedstock cultivation for evaluation of the ecosystem's productivity. Such evaluation is important to determine the nutrient level in soil, the status of soil depletion, and various other ecological parameters (Prince et al. 2001).

6.4 Biodiversity

Biodiversity indicators are useful to measure the sustainability of the agroforestry system. Bioenergy production is likely to affect biodiversity in different ways such as monoculture of energy crops, pollution from biorefineries, cultivation of invasive species, etc.; such activities can cause a significant loss of species and accelerate habitat fragmentation, degradation, or loss. On the other hand, managed perennial cropping system can improve the habitat of specific species. In general, the influence of bioenergy production on biodiversity depends on species selection, land-use change, community structure, and various other factors (Barney et al. 2008).

7. Challenges and opportunities of bioenergy economy

In order to meet the ever-increasing food, energy, as well as fuel requirements, the world's economies are swiftly preparing to move towards some strategical policies meant to transform them from a food economy to a bioenergy economy (Yadav et al. 2021). Fundamental policy considerations in the transition from a food economy to one based on bioenergy will immediately trigger a trade-off between the use of natural resources for food security, renewable energy, waste management, and biobased manufacturing due to the inevitable competition for scarce resources like land, labour, money, and water (Arndt et al. 2012). Any step toward becoming an effective bioenergy economic system needs a permanent coordinated set of policies and strategies, efficient institutional infrastructure, and human potentials in order to address this tradeoff (Hartley et al. 2018). Hence, to move towards the growth of the nation in this sector, a national development strategy is required, for aligning the bioenergy strategies with that of food security, hunger and poverty reduction, development of economy, and conservation (Resnick et al. 2012). Further, sector-wise policies engaging with energy, agricultural and forest management, renewable resources, eco-friendly technology, rural development need to be cognizant of these issues while designing the strategies for the bioenergy sector (Debnath et al. 2018). However, factors such as lack of research, institutional incompetence, and an inability to implement this harmonized set of policies make the developing economies less prepared to address the ever-growing need of becoming a bioenergy economy (Paloviita et al. 2017). Besides, there is an urgent need to identify the particular role of the bioenergy sector as well as biofuels for this larger transformation process. The bioenergy sector can flourish more if its applications and practicality are well known and studied. The dream of transforming the systems based on conventional fuels can be achieved with the appropriate implementation of applications of bioenergy (Hartley et al. 2018).

The world is going to face a large number of issues related to the environment, society as well as economy in the coming decades. According to a report by United Nations, the world's population will increase to approximately 8.5 billion by 2030, 9.7 billion by 2050, and more than 11 billion in 2100 (United Nations 2015). Hence, this jump in the world population will further lead to the generation of stress over the current available global food systems. A larger population will also show an increase in demand for commonly required natural resources that are used in the generation of food (Dobermann and Nelson 2013). Development in the bioenergy sector in the context of current economies transforming into a bioenergy economy might offer a partial solution. It is still a question of how these developing and emerging world economies will efficiently develop the policies and strategies, the developmental process, institutional infrastructure, and human capacity in order to move these natural resources and invest in them for meeting the ever-increasing demand for bio-based energy products (Von Braun 2008). Hence, the policymakers are required to keep in mind these challenges as well as the tradeoffs if they want to see their food economy getting transformed into a bioenergy economy.

In the past years, the world has faced fuel and financial crises, which has led to an increase in an interest in knowing about the implications of movement toward this enhanced reliance on bioenergy and biofuels in order to decrease the burden of the demand for fossil fuels. The developing countries framed certain policies to deal with the issue of the food crisis which ultimately led to overburdening of the available resources (Hartley et al. 2018). It is well-known fact that most of the present ecosystems worldwide have already been overexploited and has become unsustainable (Figure 4). Certain factors such as globalization and climate change could adversely impact agricultural productivity leading to an increase in the burden on the available natural resources (Baul et al. 2017). In order to follow a path of sustainable growth and for meeting the needs of food and fuel of the present and keeping in mind the needs of the future generation, sustainable options have to be explored. And this growing menace can be resolved to a large proportion by converting this

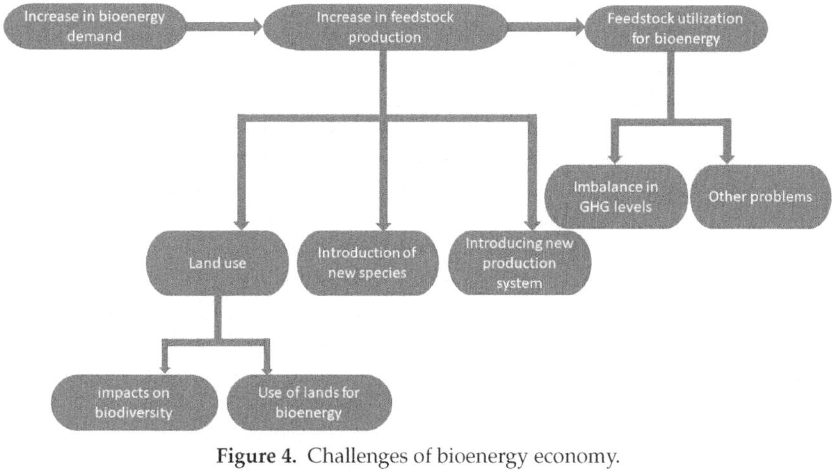

Figure 4. Challenges of bioenergy economy.

present food economy into a bioenergy economy (European Commission 2011). To transform the current food system into the bioenergy economic system, the foremost thing that we have to understand includes the vital components of a bioenergy economy.

The bioenergy economy needs to be sustainable and efficient in transforming renewable sources into food, bio-energy, and other products (Yadav et al. 2021). A bioenergy economy possesses the potential to enhance the sustainability of food, animal feed, fiber supply, ameliorate the quality of water, renewable bioenergy, improving human as well as animal health, and maintaining the biodiversity (Oliver et al. 2014). However, this potential alternative, which can lead to the sustainable growth of a nation, mostly remains underexplored. For this much-needed transformation, there is a need to have some technological innovations and viable markets, institutions, and infrastructure.

7.1.1 Increase in bioenergy demand

The IEA World Energy Outlook (2012) has suggested that there will be an increase in bioenergy demand up to the year 2035, that is, the demand for biofuels and biomass for electricity and other purposes is also expected to increase at triple the rate. These changing patterns will further impact the balance of demand and supply of bioenergy at the regional levels, resulting in an increase in trade flows and trade patterns (Matzenberger et al. 2015).

However, the current scenario of cultivating biofuel and other important bioenergy crops has invited criticism because of the land requirement for cultivating these crops that result in the encroachment of existing areas conventionally used to grow food crops. The actual tussle between land use for agricultural purposes and for biodiversity conservation is quite pertinent for the bioenergy discussions, especially if bioenergy is expected to decrease the climate changes that are going to have an impact on biodiversity as well as ecosystems (Tscharntke et al. 2012). Hence, the increasing demand for bioenergy will have certain repercussions, such as, it will lead to an increase in feedstock production, and excessive utilization of these feedstock can further negatively impact the environment.

7.1.2 Increase in Feedstock Production

The demand for ecofriendly sources has increased manifold as people are much more aware now as compared to earlier; they see bioenergy options as a potent alternative and this has actually laid more stress on the demand for increasing feedstock production. This is because the basic objectives of biodiversity conservation can restrict the availability of land area to the production of biomass (Erb et al. 2012). The discussion regarding "land sharing versus land sparing" shows various approaches for reducing the impact of agricultural activities on biodiversity as well as on ecosystem services. This debate is appropriate to bioenergy where local forests are affected by the biomass supply chain (Immerzeel et al. 2014). Also, harvesting of biomass on marginal lands for getting bioenergy in order to reduce the direct competition with food crops can lead to some economic as well as ecological issues created by low agricultural productivity (Gelfand et al. 2013). Utilizing residues such as corn stover, tree slash from agriculture and forests also has an impact on cost basis especially when these residues are crucial for soil fertility, soil erosion, as well as biodiversity conservation (Victorsson and Jonsell 2013). The bioenergy sector majorly depends on crop monocultures to be used as feedstock for the generation of biofuel, which may further show a deteriorating effect on biodiversity (Kline et al. 2015). One of the direct effects of extensive utilization of monoculture for the production of bioenergy includes the transformation or degradation of natural vegetation and habitats. Another impact of monoculture is the involvement of a combination of different plant species in order to obtain good biomass growth (Jackson et al. 2020). Other than habitat shift or loss, this rapid extension of monoculture may result in variation in species richness, abundance, composition, and distribution along with enhanced vulnerability among plants because of reduction in genetic diversity (Gonzalez-Hernandez et al. 2011). Some bioenergy crops like eucalyptus were found to have invasive potential (Kline et al. 2015). So, adding these species for bioenergy purposes can lead to some serious negative consequences. Also, expanding bioenergy production is thought to have a negative impact on land usage, the idea of producing bioenergy on degraded lands was proposed as a way to mitigate this effect. This would allow for the use of lands that are now unsuitable for agricultural or other productive purposes (Nijsen et al. 2012). For this purpose, information about the degraded lands is required, which can be achieved through various approaches to mapping (which requires expertise in this field) (Gibbs et al. 2015).

7.1.3 Other challenges

Another crucial challenge includes water consumption by bioenergy crops. Water scarcity has become a serious concern worldwide, so water consumption from bioenergy crops has become a crucial factor to evaluate. The bioenergy water footprint may vary depending upon several factors such as the crop variety used, the climatic conditions of the locality, or the agricultural practices (Mekonnen and Hoekstra 2014). Other challenges involve its effectiveness in being a negative emission renewable resource.

However, improper implementation and handling may hinder the overall goal of achieving negative emissions and can hinder the GHGs balance. Using carbon-neutral energy to process biomass as well as organic fertilizers to obtain biomass growth may decrease emissions arising from the bioenergy sector. Another important challenge in this field to address includes the transportation of biomass. Not only reducing the transportation can create a decrease in emissions, but utilizing the rail and sea route for transportation should also be considered. On the other hand, energy efficiency has become a serious shortcoming of bioenergy. No doubt, obtained biofuels usually have a smaller energy return in comparison to conventional sources. Hence, these challenges should be kept in mind while framing policies and strategies for achieving the goal of a bioenergy economy.

7.2 Opportunities

A bioenergy economy includes the generation of a variable range of goods and services that are obtained from plant and animal-based materials. It is not just a grain-based technology to obtain biofuels but it has various opportunities for social development (Figure 5). The term bioenergy economy comprises a viable substitute for everything that conventional fuel is currently being utilized for. The bioenergy economy is supposed to replace petroleum, coal, and natural gas with biomass-based products. Therefore, the biomanufacturing of products will become a crucial part of the bioenergy economy (Abhinandan et al. 2018). Hence, the bioenergy economy is the conversion of the whole economy, including the basic social structure. The fossil fuel sector is currently the dominant source of energy.

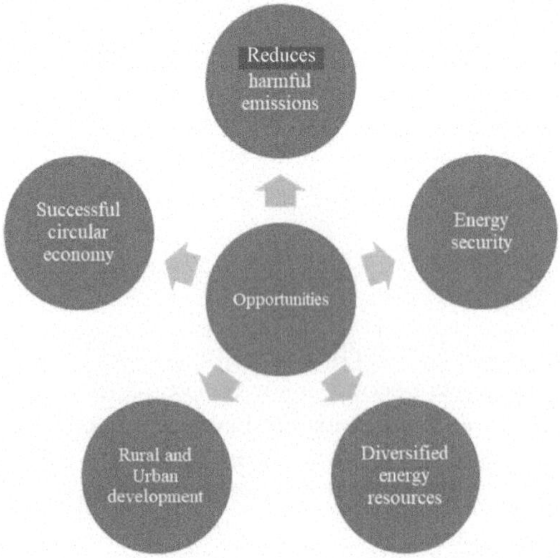

Figure 5. Opportunities of bioenergy economy.

The bioenergy sector in future will not be less than that. However, present facilities in the bioenergy sector have relatively less employment influence on the local economy, mostly due to the high capital involved. Although many studies of this sector have vastly overstated the indirect effects of biofuels (Broch et al. 2013), the major economic effects of biofuels include the enhancements in the prices of commodities at the global level along with enhanced money flow to farmer communities (McMichael 2013). In the bioenergy economy, the issues related to agricultural economies will almost be similar to that of a petrol-based economy. The major reason is that farmers will strive to generate energy nearly to an amount that can be justified on the basis of net income, the increasing bioenergy demands will be help by leading to an increase in production and lowering of the commodity prices. However, the bioenergy economy somewhat differs from that a petro-based economy in some other aspects, which include its impact on the environment, geopolitics, society, as well as technology.

One of the common raw materials playing an important role in the bioenergy economy is solar radiation. It's even spatial distribution is comparative to that of conventional fossil fuels, which show high spatially distribution as well as concentrated energy density. When solar radiations are evenly distributed, it will result in biomass being more evenly distributed, and in turn will lead to a variable spatial production, processing, as well as marketing system.

Another important factor in any economy is transportation. Transportation costs always have been a crucial aspect of economic activities and hence, any population. Transportation basically enhances the price of products to consumers resulting in a decrease of the *in situ* value of the final product to producers (Alicke et al. 2017). Rural economies are especially disturbed due to the high transportation costs as the raw material needed are mostly transported from outside the site of production and their finished products are required to be transported outside as well. Hence, rural as well as small towns, having less populations and located in some landlocked areas face double disadvantages, that is, the high cost of transporting the material inside and then transporting the finished goods outside the rural area due to a larger distance to seaports, distribution centers, or population centers.

With regard to the petro-based economy, the high transportation costs decrease the overall value of commodities and hence, leading to a reduction in the income of makers. However, in the bioenergy economy, there is a potential difference. Production of energy at the local level decreases the requirement of transportation of energy in and out of the area. When rural areas start producing their own bioenergy, it will cause saving in two directions: producers will no longer be required to pay the transportation charges. Obviously, the generation of bioenergy still needs to be economically feasible. It to the generation of bioenergy has to occur with a cost that can compete with petroleum and hence, when this will be achieved, rural producers will gain enormously.

In general, the bioenergy economy will offer sustainable growth and will lead to an improved economy and industrial sustainability. The production of bioenergy in the right way and its utilization can help in combating several problems and offer many attractive opportunities such as:

- Bioenergy can make some efficient contributions in decreasing carbon emissions, mostly from those sectors that are termed as difficult-to-decarbonize such as aviation, transportation, and industrial sectors. Hence, the bioenergy economy would contribute a significant amount in forming a green, clean, and low carbon-emission economy because of the low carbon footprint generated in bio-based products.
- It is well-known fact that there will be an energy crisis in the near future due to a shortage in the nonrenewable energy sources. Hence, bioenergy can offer some sort of energy security by biomass prioritization leading to sustainable growth.
- A successful and efficient strategy related to the bioenergy economy meant for people and in rural and urban regions will generate more employment for the people, reduce the transportation charges, and cut down the pollution level so as to make the area good for residing.
- Bioenergy economy and circular systems are regarded as economic systems that can fetch better outcomes that are sustainable also.

Hence in a similar way, many areas can be benefitted economically due to several reasons, especially cost reduction, leading to a replacement by the gains of a spatially distributed bioenergy economy. Also, the potential gains that will be obtained if the bioenergy economy will involve the safeguards required for renewable sources, mechanisms ensuring an equal distribution of the gains coming out of investing; that is, if the bioenergy economy is made to grow in an effective and optimal way. Several desired goals articulated with strategic policy can be achieved.

8. Conclusion

In conclusion, it is a well-known fact that a nation's development depends on the development of its native individuals in aspects of quality life, fulfilment of food, requirements of clothes and shelter, good finances, and accessibility and affordability of resources of the country. There is the need for development strategies promoting poverty alleviation and zero hunger by emphasising small-scale, decentralized, quick-yielding investment programs to boosted employment. Additionally, access toand more equitable distribution of arable land, credit, technology and other resources can proficiently attack poverty. Subsequently, bioenergy technology emerged as an excellent strategy having the potential to meet the energy requirements with no compromise in food production. Bioenergy production can also serve as a tool to amplify the country' economy by reducing dependence on other nations for energy needs. Additionally, bioenergy seems closely related to attaining SDGs. The SDGs also promote the deployment of bioenergy economy so as to meet energy and food security demands, deliver sustainable management of resources, lowering emissions of greenhouse gases, reducing societal inequality, promoting a healthy lifestyle, and ensuring elevated economic growth.

References

Abdeshahian, P., Lim, J.S., Ho, W.S., Hashim, H., and Lee, C.T. (2016). Potential of biogas production from farm animal waste in Malaysia. Renewable and Sustainable Energy Reviews, 60: 714–723.

Abinandan, S., Subashchandrabose, S.R., Venkateswarlu, K., and Megharaj, M. (2018). Nutrient removal and biomass production: advances in microalgal biotechnology for wastewater treatment. Critical Reviews in Biotechnology, 38(8): 1244–1260.

Adami, L., Schiavon, M., and Rada, E.C. (2020). Potential environmental benefits of direct electric heating powered by waste-to-energy processes as a replacement of solid-fuel combustion in semi-rural and remote areas. Science of The Total Environment, 740: 140078.

Ahmed, A., Ge, T., Peng, J., Yan, W.C., Tee, B.T., and You, S. (2022). Assessment of the renewable energy generation towards net-zero energy buildings: A review. Energy and Buildings, 256: 111755.

Alicke, K., Rexhausen, D., and Seyfert, A. (2017). Supply Chain 4.0 in consumer goods. Mckinsey& Company, 1–11.

Barney, J.N., and Tomaso, J.M. (2008). Nonnative species and bioenergy: Are we cultivating the next invader? BioScience, 58: 64–70 885 13.

Batini, N., Di Serio, M., Fragetta, M., Melina, G., and Waldron, A. (2022). Building back better: How big are green spending multipliers? Ecological Economics, 193: 107305.

Baul, T., Alam, A., Ikonen, A., Strandman, H., Asikainen, A., Peltola, H., and Kilpel€ainen, A. (2017). Climate change mitigation potential in boreal forests: impacts of management, harvest intensity and use of forest biomass to substitute fossil resources. Forests 8(12): 455. https://doi.org/10.3390/f8110455.

Berawi, M.A. (2019). The role of industry 4.0 in achieving sustainable development goals. International Journal of Technology, 10(4): 644–647.

Bergthorson, J.M., and Thomson, M.J. (2015). A review of the combustion and emissions properties of advanced transportation biofuels and their impact on existing and future engines. Renewable and Sustainable Energy Reviews, 42: 1393–1417.

Beyene, A., Bluffstone, R.A., Gebreegziabher, Z., Martinsson, P., Mekonnen, A., and Vieider, F.M. (2015). The improved biomass stove saves wood, but how often do people use it? Evidence from a randomized treatment trial in Ethiopia. Evidence from a Randomized Treatment Trial in Ethiopia (June 9, 2015). World Bank Policy Research Working Paper, (7297).

Bharathiraja, B., Sudharsanaa, T., Bharghavi, A., Jayamuthunagai, J., and Praveenkumar, R. (2016). Biohydrogen and Biogas–An overview on feedstocks and enhancement process. Fuel, 185: 810–828.

Bhattacharya, A., and Datta, A. (2018). Laminar burning velocity of biomass-derived fuels and its significance in combustion devices. In Sustainable Energy Technology and Policies (pp. 359–378). Springer, Singapore.

Bhuvaneshwari, S., Hettiarachchi, H., and Meegoda, J.N. (2019). Crop residue burning in India: policy challenges and potential solutions. International Journal of Environmental Research and Public Health, 16(5): 832.

Bioenergy for Sustainable Development. (2017). https://www.ieabioenergy.com/wp-content/uploads/2017/01/BIOENERGY-AND-SUSTAINABLE-DEVELOPMENT-final-20170215.pdf.

Brašanac-Bosanac, L., Ćirković-Mitrović, T., Eremija, S., Stajić, S., and Lučić, A. (2018). Improving the use of forest-based biomass for energy purposes in Serbia. Sustainable Forestry: Collection, (77-78): 113–122.

Broch, A., Hoekman, S.K., and Unnasch, S. (2013). A review of variability in indirect land use change assessment and modeling in biofuel policy. Environmental Science & Policy, 29: 147–157.

Buccola, S. (2020). Virtual Water Trade: An Economically Silent Solution to Water Scarcity Problems.

Cernev, T., and Fenner, R. (2020). The importance of achieving foundational Sustainable Development Goals in reducing global risk. Futures, 115: 102492.

Channing Arndt, Karl Pauw, and James Thurlow. (2012). Biofuels and economic development: A computable general equilibrium analysis for Tanzania, Energy Economics, 34(6): 1922–1930, ISSN 0140-9883,https://doi.org/10.1016/j.eneco.2012.07.020.

Chapple, K. (2008). Defining the green economy: A primer on green economic development. Center for Community Innovation, University of California, Berkeley.

Chowdhury, M., Islam, M., Rahman, S.M., Abubakar, I.R., Aina, Y.A., Hasan, M., and Khondaker, A.N. (2021). A review of policies and initiatives for climate change mitigation and environmental sustainability in Bangladesh. Environment, Development and Sustainability, 23(2): 1133–1161.

Clarke, L., Jiang, K., Akimoto, K., Babiker, M.G.B., Fisher-Vanden, K., Hourcade, J.C., Krey, V., Kriegler, E., Löschel, A., McCollum, D., Paltsev, S., Rose, S., Shukla, P.R., Tavoni, M., Van Der Zwaan, B. and Van Vuuren, D.P. (2014). Assessing transformation pathways. *In*: Edenhofer, O., Pichs-Madruga, R., Sokona, Y., Farahani, E., Kadner, S., Seyboth, K., Adler, A., Baum, I., Brunner, S., Eikemeier, P., Kriemann, B., Savolainen, J., Schlömer, S., Von Stechow, C., Zwickel, T., and Minx, J.C. (eds.). Climate Change 2014: Mitigation of Climate Change. Contribution of Working Group III to the Fifth assessment Report of the Intergovernmental Panel on Climate Change. Cambridge, United Kingdom and New York, NY, USA: Cambridge University press.

Debnath, D., Babu, S.C., Ghosh, P., and Helmer, M. (2018). The impact of India's food security policy on domestic and international rice market. J. Policy Model 40(2): 265–283. https://doi.org/10.1016/j. jpolmod.2017.08.006.

Destek, M.A., Sarkodie, S.A., and Asamoah, E.F. (2021). Does biomass energy drive environmental sustainability? An SDG perspective for top five biomass consuming countries. Biomass and Bioenergy, 149: 106076.

Deutch, J. (2020). Is net zero carbon 2050 possible? Joule, 4(11): 2237–2240.

Dobermann, A., and Nelson, R. (2013). Opportunities and solutions for sustainable food production. Sustainable Development Solutions Network: Paris, France.

Ebadian, M., van Dyk, S., McMillan, J.D., and Saddler, J. (2020). Biofuels policies that have encouraged their production and use: An international perspective. Energy Policy, 147: 111906.

Eras-Almeida, A.A., and Egido-Aguilera, M.A. (2020). What is still necessary for supporting the SDG7 in the Most Vulnerable Contexts? Sustainability, 12(17): 7184.

Erb, K.-H., Haberl, H., and Plutzar, C. (2012). Dependency of global primary bioenergy crop potentials in 2050 on food systems, yields, biodiversity conservation and political stability. Energy Policy, 47: 260–269.

European Commission. (2011). Communication from the Commission to the European Parliament, the Council, the European Economic and Social Committee and the Committee of the Regions Youth Opportunities Initiative.

Fava, F., Totaro, G., Diels, L., Reis, M., Duarte, J., Carioca, O.B., and Ferreira, B.S. (2015). Biowaste biorefinery in Europe: opportunities and research & development needs. New Biotechnology, 32(1): 100–108.

Gelfand, I., Sahajpal, R., Zhang, X.S., Izaurralde, R.C., Gross, K.L., and Robertson, G.P. (2013). Sustainable bioenergy production from marginal lands in the US Midwest. Nature, 493: 514–520.

Gernaat, D.E., de Boer, H.S., Daioglou, V., Yalew, S.G., Müller, C., and van Vuuren, D.P. (2021). Climate change impacts on renewable energy supply. Nature Climate Change, 11(2): 119–125.

Gibbs, H.K., and Salmon, J.M. (2015). Mapping the world's degraded lands. Applied Geography, 57: 12–21.

Gielen, D., Boshell, F., Saygin, D., Bazilian, M.D., Wagner, N., and Gorini, R. (2019). The role of renewable energy in the global energy transformation. Energy Strategy Reviews, 24: 38–50.

Gonzalez-Hernandez, J.L., Sarath, G., Stein, J.M., Owens, V., Gedye, K., and Boe, A. (2011). A multiple species approach to biomass production from native herbaceous perennial feedstocks. In: Tomes, D., Lakshmanan, P., and Songstad, D. (eds.). Biofuels. Springer, New York, NY. https://doi.org/10.1007/978-1-4419-7145-6_5.

Groom, M.J., Gray, E.M., and Townsend, P.A. (2008). Biofuels and bio886 diversity: principles for creating better policies for biofuel pro887 duction. ConservBiol., 22: 602–609.

Hartley, F., van Seventer, D., Samboko, P.C., and Arndt, C. (2019). Economy-wide implications of biofuel production in Zambia. Development Southern Africa, 36(2): 213–232.

Hartley, L. (2018). Introduction: The 'Business' of Writing Women. pp. 1830–1880. In: Hartley, L. (eds.). The History of British Women's Writing. History of British Women's Writing. Palgrave Macmillan, London. https://doi.org/10.1057/978-1-137-58465-6_1.

Hassan, S.S., Williams, G.A., and Jaiswal, A.K. (2019). Moving towards the second generation of lignocellulosic biorefineries in the EU: Drivers, challenges, and opportunities. Renewable and Sustainable Energy Reviews, 101: 590–599.

Helms, I.V., J.A. Ijelu, S.E. Wills, B.D. Landis, D.A., and Haddad, N.M. (2020). Ant biodiversity and ecosystem services in bioenergy landscapes. Agriculture, Ecosystems & Environment, 290: 106780.

High-Level Political forum on sustainable development. (2018). https://sustainabledevelopment.un.org/content/documents/195532018_background_notes_SDG_7Final1.pdf.

(https://www.sciencedirect.com/science/article/pii/S0140988312001648).

IEA (International Energy Agency). 2016. World energy model. Available from: https://www.iea.org/weo/weomodel/.

Immerzeel, D.J., Verweij, P.A., van der Hilst, F., and Faaij, A.P.C. (2014). Biodiversity impacts of bioenergy crop production: a state-of-the-art review. Glob Change Biol. Bioenergy, 6: 183–209.

Jenčík, J., Hönig, V., Obergruber, M., Hájek, J., Vráblík, A., Černý, R., and Herink, T. (2021). Advanced biofuels based on fischer–tropsch synthesis for applications in diesel engines. Materials, 14(11): 3077.

Jordan, N., Boody, G., Broussard, W., Glover, J.D., Keeney, D., McCown, B.H., McIsaac, G., Muller, M., Murray, H., Neal, J., Pansing, C., Turner, R.E., Warner, K., and Wyse, D. (2007). Environment: sustainable development of the agricultural bioeconomy. Science, 316: 1570–1571.

Keeney, D. (2008). Ethanol USA. Environ. Sci. Technol., 43: 8–11.

Klein, D., Luderer, G., Kriegler, E., Strefler, J., Bauer, N., Leimbach, M., Popp, A., Dietrich, J.P., Humpenöder, F., LotzeCampen, H., and Edenhofer, O. (2014). The value of bioenergy in low stabilization scenarios: an assessment using REMIND-MAgPIE. Clim. Chang., 123: 705–718.

Kline, K., Dale, V.H., Lee, R., and Leiby, P. (2009). In defense of biofuels, done right. Issues Sci. Technol., 25: 75–84.

Kline, K.L., Martinelli, F.S., Mayer, A.L. et al. (2015). Bioenergy and biodiversity: key lessons from the pan American Rregion. Environmental Management 56: 1377–1396. https://doi.org/10.1007/s00267-015-0559-0.

Kojima, T. (Ed.). (2019). Carbon Dioxide Problem: Integrated Energy and Environmental Policies for the 21st Century. Routledge.

Kothari, R., Vathistha, A., Singh, H.M., Pathak, V.V., Tyagi, V.V., Yadav, B.C., Ashokkumar, V., and Singh, D.P. (2020). Assessment of Indian bioenergy policy for sustainable environment and its impact for rural India: Strategic implementation and challenges. Environmental Technology & Innovation, p. 101078.

Ladu, L., and Quitzow, R. (2017). Bio-based economy: policy framework and foresight thinking. pp. 167–195. In: Food Waste Reduction and Valorisation. Springer, Cham.

Lago, C., Caldés, N., and Lechón, Y. (Eds.). (2018). The Role of Bioenergy in the Emerging Bioeconomy: Resources, Technologies, Sustainability and Policy. Academic Press.

Lin, B., and Zhu, J. (2019). Determinants of renewable energy technological innovation in China under CO2 emissions constraint. Journal of Environmental Management, 247: 662–671.

Linville, J.L., Shen, Y., Wu, M.M., and Urgun-Demirtas, M. (2015). Current state of anaerobic digestion of organic wastes in North America. Current Sustainable/Renewable Energy Reports, 2(4): 136–144.

Mathiesen, B.V., Lund, H., Connolly, D., Wenzel, H., Østergaard, P.A., Möller, B., and Hvelplund, F.K. (2015). Smart Energy Systems for coherent 100% renewable energy and transport solutions. Applied Energy, 145: 139–154.

Matzenberger, J., Kranzl, L., Tromborg, E., Junginger, M., Daioglou, V., Goh, C.S., and Keramidas, K. (2015). Future perspectives of international bioenergy trade. Renewable and Sustainable Energy Reviews, 43: 926–941.

McMichael, P. (2013). Value-chain agriculture and debt relations: Contradictory outcomes. Third World Quarterly, 34(4): 671–690.

Mekonnen, M.M., and Hoekstra, A.Y. (2014). Water footprint benchmarks for crop production: A first global assessment. Ecological Indicators, 46: 214–223.

Morton, S., Pencheon, D., and Squires, N. (2017). Sustainable Development Goals (SDGs), and their implementation A national global framework for health, development and equity needs a systems approach at every level. British medical bulletin, 1–10. https://www.un.org/sustainabledevelopment/.

Nations, U. (2015). World population prospects: The 2015 revision. United Nations Econ. Soc. Aff., 33(2): 1–66.

Nazari, L., Xu, C.C., and Ray, M.B. (2021). Resource Utilization of Agricultural/Forestry Residues via Fractionation into Cellulose, Hemicellulose and Lignin. Advanced and Emerging Technologies for Resource Recovery from Wastes, 179–204.

Nijsen, M., Smeets, E., Stehfest, E., and van Vuuren, D.P. (2012). An evaluation of the global potential of bioenergy production on degraded lands. Gcb Bioenergy, 4(2): 130–147.

Nunes, L.J.R., Matias, J.C.O., and Catalão, J.P.S. (2014). A review on torrefied biomass pellets as a sustainable alternative to coal in power generation. Renewable and Sustainable Energy Reviews, 40: 153–160.

Oliver, C.D., Nassar, N.T., Lippke, B.R., and McCarter, J.B. (2014). Carbon, fossil fuel, and biodiversity mitigation with wood and forests. J. Sustain. For. 33(3): 248–275. https:// doi.org/ 10.1080/10549811.2013.839386.

Omisore, A.G. (2018). Attaining sustainable development goals in sub-Saharan Africa; The need to address environmental challenges. Environmental Development, 25: 138–145.

Owusu, P.A., and Asumadu-Sarkodie, S. (2016). A review of renewable energy sources, sustainability issues and climate change mitigation. Cogent Engineering, 3(1): 1167990.

Paloviita, A., Kortetmäki, T., Puupponen, A., and Silvasti, T. (2017). Insights into food system exposure, coping capacity and adaptive capacity. British Food Journal.

Pastor, A.V., Palazzo, A., Havlik, P., Biemans, H., Wada, Y., Obersteiner, M., and Ludwig, F. (2019). The global nexus of food–trade–water sustaining environmental flows by 2050. Nature Sustainability, 2(6): 499–507.

Popp, J., Lakner, Z., Harangi-Rakos, M., and Fari, M. (2014). The effect of bioenergy expansion: Food, energy, and environment. Renewable and Sustainable Energy Reviews, 32: 559–578.

Prince, S.D., Haskett, J., Steininger, M., Strand, H., and Wright, R. (2001). Net primary production of U.S. Midwest croplands from agricultural harvest yield data. Ecol. Appl., 11: 1194–1205.

Resnick, D., Tarp, F., and Thurlow, J. (2012). The political economy of green growth: Cases from Southern Africa. Public Administration and Development, 32(3): 215–228.

Rose, S.K., Kriegler, E., Bibas, R., Calvin, K., Popp, A., van Vuuren, D.P., and Weyant, J. (2014). Bioenergy in energy transformation and climate management. Climatic Change, 123(3): 477–493.

Roy, J., Tscharket, P., Waisman, H., Abdul Halim, S., Antwi-Agyei, P., Dasgupta, P., and Suarez Rodriguez, A.G. (2018). Sustainable Development, Poverty Eradication and Reducing Inequalities.

Scarlat, N., Dallemand, J.F., Monforti-Ferrario, F., and Nita, V. (2015). The role of biomass and bioenergy in a future bioeconomy: Policies and facts. Environmental Development, 15: 3–34.

Segerstedt, E., and Abrahamsson, L. (2019). Diversity of livelihoods and social sustainability in established mining communities. The Extractive Industries and Society, 6(2): 610–619.

Seinfeld, J.H., and Pankow, J.F. (2003). Organic atmospheric particulate material. Annu. Rev. Phys. Chem. 54: 121–140.

Shahsavari, A., and Akbari, M. (2018). Potential of solar energy in developing countries for reducing energy-related emissions. Renewable and Sustainable Energy Reviews, 90: 275–291.

Sharma, Y.C., and Singh, V. (2017). Microalgal biodiesel: a possible solution for India's energy security. Renewable and Sustainable Energy Reviews, 67: 72–88.

Simonyan, K.J., and Fasina, O. (2013). Biomass resources and bioenergy potentials in Nigeria. African Journal of Agricultural Research, 8(40): 4975–4989.

Sims, R.E. Mabee, W. Saddler, J.N., and Taylor, M. (2010). An overview of second generation biofuel technologies. Bioresource Technology, 101(6): 1570–1580.

Singh, H.M., Kothari, R., Gupta, R., and Tyagi, V.V. (2019). Bio-fixation of flue gas from thermal power plants with algal biomass: Overview and research perspectives. Journal of Environmental Management, 245: 519–539.

Sinha, S.K., Subramanian, K.A., Singh, H.M., Tyagi, V.V., and Mishra, A. (2019). Progressive trends in bio-fuel policies in India: targets and implementation strategy. Biofuels, 10(1): 155–166.

Siwal, S.S., Zhang, Q., Devi, N., Saini, A.K., Saini, V., Pareek, B., and Thakur, V.K. (2021). Recovery processes of sustainable energy using different biomass and wastes. Renewable and Sustainable Energy Reviews, 150: 111483.

Sparrevik, M., Adam, C., Martinsen, V., and Cornelissen, G. (2015). Emissions of gases and particles from charcoal/biochar production in rural areas using medium-sized traditional and improved "retort" kilns. Biomass and Bioenergy, 72: 65–73.

Target and Indicators.https://sdgs.un.org/goals/goal7.

Tscharntke, T., Clough, Y., Wanger, T.C., Jackson, L., Motzke, I., Perfecto, I., Vandermeer, J., and Whitbread, A. (2012). Global food security, biodiversity conservation and the future of agricultural intensification. Biol. Conserv., 151: 53–59.

Tucho, G.T., and Kumsa, D.M. (2020). Challenges of achieving sustainable development goal 7 from the perspectives of access to modern cooking energy in developing countries. Frontiers in Energy Research, p. 286.

Victorsson, J., and Jonsell, M. (2013). Ecological traps and habitat loss, stump extraction and its effects on saproxylic beetles. For EcolManag., 290: 22–29.

von Braun, J. (2012). The role of Science and research for development policy and the millennium development goals.

Wu, M., Mintz, M., Wang, M., and Arora, S. (2009). Water consumption in the production of ethanol and petroleum gasoline. Environ. Manage., 44: 981–997.

Yadav, B., Atmakuri, A., Chavan, S., Tyagi, R.D., Drogui, P., and Pilli, S. (2021). Role of Bioeconomy in Circular Economy. In Biomass, Biofuels, Biochemicals (pp. 163–195). Elsevier.

Zhenmin, L., and Espinosa, P. (2019). Tackling climate change to accelerate sustainable development. Nature Climate Change, 9(7): 494–496.

3

Bioeconomy:
Current Status and Challenges

Shamshad Ahmad,[1,*] *Anu Bharti,*[2] *Mohd Islahul Haq*[3] and *Richa Kothari*[2]

1. Introduction

The world's delicate ecosystems are becoming increasingly threatened by the ever-growing human population. By the end of 2050, the population is expected to exceed nine billion due to the current, rising pace of human growth. The needs of society are changing quickly along with the growth of human civilisation. The only source that can satisfy a human desire is nature. Therefore, due to growing population, urbanization, deforestation, industrialization, need more energy with the technological advancement majority of sectors to satisfy our growing need and greed, humans are primarily to blame for ecological imbalance and environmental degradation. The bioeconomy is the product of the interaction between economists and biotechnology inventors, and as a consequence, there is a private economic interest that frequently ignores environmental repercussions, particularly at the societal level and the demands of future generations (Aguilar et al. 2019). Political conditions, business interests and society's level of social responsibility, existing economic development routes, bioresource supply and demand, and other factors all influence

[1] CSIR-National Environmental Engineering Research Institute Nehru Marg, Nagpur, Maharashtra, India-440 020.
[2] Department of Environmental Sciences, Central University of Jammu, Rahya-Suchani, Bagla, Samba, Jammu and Kashmir, India- 181143.
[3] Dr. Rammanohar Lohia Avadh University, Ayodhya, U.P. India.
* Corresponding author: shamshadahmad93@gmail.com

the dangers of bioeconomic transformation (Angouria-Tsorochidou et al. 2021). Such factors must be taken into consideration while framing a plan for a bioeconomy. The plan needs to be circular in nature and implemented in places where the most important biomaterials are discovered (Stegmann et al. 2020). Hence, a firm optimism is required with respect to bioeconomy if a sustainable world is desired.

The concept of bioeconomy still requires a practical solution at local as well as global level. Presently, most of the nations in the world are dependent upon agriculture for their living. It not only ensures food security but also encourages enterpreneurship, generates employment, and considerably improves society as a whole. The bioeconomy has a stern emphasis on the efficient and sustainable utilization of all kinds of biological sources that are regenerative in nature as show in the Table 1 (Dantas et al. 2021). As a result, production may continue despite natural resource losses or additional environmental costs, and moreover certain bioeconomy-produced objects are not always biodegradable (Hetemäki et al. 2017). Thus, the notion that the growing bioeconomy will lead to an economy that is environmentally friendly as well as sustainable is unfounded, because it is not necessary that technologies become environmentally sustainable simply by switching to renewable resources (Maksymiv et al. 2021). Because a firm idea for environmental ethics states that no technical intercession may be forced over the nature exceeding its available receptive capacity, the solution to this conceptual paradox can be found in bioethics (Székács 2017). While the meaning remains less or more the same, the exact definition of bioeconomy with respect to bioeconomy strategies differs in its specificity. These definitions are discussed in brief in the Table 1.

Scientists explore policy approaches aimed at supporting and expanding the bioeconomy. When analyzing efforts with respect to Sustainable Development Goals at the global level, the argument relies on a strong governance structure, which is required for path to sustainability (Dietz et al. 2018). Each country's strategy identifies the governance conflicts that must be restrained in order to build a sustainable bioeconomy. It also specifies the political tools that each strategy use to control these conflicts and lessen the dangers that ensue. Countries use Enabling governance in their political policies to get around issues with route dependencies in the creation of a sustainable bioeconomy (enabling and constraining governance) (Heimann et al. 2019). Another challenge, we feel, which is important to address is the effective management of business processes along with implementing the SDGs by each organization whose operations are affected by bioeconomy characteristics (Maksymiv et al. 2021). Understanding the needs of various countries, as well as their

Table 1. Concept of bioeconomy in bioeconomy strategies put forth by several agencies.

European Commission 2012	The production of value-added things like food, feed, biological by-products, as well as bioenergy from renewable biological resources is referred to as bioeconomy. The term "bioeconomy" refers to the process of converting biomass into high-value things such as food, feed, biobased products, and bioenergy. Agriculture, forestry, fishing, food, pulp and paper manufacturing, as well as chemical, biotechnology, and energy sectors are all included.
European Commission 2018	All industries as well as systems that depend upon biobased resources such as animals, plants, microorganisms, and derived biomass, including organic waste, are included in the bioeconomy. It links terrestrial as well as marine ecosystems, along with all primary producing sectors such as agriculture, forestry, pisciculture, and aquaculture that utilize and produce biological resources, as well as all sectors (economic as well as industrial) that make use of the biological resources and processes in order to obtain food, feed, bio-based products, energy, and other services.
Organisation for Economic Co-operation and Development (OECD) 2009	In the bioeconomy, biotechnology is crucial in primary agriculture and industry, especially when advanced life sciences are employed to transform biomass into minerals, chemicals, and fuels.
White House 2012	A bioeconomy is one in which biological science research and innovation create economic activity as well as societal gain. To cite a few bioeconomy examples from the United States, there are novel drugs and diagnostics for enhanced human well-being, high yield of food crops, biofuels being developed to be less reliant on oil, and biobased chemical components all around.
Federal Ministry of Food and Agriculture 2013	The "biobased economy" is based on a structural shift away from an economy that is predominantly reliant on limited fossil resources, particularly petroleum, and toward one that is growing more reliant on renewable resources. By exploiting biological operations as well as resources, enhancing them, and so increasing their activity potential while also making their usage more effective as well as sustainable, the bioeconomy blends technology, economics, and environmental issues. The bioeconomy replaces raw materials derived from fossil fuels, along with creating innovative goods and processes.

expectations with respect to bioeconomy is critical for reaching the UN's sustainable development objectives, although these expectations can be unduly optimistic at times (Kurki et al. 2021).

The German Bioeconomy Council conducted an online survey of 345 experts from 46 countries to look into what the global bioeconomy would look like in the next 20 years (Issa et al. 2019). They found that, first and foremost, the future bioeconomy must handle humanity's energy, agriculture, and food needs.

Additionally, one-of-a-kind goods manufactured from renewable resources are expected to play an important role. Although all UN Sustainable Development Goals will have an impact on future bioeconomy success stories, five were chosen for the study: SDG 12 (responsible consumption and production), SDG 9 (industry, innovation, and infrastructure), SDG 13 (climate action), SDG 7 (affordable and clean energy), and SDG 11 (sustainable development) (sustainable cities and communities). The International Council for Science investigated the nature of SDG links and found that the four SDGs studied in-depth (SDGs 2, 3, 7, and 14) are usually synergistic (Biber-Freudenberger et al. 2020). This basically rests on the idea that a science-based study of the linkages across SDG sectors (which is presently lacking), might aid in more coherent and effective decision-making, as well as enhanced follow-up and progress monitoring. Getting to know about the possible trade-offs and synergistic interactions across the SDGs, as indicated in Table 2, are crucial for long-term sustainable development goals.

2. Changing perspectives on the bioeconomy

The bioeconomy concept was influenced by two views, as previously stated: (1) resource substitution, and (2) biotechnology innovation. Resource substitution became more significant in the initial phase of the twenty-first century, despite the fact that biotechnology innovation has long been considered as a bioeconomy potential (Fader et al. 2018). The idea of "peak oil", that predicts oil extraction rates have peaked and will begin to drop in the future while oil prices continue to rise, was a key component of the resource substitution strategy (Kuhns et al. 2018). The comparative value of utilizing biomass for energy and material consumption grows as the price of oil rises. This line of inquiry reinforced the resource substitution stance of a bioeconomy (Hagemann et al. 2018). The bioeconomy encompasses the biomass production in a variety of forms, with respect to its conditioning as well as conversion via various methods, as well as the production along with the sale of food, feedstock, fiber, fuel, and other items (Ghodake et al. 2021). Hence, the concept of bioeconomy is much more than merely the biomass production process.

The crisis over oil prices that occurred in 2007–2008 bolstered the "peak oil" theory. Also, an increase in food crop utilization for production of biofuel during the oil price crisis contributed to a rise in food costs. The rise in oil prices played a significant effect in this development. Policies, like biofuel subsidies along with laws requiring biofuel to be mixed with ordinary gasoline, have come under scrutiny after studies indicated their potential impact on food prices (Noh et al. 2016). The bioeconomy was affected in two ways as a result of these developments: First, as stated

Table 2. Highlights of a synergistic interconnection between a sustainable bioeconomy and the SDGs.

2 ZERO HUNGER	**SDG 2: End hunger, achieve food security and improved nutrition and promote sustainable agriculture**	Connecting smallholder farmers to markets, value chains, and agro-bioeconomy prospects is critical to increasing agricultural output, reducing poverty, and enhancing rural living conditions. A large proportion of organic waste is created even today across food supply chains, that may be kept away or efficiently utilized in the bioeconomy for the well-being of people and the environment. Furthermore, another most exciting aspect of the bioeconomy is its ability to provide long-term protein supplies for human and animal use. Microorganism-based food applications, such as microbiomes-based solutions, are also being developed.
3 GOOD HEALTH AND WELL-BEING	**SDG 3: Ensure healthy lives and promote well-being for all at all ages**	Conventional biomass as an energy source has poor health consequences because of the air pollution, especially indoor type. Switching to alternative sources of energy like biogas, may assist developing nations in reducing their reliance on fuel wood. In the bioeconomy, there are opportunities in the medical and pharmaceutical areas, as well as in the health domain. The possible health consequences of the bioeconomy, on the other hand, must be better recognized and disclosed to the general population.
7 AFFORDABLE AND CLEAN ENERGY	**SDG 7: Ensure access to affordable, reliable, sustainable and modern energy for all**	SDG 7 is crucial, with 1.1 billion people lacking access to electricity and another three billion people depending upon conventional biomass for home energy (fuelwood, farm wastes, animal dung, and charcoal). Increased energy availability, as well as decreased or eliminated use of conventional biomass, help in achieving a number of SDGs. Poor and vulnerable individuals commonly waste time (gathering fuel) or money on poor-quality energy suppliers. Enhanced energy access also enhances adaptability.
9 INDUSTRY, INNOVATION AND INFRASTRUCTURE	**SDG 9: Build resilient infrastructure, promote inclusive and sustainable industrialization and foster innovation**	The bioeconomy has the potential to develop new value chains that connect agriculture, fisheries, and bio-based companies. Life sciences' fast expansion, as well as digitalization and convergence of key technologies in applications, is a primary driver of bioeconomy innovation. Promising improvements have been made in genetics, big data analysis, artificial intelligence, biotechnology, neurotechnology, and nanotechnology, to name a few. These high-tech applications offer enormous long-term growth and bioeconomy potential.

Table 2 contd. ...

...Table 2 contd.

10 REDUCED INEQUALITIES	**SDG 10: Reduce inequality within and among countries**	Many bioresource-rich countries lack scientific and technological investments that would allow them to participate in technological advances, whereas biotechnology as well as concerned advanced developments are commanded by a very few centers of excellence, the majority of which are located in developed countries. Many developing nations lack technology and commercial incubation facilities, making it challenging to bring bioscience concepts to the market. An enabling environment, on the other hand, can aid in the reduction of global disparities by promoting the approach of science and technology to maximize the utilization as well as value of bio-based resources in poorer nations where they exist in abundance. Encouragement of innovative value-networks covering agriculture, fisheries, and bio-based businesses may assist in alleviating internal imbalances by encouraging growth in rural and coastal areas.
11 SUSTAINABLE CITIES AND COMMUNITIES	**SDG 11: Make cities and human settlements inclusive, safe, resilient and sustainable**	Integrating biological concepts into urban planning and management has become a critical component in constructing "greener cities" that provide a higher quality of life while reducing GHG emissions. Local bio-product manufacturing, vertical farms, recycling process, and efficient management of the waste are all critical components for the creativeness, and sustainability of communities. Building "greener cities" that offer a greater quality of life while lowering GHG emissions has grown dependent on integrating biological notions into urban planning and management. Local bioproduct production, vertical farming, recycling, and effective waste management are all essential for the sustainability and creativity of communities. In the modern world, building designs and construction techniques based on biological principles and renewable resources are gaining popularity.

12 RESPONSIBLE CONSUMPTION AND PRODUCTION	**SDG 12: Ensure sustainable consumption and production patterns**	Management of generation, circular processes, as well as functional resource utilization are examples of bioeconomy concepts that should be adopted more generally, particularly in the domain of primary sector. Improving process of consumption as well as production, especially by avoiding and/or lowering losses (e.g., food loss and waste) and making better use of agricultural and fisheries inputs, is of worldwide relevance and critical for resource efficiency. Furthermore, the bioeconomy has the potential of encouraging the creation of sustainable products, such as food packaging substances obtained through bio-based components such as agri-based food wastes rather than fossil-based plastics, which can help to combat plastic pollution.
13 CLIMATE ACTION	**SDG 13: Take urgent action to combat climate change and its impacts**	Climate change and its negative implications must be recognized as a key future obstacle to sustainable development, as must the requirement for undertaking the GHG reduction techniques while guaranteeing energy security via expanded variability of energy sources, composing of more sustainable biomass applications.

further below, the possible conflict between guaranteeing food supply and burning biomass for energy purposes has become a hot subject in the public policy discussion around the bioeconomy (Vogelpohl et al. 2021). Second, the need to boost biomass production productivity and create biomass production and use alternatives that do not clash with food supply has gotten more attention (Popp et al. 2014). Second-generation methods as well as utilization of by-products and waste materials for production of bioenergy are examples of such solutions. Both energy as well as food costs fell sharply after 2010, and that resulted in more variability in the oil prices than in the 1990s (Sillanpää et al. 2017). Although, the future of oil prices is hard to foresee, oil scarcity is not a compelling reason for the substitution of resources (Baumeister and Kilian 2016).

The use of renewable energy sources rather than fossil fuels has become increasingly important as a result of climate change (Brockway et al. 2019). The United Nations Framework Convention on Climate Change's Paris Agreement (2019) formed a fundamental foundation for resource substitution. While substituting the resources is very vital, bioeconomies has switched to focusing on the innovations in the field of biotechnology (Wydra et al. 2019). As a result, in recent years, the bioeconomy's ability to commercialize biotechnology and, more broadly, life sciences innovations has emerged as a critical aspect (Issa et al. 2019).

3. Rise of the bioeconomy: A global concept

The concept of bioeconomy is being promoted worldwide, including by the European Union, since the early 2000s. As an official bioeconomy policy, the Obama administration produced the "National Bioeconomy Blueprint" in 2012 (Frisvold et al. 2021). This notion also integrates the two previously mentioned perspectives on the bioeconomy: biotechnology innovation and resource substitution (Bröring et al. 2020). Several governments, in both developed and developing countries, have provided bioeconomy-related blueprints and programmes during the beginning of the twenty-first century. Malaysia launched a "Bioeconomy Transformation Program" in 2012, while South Africa announced a bioeconomy strategy in 2013 (Hetemäki et al. 2017). While the number of countries with specialized bioeconomy policies is minimal, several governments use biotechnology and/or renewable resource programmes (Meyer et al. 2017). In December 2015, Berlin hosted the first Global Bioeconomy Summit. The council was assisted by a global advisory group. There were around 700 bioeconomy practitioners in attendance from over 80 nations (Gerdes et al. 2018). The expanding number of countries with bioeconomy-related goals and policies, as well as the scientific literature, show that the bioeconomy is becoming a global concept.

4. Bioeconomy and sustainability

The evolving definition of the bioeconomy reflects rising concerns about its long-term viability. The following statement occurs in the *Making Bioeconomy Work for Sustainable Development* communique from the Global Bioeconomy Summit 2015 (Gerdes et al. 2018). The word "sustainable development" can be used to refer to the larger societal goal of "sustainability". In the 1980s, this idea had already found its way into the international policy agenda. The definition was given by The United Nations Commission on Environment and Development for "sustainable development" in its report *Our Common Future* (Borowy et al. 2013). The Brundtland Commission is named after the name of Norway's prime minister at the time, Gro Harlem Brundtland, and he was the country's only political leader to occupy this position after serving as an environment minister (Haack et al. 2022). In its report, *Our Common Future*, the United Nations Commission on Environment and Development defined "sustainable development" (Borowy et al. 2013). The phrase "sustainable development" refers to the goal of tackling these two issues together. During the 1992 *International Conference on Environment and Development*, often referred to as the "Rio de Janeiro Earth Summit", the concept of sustainable development was confirmed. Nearly 170 nations signed "Agenda 21", a major global action plan with four programme domains that included social as well as economic challenges, resource conservation along with its management, major group strengthening, civil society organizations, and implementation tactics (UN 1992). The notion of "sustainable development" was backed by Agenda 21 as having three major areas: economic, social, and environmental (Li 2022). As a result, a sustainable bioeconomy incorporates environmental along with economic as well as social components.

5. Bioeconomy as a path to "Massive Societal Transformation"

As the preceding explanations demonstrate, the developmental idea regarding bioeconomy was first defined by attention on the "supply side", that is, the provision of commodities and services on bio-based resources as well as biotechnological processes. During the past few years, a greater emphasis has been laid upon the demand and functioning of bioeconomy in society as a whole (Eversberg et al. 2022). A group belonging to the University of Hohenheim started foundation course for master's programme "Bioeconomy", that began in 2014. People's tastes and values, which gets transformed into wants as well as desires for innovative goods that are bio-based, hold enormous importance for the bioeconomy due to the creation of those value-added products. A transdisciplinary systems analysis is required for this complete understanding of the bioeconomy (Perišić et al. 2022).

Taking the bioeconomy's societal embeddedness, a step further, the bioeconomy may be seen as a component of a societal transformation process that will eventually be necessary to transition the existing economic setup into a system which is economical, ecological, as well as socially sustainable (Ho et al. 2022). The understanding of the problems involved in this change has led to the idea that creating economic incentives and implementing environmentally friendly regulations would not be enough. Finally, "a massive societal transformation" is necessary, which "includes substantial variability to infrastructures, generation processes, regulatory setup, along with different lifestyles, as well as a latest sort of interplay among the policies, social life, science, as well as the economy" (Allain et al. 2022). According to this viewpoint, the bioeconomy is an essential component of the current epoch that will finally bring back the benefits of the well-established industrial civilization (Bijon et al. 2022). As previously stated, the phrase "bioeconomy strategies" refers to policy blueprints or strategic documents which are made public by the concerned governments or legislatures. To better understand government bioeconomy policies, evaluate a country's competitive advantage in developing various components of the bioeconomy (Näyhä et al. 2019). Each industrial cluster has unique microeconomic environment that is described by the Porter diamond model, along with the different elements that influences the feasibility and existence of clusters (Porter 1990). It also emphasises on the fact that governments may make a country more competitive on the globe and provides recommendations for how to do so (Adetoyinbo et al. 2022). This conceptual framework can at some point of time serve in the framing of a country's own bioeconomy policies or blueprints.

6. Bioeconomy: Vision and opportunities

A bioresource (substitution) vision: In order to decarbonize development, bio-based renewable resources must be utilized in place of fossil fuels and other fossil-based resources.

Innovative vision in relation with Biotechnology: Promotion of new techniques, social, as well as institutional restructuring such as a greater number of new sustainable commodities, effective and plunged utilization of bio-based resources including biomass, recycling, enhancing longevity as well as mending, bringing in second- as well as third-generation bio-based sources such as lignin, algae, etc., and effective desegregated generation as well as consumption throughout the worldwide value chains or nets, covering the entire range of bio-products into account

(Schmidt et al. 2012). All of these steps can help in achieving sustainable goals with an innovative set of policies and framework.

Agricultural innovation in compliance with rural development: Altering, revitalizing, along with modernizing existing agricultural, forestry, as well as production of biomass in European countries along with other adjacent areas through a broader range of highly effective crops that are capable of adaptation, improvised and multi-faceted generation systems, and sustainable escalation (Hoff et al. 2018), that makes the land present at the margins more high-yielding and efficiently connects farmers to merchandise, while also offering new agriculture-based and bioresources processing employment leading to an enhanced living standard.

International Cooperation for Developmental Vision: Creating new offerings for the world, such as knowledge, technology, and innovation transfers, improved education and skills, and a stronger focus on the strengths and advantages of areas along with increased value-added products, and employment moved "upstream" in global value nets that are close to the production of primary biomass (globally), and an even and fair distribution of wealth (Lokesh et al. 2018). The vision for development can hence be merged with an integrated international cooperation to bring into the reality the basic goal of sustainable development.

7. A Bio-ecological vision

This vision supporting the transition to a bioeconomy through an approach towards the topography and ecosystem, rehabilitation of deteriorated lands for the purpose of biomass generation as well as other ecosystem services, protection of biodiversity, reduction of waste, addressing handling of demand and replacing the critical products such as plastics, removing the existing patterns responsible for the overexploitation of available natural resources, which in turn are causes deterioration of the present environment (Hurlbert et al. 2019). Hence, a transition of any economy into bioeconomy can be made swiftly through following a bio-ecological pattern, thus safeguarding the existing natural resources.

8. Bioeconomy: Challenges and risks

Bioeconomy approaches provide challenges, choices, and potential conflicts to Europe and beyond, including sustainable biomass and bioresources production, processing, and use, responsible ecological as well as environmental management, and comprehensive development (Hoff et al. 2018). In addition to present and future carbon sequestration needs for climate protection, changes in the bioeconomy may put greater demand on bioresources, and hence on increased biomass output (Yadav et al. 2021). The full implementation of the vast array of new bioeconomy efforts will very certainly raise demand and competition, demanding new supply and demand limitations (Wydra et al. 2021). Unless bioeconomy transitions are followed by technological and other innovative measures such as enlarging the genetic well as a feedstock resource, plunged utilization, altering the consumption trends, issues such as resource overexploitation can occur, which can further pose threats to the biodiversity and its related functions as well as services (Mohan et al. 2019). Some of the negative consequences of bio-based commodities may be stronger per unit than those of fossil-based goods, depending upon the route followed to become a bioeconomy.

Close coordination of the various national bioeconomies is also required to avoid these potential risks and achieve the aforementioned goals, which include best matching bioresources generation and consumption trends; hence, ensuring bioresources sourcing and allotting is sustainable in nature and has beneficial uses (Devaney and Iles 2019). The sustainable development goals (SDGs) need major bioresources production, processing, and consumption improvements, as well as socioeconomic and institutional innovations as important components of bioeconomy transition paths (Ray et al. 2021). This demonstrates that certain economies, particularly those in underdeveloped nations, are heading towards the new bioeconomies without having to pass via the phase of fossil fuels or other developmental curves that are not sustainable in nature (Hoff et al. 2018). Science could be able to help us better comprehend the dangers, minimizes tradeoffs, and build knowledge-based long-term bioeconomy strategies. For example, studies have demonstrated that by combining the biomass sourcing through a sustainable mode with agricultural intensification that too via sustainable mode, backed by consistent strategies, it can enhance the availability of bio-based sources and ensuring food security along with minimizing the problem of climate change as well as the present negative impact on biodiversity (Albrecht and Ettling 2014). Therefore, these things should be kept in mind to have a lucid vision regarding bioeconomy.

There are several risks available with respect to the bioeconomy transitions. One such risk includes the maintenance of resource-intensive consumption trend followed by developed countries, such as Europe, which is based on enhancing the overall imports of raw materials that are bio-based, thus making the bioeconomy a justification to continue doing business like they used to do before, while some countries in the south continue to be the exclusive suppliers of these bio-based resources that too by ignoring their own progress towards become modern bioeconomies, which involve value-added byproducts (Bracco et al.

2018). More international collaboration and a more equal sharing of the advantages related to the bioeconomy transition at a global level across the whole supply chain are required to offset such issues. Firm approaches meant to analyze the cost-benefit distribution, on the other hand, are mainly nonexistent, at present (Lodge et al. 2016). The target to achieve a bioeconomy friendly society involves minimizing all types of risk which is applicable in bioeconomy process.

9. Bioeconomy policy: A target for global sustainability

Bioeconomy-related policies can be recognized as those having a firm tilt towards bioeconomic development, mainly in biotechnological domain, bioenergy and bioeconomy, or industry (Prochaska and Schiller 2021). Several political policies within the conventional bioeconomic fields, including the production sector such as agriculture, forestry, or marine along with policies related to research and innovation areas were addressed only if they hierarchized bioeconomy; please see Figure 1. A very similar concept was used to frame policies and strategies covering the overriding targets meant for sustainable, green and blue growth, along with a circular economy (Chowdhury et al. 2020). Thus, in order to have a sustainable society, potential concepts should be kept in mind before framing the policy blueprints.

Strategies related to bioeconomy are meant to have some characteristics such as high technology, vital role in research and innovation, bioenergy, blue economy (policies covering the ocean or marine bioeconomy), green economy (policy addressing the biobased innovations), bio-based economy (policies addressing the bio-based economy along with

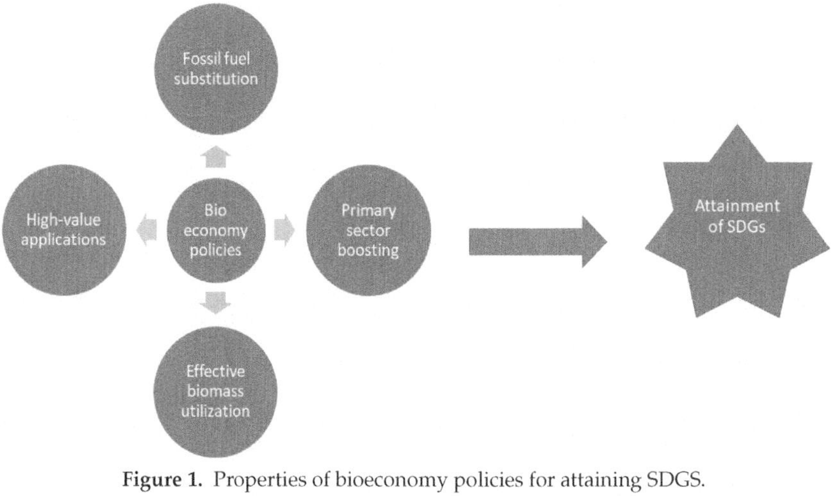

Figure 1. Properties of bioeconomy policies for attaining SDGS.

focusing on the bioeconomic development), circular bioeconomy (policies firmly focusing on the circular utilization of bio-based resources), forest bioeconomy (policies focusing on the conversion of resources from forests into valuable products), and the main aspects of bioeconomy (include policies that broadly focus on bioeconomy development, bioeconomy documents other than policies, and regional bioeconomy strategies) (Fava et al. 2021). Although, at the global level, the bioeconomy term has been accepted to a large extent for framing policy papers and strategies. However, there exists no general definition that can define the political concept of bioeconomy. Moreover, its meaning is evolving continuously.

Bioeconomy policies can prove to be a crucial component in bringing the SDGs into reality. Figure 1 depicts the properties of bioeconomy policies that can ultimately lead towards the path of attaining SDGs (Dietz et al. 2018). For example, introduction of some productive models that take advantage of R&D to utilize biological sources in a sustainable and efficient way to make fossil fuel substitutes such as bioenergy, biofertilizers, etc., and addressing some SDGs like, affordable and clean energy (SDG 7) and industry and innovation (SDG 9). Another example includes the boosting of the primary sector by making a productive use of biomass and covering SDGs such as climate action (SDG 13) and responsible consumption and production (SDG 12). Good bioeconomic policies can also ensure the production of high-value products, processes, and systems covering SDGs such as new sources of decent work and sustainable economic growth (SDG 8).

The term "bioeconomy" acquired importance when the European Union announced it to give attention to bioeconomy strategies and policies in 2012; it was reviewed again in 2018 for its relevancy (Patermann and Aguilar 2018). According to the reviewed strategy by EU, bioeconomy needs to cover almost every sector and system that depends upon the biological resources such as animals, plants, microorganisms, and biomass, their functions and principles (Toteva et al. 2020). It involves and integrates the land as well as aquatic ecosystems along with the functions and services they offer, all sectors that utilize or produce biological products such as agriculture, forestry, fisheries, and aquaculture, and all economic and industrial sectors that utilize the biological sources and related processes in order to obtain food, feedstock, biological products, energy, and other services. The European bioeconomy policy is projected to have a sustainable as well as circular economy in order to drive the industries, modernizing the primary production systems, and protecting the environment and biodiversity. At present, the world is moving its interest towards bioeconomy (Aguilar et al. 2019). The concept of bioeconomy is gaining ever-increasing attention from developed as well as developing governments and policy makers. Countries look up to bioeconomy as the basis in addressing the issues related to fossil fuels consumption, changing climate, food and nutrition, and moving towards to create a healthy and ecofriendly world.

The *India Bioeconomy Report 2019* is an effort that brings together a multi-faceted approach along with suitable dimensions to the forefront. That is, the major goals of the Indian bioeconomy include a $100 billion Bioeconomy by 2025, 10 million jobs meant to drive the bioeconomy based upon products and services, offering easy access to the global healthcare system, developing effective medical treatments such as gene therapy, cell therapy, and stem cells therapy, manufacturing high-yielding agricultural products which are environmentally sustainable, utilizing bioremediation, and other related methods along with bio-fuels in order to decrease the carbon footprint and pave a way for a green future for India (Mohan et al. 2019).

The basic perception of the bioeconomy remains to be a diversified field which simply depicts a picture that there exists not one bioeconomy but many. A bioeconomy policy as well as its definition differs according to the technological, natural resource base and economic capability of a country.

Despite the fact that the bioeconomy is now viewed as global policy processes as opposed to those in 2010, several gaps in the international institutionalisation still exist (Bastos Lima 2021). Hence, in order to avoid such replication of attempts, current macroregional as well as global bioeconomy initiatives need to be merged in an efficient manner to obtain one single global platform, which can offer advanced regional collaboration. A global platform can aid in the global interconnection of various bio economies, with special reference to biomass energy, management of value-added products, and efficient technologies having an equal division of labor. For bioeconomics to be effective and sustainable, there has to be an evaluation of bioeconomic advancements that are better linked globally. In order to make bioeconomics a successful and sustainable domain, a proper assessment of bioeconomic advances aligned in a better way at the international level is required.

10. Conclusion

Globally, an increasing number of developed and developing nations are pursuing national bioeconomy goals and stimulating the development of bio-based energy, goods, materials, and chemicals. Bioeconomies possess enormous potential capable of achieving the SDGs by moving economic development toward low-emission paths that are climate resilient and are in compliance with the Paris Agreement. Increased biomass usage, on the other hand, has the potential to harm ecosystems and communities if not framed in a way which is sustainable as well as responsible.

References

Adetoyinbo, A., Gupta, S., Okoruwa, V.O., and Birner, R. (2022). The role of institutions in sustaining competitive bioeconomy growth in Africa–Insights from the Nigerian maize biomass value-web. Sustainable Production and Consumption, 30: 186–203.

Aguilar, A., Twardowski, T., and Wohlgemuth, R. (2019). Bioeconomy for sustainable development. Biotechnology Journal, 14(8): 1800638.

Albrecht, K., and Ettling, S. (2014). Bioeconomy strategies across the globe. Rural, 21(48): 3.

Allain, S., Ruault, J.F., Moraine, M., and Madelrieux, S. (2022). The 'bioeconomics vs bioeconomy' debate: Beyond criticism, advancing research fronts. Environmental Innovation and Societal Transitions, 42: 58–73.

Angouria-Tsorochidou, E., Teigiserova, D.A., and Thomsen, M. (2021). Limits to circular bioeconomy in the transition towards decentralized biowaste management systems. Resources, Conservation and Recycling, 164: 105207.

Bastos Lima, M.G. (2021). International Bioeconomy Governance: Unveiling the Initial Patterns. In The Politics of Bioeconomy and Sustainability (pp. 67–86). Springer, Cham.

Bastos Lima, M.G. (2021). Corporate power in the bioeconomy transition: The policies and politics of conservative ecological modernization in Brazil. Sustainability, 13(12): 6952.

Baumeister, C., and Kilian, L. (2016). Forty years of oil price fluctuations: Why the price of oil may still surprise us. Journal of Economic Perspectives, 30(1): 139–60.

Biber-Freudenberger, L., Ergeneman, C., Förster, J.J., Dietz, T., and Börner, J. (2020). Bioeconomy futures: Expectation patterns of scientists and practitioners on the sustainability of bio-based transformation. Sustainable Development, 28(5): 1220–1235.

Bijon, N., Wassenaar, T., Junqua, G., and Dechesne, M. (2022). Towards a sustainable bioeconomy through industrial symbiosis: Current situation and perspectives. Sustainability, 14(3): 1605.

Borowy, I. (2013). Defining sustainable development for our common future: A history of the World Commission on Environment and Development (Brundtland Commission). Routledge.

Bracco, S., Calicioglu, O., Gomez San Juan, M., and Flammini, A. (2018). Assessing the contribution of bioeconomy to the total economy: A review of national frameworks. Sustainability, 10(6): 1698.

Brockway, P.E., Owen, A., Brand-Correa, L.I., and Hardt, L. (2019). Estimation of global final-stage energy-return-on-investment for fossil fuels with comparison to renewable energy sources. Nature Energy, 4(7): 612–621.

Bröring, S., Laibach, N., and Wustmans, M. (2020). Innovation types in the bioeconomy. Journal of Cleaner Production, 266: 121939.

Chowdhury, S., Kain, J.H., Adelfio, M., Volchko, Y., and Norrman, J. (2020). Greening the Browns: A bio-based land use framework for analysing the potential of urban brownfields in an urban circular economy. Sustainability, 12(15): 6278.

Dantas, T.E., De-Souza, E.D., Destro, I.R., Hammes, G., Rodriguez, C.M.T., and Soares, S.R. (2021). How the combination of Circular Economy and Industry 4.0 can contribute towards achieving the Sustainable Development Goals. Sustainable Production and Consumption, 26: 213–227.

Devaney, L., and Iles, A. (2019). Scales of progress, power and potential in the US bioeconomy. Journal of Cleaner Production, 233: 379–389.

Dietz, T., Börner, J., Förster, J.J., and Von Braun, J. (2018). Governance of the bioeconomy: A global comparative study of national bioeconomy strategies. Sustainability, 10(9): 3190.

European Commission (2012). Innovating for sustainable growth: A bioeconomy for Europe. Brussels: EuropeanCommission.https://publications.europa.eu/de/publication-detail/-/publication/1f0d8515-8dc0-4435-ba53-9570e47dbd51.

European Commission. (2018). A Clean Planet for all: A European strategic long-term vision for a prosperous, modern, competitive and climate neutral economy. Available at https://eur-lex.europa.eu/legal-content/EN/TXT/?uri=CELEX%3A52018 DC0773 [11.09.20]

European Commission. (2018). A sustainable bioeconomy for Europe: strengthening the connection between economy, society and the environment: Updated Bioeconomy Strategy. Available athttps://ec.europa.eu/research/bioeconomy/pdf/ec_bioeconomy _strategy_2018.pdf [11.09.20].

Eversberg, D., and Fritz, M. (2022). Bioeconomy as a societal transformation: Mentalities, conflicts and social practices. Sustainable Production and Consumption.

Fader, M., Cranmer, C., Lawford, R., and Engel-Cox, J. (2018). Toward an understanding of synergies and trade-offs between water, energy, and food SDG targets. Frontiers in Environmental Science, 6: 112.

Federal Ministry of Education and Research and Federal Ministry of Food and Agriculture.2020. National Bioeconomy Strategy—Summary. https://www.bmbf. de/files/2020_1501_.

Fava, F., Gardossi, L., Brigidi, P., Morone, P., Carosi, D.A., and Lenzi, A. (2021). The bioeconomy in Italy and the new national strategy for a more competitive and sustainable country. New Biotechnology, 61: 124–136.

Frisvold, G.B., Moss, S.M., Hodgson, A., and Maxon, M.E. (2021). Understanding the US bioeconomy: A new definition and landscape. Sustainability, 13(4): 1627.

Gerdes, H., Kiresiewa, Z., Beekman, V., Bianchini, C., Davies, S., Griestop, L., and Vale, M. (2018). Engaging stakeholders and citizens in the bioeconomy: Lessons learned from BioSTEP and recommendations for future research. Ecologic Institute: Berlin, Germany.

Ghodake, G.S., Shinde, S.K., Kadam, A.A., Saratale, R.G., Saratale, G.D., Kumar, M., and Kim, D.Y. (2021). Review on biomass feedstocks, pyrolysis mechanism and physicochemical properties of biochar: State-of-the-art framework to speed up vision of circular bioeconomy. Journal of Cleaner Production, 297: 126645.

Haack, K. (2022). Explaining access to executive leadership in UN agencies. In Women's Access, Representation and Leadership in the United Nations (pp. 99–140). Palgrave Macmillan, Cham.

Hagemann, N., Gawel, E., Purkus, A., Pannicke, N., and Hauck, J. (2016). Possible futures towards a wood-based bioeconomy: A scenario analysis for Germany. Sustainability, 8(1): 98.

Heimann, T. (2019). Bioeconomy and SDGs: Does the bioeconomy support the achievement of the SDGs? Earth's Future, 7(1): 43–57.

Hetemäki, L., Hanewinkel, M., Muys, B., Ollikainen, M., Palahí, M., Trasobares, A., and Potočnik, J. (2017). Leading the way to a European circular bioeconomy strategy (Vol. 5, p. 52). Joensuu, Finland: European Forest Institute.

Ho, C.H., Böhm, S., and Monciardini, D. (2022). The collaborative and contested interplay between business and civil society in circular economy transitions. Business Strategy and the Environment.

Hoff, H., Johnson, F.X., Allen, B., Biber-Freudenberger, L., and Förster, J.J. (2018). Sustainable bio-resource pathways towards a fossil-free world: The European Bioeconomy in a global development context. In Policy paper produced for the IEEP Think 2030 conference. Institute for European Environmental Policy (IEEP), Brussels.

Hurlbert, M., Krishnaswamy, J., Johnson, F.X., Rodríguez-Morales, J.E., and Zommers, Z. (2019). Risk Management and Decision Making in Relation to Sustainable Development.

Issa, I., Delbrück, S., and Hamm, U. (2019). Bioeconomy from experts' perspectives–Results of a global expert survey. PloS one, 14(5): e0215917.

Kuhns, R.J., Shaw, G.H. (2018). Peak oil and petroleum energy resources. In Navigating the Energy Maze (pp. 53–63). Springer, Cham.

Kurki, S., and Ahola-Launonen, J. (2021). Bioeconomy in maturation: a pathway towards a "Good" bioeconomy or distorting silence on crucial matters? In Bio# Futures (pp. 165–199). Springer, Cham.

Li, D. (2022). The UN's 'Big Agenda': Building an Inclusive Global Cooperation Framework. In China's Role in Global Governance (pp. 129–158). Palgrave Macmillan, Singapore.

Lodge, D.M., Simonin, P.W., Burgiel, S.W., Keller, R.P., Bossenbroek, J.M., Jerde, C.L., and Zhang, H. (2016). Risk analysis and bioeconomics of invasive species to inform policy and management. Annual Review of Environment and Resources, 41: 453–488.

Lokesh, K., Ladu, L., and Summerton, L. (2018). Bridging the gaps for a 'circular' bioeconomy: selection criteria, bio-based value chain and stakeholder mapping. Sustainability, 10(6): 1695.

Maksymiv, Y., Yakubiv, V., Pylypiv, N., Hryhoruk, I., Piatnychuk, I., and Popadynets, N. (2021). Strategic challenges for sustainable governance of the bioeconomy: preventing conflict between SDGs. Sustainability, 13(15): 8308.

Meyer, R. (2017). Bioeconomy strategies: Contexts, visions, guiding implementation principles and resulting debates. Sustainability, 9(6): 1031.

Mohan, S.V., Dahiya, S., Amulya, K., Katakojwala, R., and Vanitha, T.K. (2019). Can circular bioeconomy be fueled by waste biorefineries—A closer look. Bioresource Technology Reports, 7: 100277.

National-Bioeconomy-Strategy_Summary_accessible.pdf (accessed January 15, 2020).

Näyhä, A. (2019). Transition in the Finnish forest-based sector: Company perspectives on the bioeconomy, circular economy and sustainability. Journal of Cleaner Production, 209: 1294–1306.

Noh, H.M., Benito, A., and Alonso, G. (2016). Study of the current incentive rules and mechanisms to promote biofuel use in the EU and their possible application to the civil aviation sector. Transportation Research Part D: Transport and Environment, 46: 298–316.

Organisation for Economic Co-operation and Development. (2009). The bioeconomy to 2030—Designing a policy agenda. Paris: OECD. http://www.oecd.org/futures/bioeconomy/2030.

Patermann, C., and Aguilar, A. (2018). The origins of the bioeconomy in the European Union. New biotechnology, 40: 20–24.

Perišić, M., Barceló, E., Dimic-Misic, K., Imani, M., and SpasojevićBrkić, V. (2022). The role of bioeconomy in the future energy scenario: A state-of-the-art review. Sustainability, 14(1): 560.

Popp, J., Lakner, Z., Harangi-Rakos, M., and Fari, M. (2014). The effect of bioenergy expansion: Food, energy, and environment. Renewable and Sustainable Energy Reviews, 32: 559–578.

Porter, M.E. 1990. The Competitive Advantage of Nations. New York: The Free Press.

Prochaska, L., and Schiller, D. (2021). An evolutionary perspective on the emergence and implementation of mission-oriented innovation policy: the example of the change of the leitmotif from biotechnology to bioeconomy. Review of Evolutionary Political Economy, 2(1): 141–249.

Ray, L., Pattnaik, R., Singh, P.K., Mishra, S., and Adhya, T.K. (2021). Environmental impact assessment of wastewater based biorefinery for the recovery of energy and valuable bio-based chemicals in a circular bioeconomy. In Waste Biorefinery (pp. 67–101). Elsevier.

Schmidt, O., Padel, S., and Levidow, L. (2012). The bio-economy concept and knowledge base in a public goods and farmer perspective. Bio-based and Applied Economics, 1(1): 47–63.

Shrestha, S., Karky, B.S., and Karki, S. (2014). Case study report: REDD+ pilot project in community forests in three watersheds of Nepal. Forests, 5(10): 2425–2439.

Sillanpää, M., and Ncibi, C. (2017). Bioeconomy: Multidimensional impacts and challenges. In A Sustainable Bioeconomy (pp. 317–343). Springer, Cham.

Staniškis, J.K. (2022). Sustainability challenges in an business organisation. In Transformation of Business Organization towards Sustainability (pp. 3–14). Springer, Cham.

Stegmann, P., Londo, M., and Junginger, M. (2020). The circular bioeconomy: Its elements and role in European bioeconomy clusters. Resources, Conservation & Recycling: X, 6: 100029.

Székács, A. (2017). Environmental and ecological aspects in the overall assessment of bioeconomy. Journal of Agricultural and Environmental Ethics, 30(1): 153–170.

Toteva, D., Popov, R., and Marinov, P. (2020). Establishment of a system of indicators for transforming the regional farm of bulgaria towards the bioeconomy in the context of biomass. Trakia Journal of Sciences, 18.

UN (1992) Agenda 21. United Nations Conference on Environment and Development Rio de Janeiro Brazil 3 to 14 June 1992. https://sustainabledevelopment.un.

Vogelpohl, T., Beer, K., Ewert, B., Perbandt, D., Töller, A.E., and Böcher, M. (2021). Patterns of European bioeconomy policy. Insights from a cross-case study of three policy areas. Environmental Politics, 1–21.

White House. (2012). National Bioeconomy Blueprint. Washington, DC.

Wydra, S. (2019). Value chains for industrial biotechnology in the bioeconomy-innovation system analysis. Sustainability, 11(8): 2435.

Wydra, S., Hüsing, B., Köhler, J., Schwarz, A., Schirrmeister, E., and Voglhuber-Slavinsky, A. (2021). Transition to the bioeconomy–Analysis and scenarios for selected niches. Journal of Cleaner Production, 294: 126092.

Yadav, B., Atmakuri, A., Chavan, S., Tyagi, R.D., Drogui, P., and Pilli, S. (2021). Role of bioeconomy in circular economy. In Biomass, Biofuels, Biochemicals (pp. 163–195). Elsevier.

4

Butanol Biofuels:

Current Status and Challenges

Sonika Kumari,[1,2] *Pankaj Kumar,*[1]
Veeramuthu Ashokkumar,[3] *Richa Kothari,*[2]
Sheetal Rani,[1] *Jogendra Singh*[1] and *Vinod Kumar*[1,*]

1. Introduction to butanol biofuels: A viable concept of green energy

The recent increase in industrialization and globalization has also increased energy demands tremendously. For fulfilling these energy demands, fossil fuels are being exploited recklessly. Hence, the extensive utilization of fossil fuels has contributed to negative impacts on the environment, such as global warming, greenhouse gas emission, and ozone hole depletion (Ruan et al. 2019). Looking at the impact caused by fossil fuels on the environment, the research for alternative energy has become the priority. This is the scenario that has urged the world to get introduced to bioenergy, that is, biomass-based energy. Bioenergy comprises biofuels, which are greener energy fuels derived from renewable substrates. The fuels derived from biological sources are called biofuels. These biofuels comprise

[1] Agro-ecology and Pollution Research Laboratory, Department of Zoology and Environmental Science, Gurukula Kangri (Deemed to be University), Haridwar 249404, Uttarakhand, India.
[2] Department of Environmental Sciences, Central University of Jammu, RahyaSuchani, Bagla, Samba 181143, Jammu and Kashmir, India, Email: rvashok2008@gmail.com
[3] Center of Excellence in Catalysis for Bioenergy and Renewable Chemicals (CBRC), Faculty of Science, Chulalongkorn University, Pathum Wan, Bangkok 10330, Thailand.
Emails: sonikakumari02gkv@gmail.com; kumarpankajgkv@gmail.com; kothariricha21@gmail.com; sheetalrani340@gmail.com; jogendrasinghpatil@gmail.com
* Corresponding author: drvksorwal@gkv.ac.in

biobutanol, bioethanol, biomethanol, biodiesel, biogas, and biohydrogen, among which biobutanol is attracting more attention because of having gasoline-like energy properties (Liu 2021). Biofuels can be produced from different feedstocks such as agricultural residues, microalgae, edible-oil crops, non-edible oil crops, waste-cooking oil, animal fat, etc. Recently, numerous studies on biofuel have shown that microalgae have immense potential for biofuel production (Satlewal et al. 2018, Milano et al. 2016). In a study, Callegari et al. (2020) suggested that the excessive utilization of fossil fuels has resulted in carbon emission issues, which have led to discovering renewable energy sources like biofuels. Another study by Satlewal et al. (2018) concluded that rice straw can be efficiently managed by using it as a feedstock for the production of biofuels.

Biobutanol is a colorless, four-carboned alcohol-based fuel having a banana-like smell. High-energy density, less corrosiveness and water solubility, high flashpoint, low vapor pressure, octane-enhancement property, etc., makes butanol a better choice over the other alcohol-based fuels (Obergruber et al. 2021). Being gasoline-like, biobutanol is compatible with the present gasoline-based vehicles without any modification. Butanol can be used in pure form or as a blend with gasoline. Despite the other organic feedstocks used for biofuel production, butanol is also produced from syngas, glycerol, and ethanol (Veza et al. 2021). Moreover, biofuels help boost the global economy, reduce dependency on fossil fuels, mitigate the impacts of climate change, etc. (Mishra 2017).

Various researchers have studied the potential of biofuels. Among which Skaggs et al. (2018) emphasized the use of waste materials for biofuel production. It was concluded that using organic wastes such as animal manure, food processing waste, fats, oils, grease, and municipal wastewater sludge can produce up to 22.3 GL/y (5.9 B gal/y) of a bio-crude oil, meeting the country's 23.9% demand for aviation kerosene. Through a study, Alavijeh and Karimi (Alavijeh et al. 2019) demonstrated that the country has the potential of utilizing corn stoves for butanol production. About 84 million tonnes of corn stover can be collected which can produce 10.48 gigalitres (Gl) of biobutanol. This can fulfill the 11.8% of domestic gasoline requirement of the country. Therefore, keeping in view the present perspectives, this chapter deals with current advances and challenges in the production of butanol biofuels.

2. Chemical *vs.* biological synthesis pathways

Biobutanol is a commercially important chemical having wide utility as a solvent for paints, waxes, shellac, rubbers, fuel additives, etc. It can be synthesized by chemical and biological pathways (Figure 1). Butanol synthesized by both pathways has identical properties. However, the

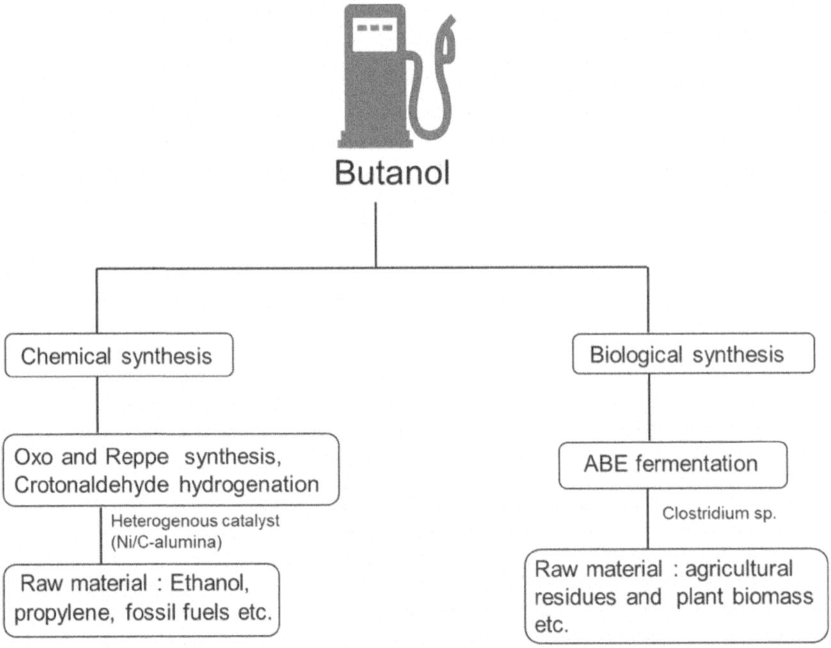

Figure 1. Methods of butanol production.

chemical method of butanol production is less time-consuming but not environmentally friendly (Zhen et al. 2020).

2.1 Chemical synthesis methods

The chemical synthesis of butanol uses chemical-based substrates such as ethanol, propylene, fossil fuels, and gases in the presence of a suitable catalyst. Chemically butanol is synthesized by three processes: oxo synthesis, Crotonaldehyde hydrogenation, and Reppe synthesis (Kazemi et al. 2019). All the three processes of butanol synthesis involve three reactions along with the presence of a catalyst, which react in different manners as given below:

Dehydrogenation: In this reaction, the removal of a hydrogen molecule from the chemical substrate such as ethanol takes place leading to the production of ethyl acetate and acetaldehydes.

Aldol condensation: Aldol condensation is also known as the Guebert reaction, first studied in the 1890s by Marcel Guerbet. In this reaction, enol or an enolate ion reacts with a carbonyl compound followed by dehydrogenation, resulting in the formation of aldehyde in the presence of a heterogeneous catalyst such as a mixture of Mg and Al oxides, Na/ZrO_2, Rh or Ru, etc. (Uyttebroek 2015).

Hydrogenation: After aldol condensation, the hydrogenation of the aldol-adducts takes place to increase its solubility in the aqueous phase. Moreover, in a reaction involving unsaturated aldehydes, the hydrogenation of the C = C bond is thermodynamically favored over the C = O bond.

In their work, Santacesaria et al. (2012) demonstrated that for the production of butanol from the dehydrogenation of ethanol, aldol condensation of acetaldehyde, and hydrogenation of aldehyde are carried out in successive order. Different catalysts are used for butanol production, based upon selectivity and butanol yield. Another study by Perrone et al. (2018) suggested that In-CuMgAl showed the best butanol selectivity of 65% with only 3% of ethanol production up to 443K temperature.

2.2 Biological synthesis

The biological synthesis of butanol takes place via acetone butanol ethanol (ABE) fermentation (Huzir et al. 2018). In this synthesis, organic materials such as sugars, starch, agricultural residue, and algae are used as the substrates for butanol production in the presence of microorganisms. It has been reported that solventogenic *Clostridium* sp. such as *C. acetobutylicum* ATCC824, *C. saccharobutylicum*, *C. saccharoperbutylacetonium*, and *C. beijerinckii*are widely used for biological butanol synthesis (Huang et al. 2018).

During butanol production through biological synthesis, pretreatment is an important step before fermentation. Pretreatment increases the speed of the fermentation process by enabling the breakdown of lignin and hemicellulose into simpler carbohydrates. There are various pretreatment techniques, such as physical (milling), chemical (acid, alkaline hydrolysis), physicochemical (steam explosion), and biological (enzymatic hydrolysis) pretreatment techniques that can be used depending upon the type of feedstock used (Karimi et al. 2015).

In ABE fermentation, carbohydrates such as glucose get converted into acetone, butanol, and ethanol in the presence of microorganisms followed by separation and purification. The conventional ABE process produces low butanol yield. To increase the butanol production via ABE fermentation, metabolic engineering of microorganisms is being carried out to improve the yield. It has been seen that using *Bacillus* 15 sp. results in greater production of butanol than that from the solventogenic *Clostridium* species (Ng et al. 2016). In a study carried out by Procentese et al. (2017), it was observed that *Lactuca sativa* leaves pretreated with 80 kg/m^3 of NaOH produced 1.1 g/L butanol via ABE fermentation using *Clostridium acetobutylicum* DSMZ 792. Another experiment by Cai et al. (2016) on biobutanol production using sweet sorghum juice as substrate showed that the productivity of ABE fermentation can be improved by

Table 1. Different microbes used for the biosynthesis of butanol biofuels.

Organism	Substrate	Butanol Efficiency	References
Clostridium saccharoperbutylacetonicum N1-4	Palm kernel cake	3.27 g/L	(Shukor et al. 2016)
Clostridium beijerinckii DSM-6422	Seaweed (*Laminaria digitate*)	7.16 g/L	(Hou et al. 2017)
Nesterenkonia sp. strain F	Glucose rich media	66–105 mg/L	(Amiri et al. 2016)
C. saccharobutylicum DSM 13864	corn stover	9.02 ± 0.11 g/L	(Hijosa-Valsero et al. 2020)
C. acetobutylicum NCIM 2877	Orange peel waste	19.5 g/L	(Joshi et al. 2015)
C. pasteurianum	*Dunaliella* sp. and glycerol	14–16 g/L	(Jang et al. 2012)
Enterococcus hirae	Sago effluent and oil cakes	6.95 g/L	(Neethu and Murugan 2018)
Saccharomyces cerevisiae	Sunflower and poultry offal meals	59.94 mg/L	(Santos et al. 2020)
Coculture of *Bacillus subtilis* and *Clostridium acetobutylicum* ATCC 824	Agave hydrolysates	8.28 g/L	(Oliva-Rodríguez et al. 2019)
Coculture of *C. beijerinckii* 10132 with *C. cellulovorans* 35296	Wheat straw	14.20 g/L	(Valdez-Vazquez 2015)
Coculture of *C. saccharoperbutylacetonicum* and *Phlebia* sp. MG33 60-P2	Unbleached hardwood kraft pulp	2.5–3.2 g/L	(Tri et al. 2020)

integration of gas stripping, resulting in the production of 112.9, 44.1, 9.5 g/L butanol, acetone, ethanol, respectively in 312 hr of fermentation. Table 1 shows the butanol production potential of different microbes.

3. Historical progression of butanol biofuels

Butanol production is one of the oldest industrial fermentation processes that dates back to the early twentith century (Figure 2). In the year 1861, Louis Pasteur discovered microbial butanol fermentation, and then Auguste Fernbach in the year 1911 isolated a microbial strain capable of producing butanol by fermentation of potato starch but the fermentation of corn starch was not conceivable by the strain (Moon et al. 2016). Between the years 1912 to 1914, Chaim Weizmann studied several microorganisms and isolated *Clostridium acetobutylicum,* which could produce butanol from different starchy substances, as compared to

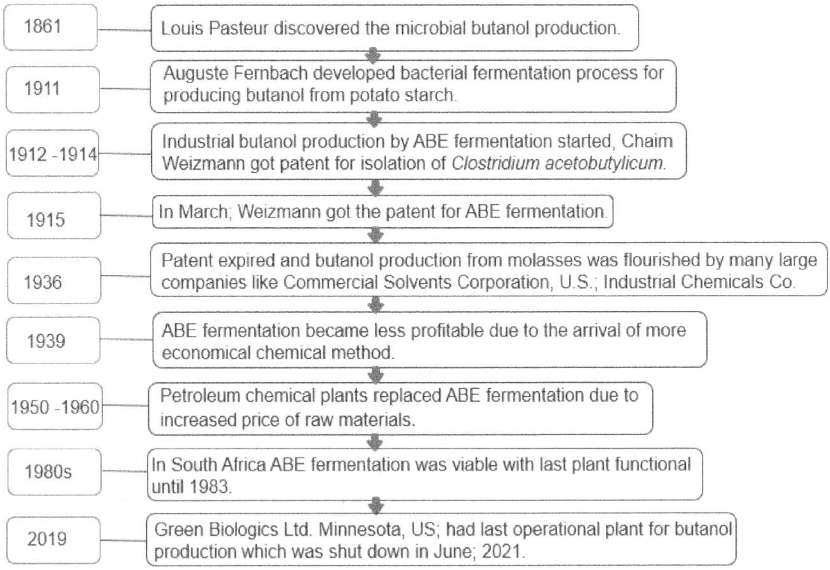

Figure 2. Historical developments in the advancement of butanol biofuel production.

the strain isolated by Fernbach (Sauer 2016). In the year 1915, Weizmann isolated *Clostridium acetobutylicum* and got a patent for ABE fermentation for butanol production using *Clostridium acetobutylicum*. This led to a boost in industrial butanol production and a high quantity of butanol was produced by ABE fermentation.

Although from 1916 to the late 1930s, ABE fermentation was the major butanol producing commercial process, Weizmann's patent expired in 1936 (Nanda et al. 2017). More research was carried out on the fermentation process, which resulted in making the butanol production more economical, by decreasing the fermentation temperature of 31°C and using a variety of materials like molasses and industrial sugars as carbon sources. From 1936 onwards, numerous fermentation plants were set up in different parts of the world and butanol production from molasses progresses rapidly began involving many large producers such as Commercial Solvents Corp., U.S. Industrial Chemicals Co., up to the 1950s (Alavijeh 2019).

It is important to note that the increased cost of raw materials as compared to the butanol yield makes the process expensive. Also, the arrival of the chemical method of butanol production resulted in a decline in the ABE fermentation process and the closing of many ABE fermentation units after 1950. On the other hand, the petrochemical plants took up butanol production to the next level due to the availability of cost-efficient chemical-based raw materials between 1950 and 1960, which eventually replaced ABE fermentation (Qureshi 2009). However, until the 1980s, the ABE fermentation remained viable in some countries like South Africa, the former Soviet Union, Egypt, China, with South Africa having the last ABE fermentation plant functional up to 1983 (Patakova et al. 2011). Moreover, Green Biologics Ltd. Company in Minnesota, US had the last operational plant for butanol production, which was shut down in June 2021. This scenario provides insight into how the butanol production sector needs more attention.

4. Current advances in the production of butanol biofuels

Recently, oxides of carbon are being used as the substrate for the production of biofuels by utilizing microbes such as *Synechococcus* sp. PCC 7942N and *Clostridium* sp. MTButOH 1365. These microbes harbor the Calvin cycle and Wood-Ljungdahl pathway. They are being genetically engineered for enhancing butanol production (Woo et al. 2019). Traditionally butanol production involves the use of solventogenic *Clostridium* via ABE fermentation but due to its drawbacks such as slow growth rate and poor butanol tolerance some potential non-solvent industrial strains such as *Saccharomyces cerevisiae*, and *Escherichia coli* are being explored (Ly et al. 2021). During the butanol production process, bacteriophages create problems with the yield, but with the help of microbiology and biotechnology, bacteriophage-resistant strains have been developed to handle the issue of infection and get better butanol yield (Generoso et al. 2015). It has been reported that electron transfer and redox metabolism play crucial roles in butanol bioconversion. So, butanol production can be enhanced by shifting from bio-catalysis to bio-electrocatalysis (Liu et al. 2021). The use of catalysts in butanol production is attracting industrial attention. Both homogenous (Iridium and Ruthenium complex) and heterogeneous (ZrO_2 and zeolite) catalysts can be used in butanol production. However, heterogeneous catalytic conversion is preferred because of its advantages over homogenous catalytic conversion (Wu et al. 2018). Butanol production can be increased by using a combination of endogenous and exogenous enzymes along with the elimination of competing pathways (Shi et al. 2016). Nowadays, mathematical modeling methods are also being widely used to optimize the butanol production conditions and the biochemical and physiological requirements of the strain used (Gürgen et al. 2018).

Previously, numerous studies have been done on the advancement in butanol production. In a study, Al-Shorgani et al. (Al-Shorgani et al. 2018) demonstrated that the response surface method was helpful for optimized butanol production using *C. acetobutylicum* YM1 and controlling important factors. From their results, it was concluded that glucose and butyric acid concentrations were the most important factors in butanol production. The model was successfully able to optimize the C/N ratio at 65 and glucose and butyric acid at 50 and 8.7 g/L respectively for maximum butanol production of 13.87 g/L. Another study by Wen et al. (2019) explored the putative potential of *C. cellulovorans* DSM 743B using a combination of metabolic and evolutionary engineering. It was observed that n-butanol production from lignocellulosic material, that is, deshelled corn cobs (AECC) increased from 0.025 to 3.47 g/L using genetically modified *C. cellulovorans*. In their work, He et al. (2016) reported that butanol productivity was increased by 60.3% and acetone and hydrogen production was reduced when the microbial cells increased the ATP production and NADH availability in *C. beijerinckii* IB4.

5. Serviceability of butanol as an alternative biofuel

Butanol is one of the alcohol-based ecofriendly and economical alternative fuels due to its similarity to gasoline in terms of heat of vaporization and energy density (Wei et al. 2016). A wide variety of feedstock such as agricultural waste (corn-fiber, rice, and wheat straw), forest residues, food industry waste, municipal waste, sewage waste, etc. (Figure 3) can be utilized for butanol production (Procentese et al. 2017). Table 2 shows recent studies which evaluated the butanol yield from various feedstocks. Butanol is being used as vehicular fuel in combustion engines without any modification due to its compatibility with the present gasoline-based engines. However, there are very limited studies available on the use of

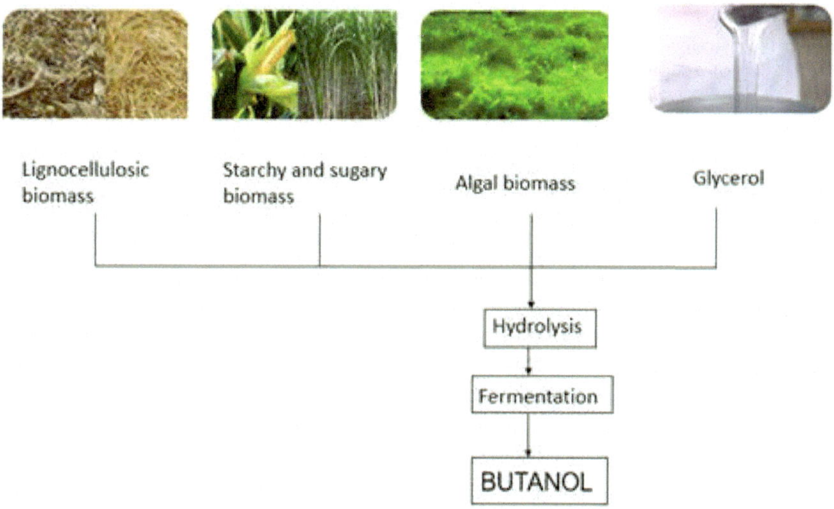

Figure 3. Butanol production from different feedstocks.

Table 2. Recent studies on butanol biofuels production using different wastes.

Feedstock	Method	Butanol yield	References
Biodegradable fraction of municipal solid waste	Acetone butanol ethanol (ABE) Fermentation	83.9 g/Kg	(Farmanbordar et al. 2018)
Wheat starch wastewater	ABE Fermentation	9.41 g/L	(Luo et al. 2018)
Pineapple waste juice	ABE Fermentation	0.08 g/g	(Sanguanchaipaiwong et al. 2018)
Corn processing industry waste	ABE Fermentation	11.92 g/L	(Zhang and Baolei 2018)
Agricultural waste (Cassava stems, banana crop residue)	Enzymatic hydrolysis, fermentation	11.68 ± 0.31 g/L and 14 g/L respectively.	(Saekhow et al. 2020, Reddy et al. 2020)
Coffee silverskin	ABE Fermentation	7.02 ± 0.27 g/L	(Hijosa-Valsero et al. 2018)
Algal biomass (*Laminaria digitate*, *Chlorella* sp.)	ABE Fermentation	7.16 g/L,6.23 ± 0.19 g/L respectively	(Hou et al. 2017, Onay 2018)
Ethanol	Catalytic conversion	34%	(Kulkarni et al. 2018)

pure butanol as engine fuel. Thus, butanol is blended with gasoline or petroleum-based diesel fuel (Han et al. 2017). Although, butanol produced from the same feedstock has several advantages over the other bio-alcohol fuels such as ethanol and methanol. In comparison to ethanol, the engine running on a butanol blend gives better mileage with lesser fuel consumption. It has been found that the NO_x emission increases up to 50% butanol (B50) as the amount of butanol increases above 50%, and the NO_x emission starts to decrease (Iliev 2021). Also, the use of butanol in fuel blends reduces soot emission.

The less corrosive nature of butanol makes it a "drop-in" fuel for the present fuel-distributing infrastructure (Han et al. 2017). Butanol can be used at a wider range of temperatures, has better lubricating properties, and is miscible with diesel fuel in comparison to other alcohol fuels like ethanol (Kumar et al. 2013). In their experiment, Varol et al. (2014) concluded that with the increase in the concentration of alcohols (methanol, ethanol, butanol) in the blends, there is a reduction in CO emission as compared to that of net gasoline, which is due to the presence of higher oxygen. In a study (Yu et al. 2018) investigated that an n-butanol + gasoline blend results in a reduction of gaseous emission from a compression ignition (CI) engine since n-butanol reduces the accumulation of particles. The minimum particulate emission was recorded with 20% n-butanol. In their work, Yang et al. (2015) showed that the addition of butanol to biodiesel or diesel blends caused a reduction in the formation of primary and secondary carbonyl compounds, including formaldehyde (FOR) and acetaldehyde (ACE). Besides, the fuel utility butanol has several other uses also, such as it can be used as an extractant by the food and pharmaceutical industries, in drugs, cosmetics, detergent formulations, 2-butoxyethanol synthesis, etc. (Dewoolkar et al. 2017). as chemical precursor in the production of plastics, polymers, paint thinners.

6. Challenges and future perspectives in the adoption of butanol biofuels

Despite the certain advantages of butanol, its production methods have their limitations and compensations. However, the biological synthesis of butanol is the most preferred one due to safety reasons. Below are some major challenges which are faced by the butanol biofuel production sectors (Table 3).

Cost-intensive: The production of butanol is cost demanding since the feedstock for butanol production is one of the major obstacles in butanol production. It has been seen that the cost of the raw material itself is about 60% of the total production cost (Gao et al. 2014). However, the adoption of low-cost sustainable biomasses as a substrate, such as waste materials (agricultural and forest residues) can help in resolving this issue (Cheng et al. 2019). Also, the distillation process involved in butanol production demands a high energy input as compared to the butanol concentration, which makes it a cost-intensive process (Nanda et al. 2017). In a study, Xue et al. (2014) showed that the gas stripping and concentration processes like distillation required nearly 14–31 MJ/kg of energy which is quite high as compared to the amount of fuel recovered, making the process energy and cost-intensive.

Wastewater generation: During butanol production, a huge volume of water is used in the pretreatment process, thereby contributing to the wastewater being generated (Xue et al. 2017). As compared to ethanol, the production of butanol generates more wastewater, leading to adverse environmental impacts (Mariano et al. 2013). From their work, Xiao et al. (2019) suggested that using hybrid pretreatment in butanol production from lignocellulosic biomass, wastewater generation reduced from (2.11–3.46 L/g-butanol) to 0.83 L/g-butanol than that of conventional dilute sulfuric acid (DA) and aqueous ammonia (AA) pretreatment.

Butanol toxicity: During butanol production by ABE fermentation, the major problem is butanol toxicity (Ibrahim 2017). *Clostridium* sp. is the

Table 3. Advantages and challenges in adoption of butanol biofuels.

S. No.	Advantage	Description
1.	High energy density	An engine running on butanol gives more mileage than that running on conventional fuels.
2.	Low hygroscopicity	Makes fuel corrosive therefore compatible with present pipeline infrastructure.
3.	Low vapor pressure and high flashpoint	Makes fuel safer for handling and transportation; a high temperature is needed to explode.
4.	Fuel additive	The calorific value and octane number of butanol are very similar to gasoline showing similar engine performance therefore, it can be used as a fuel additive.
5.	Variety feedstock	Butanol can be produced from a wide variety of feedstock such as industrial, municipal, and agricultural wastes.
6.	Renewable and greener energy source	Biodegradable in nature reduces vehicular emission and dependency on nonrenewable energy sources and improves atmospheric conditions.
	Challenge	**Proposed solution**
1.	Cost-intensive	Renewable feedstocks (waste materials), optimization of media, and cost-efficient pre-treatment methods.
2.	Low concentration and butanol toxicity	Genetically modified strains can solve the low concentration and toxicity level issue.
3.	Bacteriophage infection	Bacteriophage-resistant strains can be generated or infection can be checked at the initial stage by adding antibiotics, non-ionic detergents, and chelating agents.
4.	Wastewater generation	Wastewater generation during butanol production can be reduced by using hybrid pre-treatment.

most commonly used organism for biological butanol production via ABE fermentation. It has been reported that *Clostridium* sp. is rarely able to tolerate more than 2% butanol (Cai et al. 2013). Through a study by Garcia et al. (2011), it is evident that a wild-type of *Clostridium* strain can produce a maximum of up to 12–13 g/L of butanol following the ABE fermentation process. Another experiment by (Qureshi et al. 2014) showed that the (15–20 g/L) concentration of biobutanol is toxic to the microbial cell and thus resulted in retarding the further glucose uptake by the microorganisms.

Bacteriophage infection: During butanol production, bacteriophage infection is one of the common challenges. During the ABE fermentation, opportunistic bacteriophages infect the *Clostridium* cells, thereby hampering the butanol production (Survase et al. 2012). It has been found that *Siphoviridae* and *Podoviridae* bacteriophage infect *Clostridium madisonii* and *C. beijerinckii* P260, respectively, resulting in retarded growth and reduced butanol production (Nanda et al. 2017).

As stated above, there is an urgent need for a better understanding of the physiology of strains used for butanol production and their induced genomic regulation. Moreover, genetic engineering could help in the modification of the bacterial genome which can tolerate high butanol concentration to improve the butanol productivity and yield. An adequate solvent recovery system is necessary for reducing the product toxicity and the cost of the downstream process. In the chemical synthesis of butanol, the nanocatalysts having good selectivity for the raw material like ethanol even under mild conditions can help in increasing productivity. In the present time, when fossil fuel depletion is forthcoming, renewable fuels are becoming the ray of hope, to meet the future energy demands. Therefore, butanol is one such fuel that has the potential to replace gasoline in the near future due to its similar energy properties.

7. Conclusion and recommendations

Butanol biofuels have appeared as an alternative green energy source that could contribute to achieving sustainable development goals of clean and affordable energy for all. However, as stated in the current chapter, the status of the butanol biofuel sector is not good at all which needs proper attention from both research and industrial sectors. The present chapter focused on current advances and challenges associated with mainstreaming of butanol biofuels around the world. The feasibility of butanol biofuels and their industries should be properly increased by implementing more efficient decision-making systems. Keeping this in view, policymakers are suggested to take measures for addressing the current challenges associated with the butanol biofuel sector.

Acknowledgments

Authors are highly grateful to the Department of Zoology and Environmental Science, Gurukul Kangri (Deemed to be University), Haridwar, Uttarakhand, India, and Department of Environmental Sciences, Central University of Jammu, Samba, Jammu and Kashmir, India for academically supporting the work.

References

Alavijeh, Masih Karimi, and Keikhosro Karimi. (2019). Biobutanol production from corn stover in the US. Industrial Crops and Products, 129: 641–53.

Al-Shorgani, Najeeb Kaid Nasser, HafizaShukor, Peyman Abdeshahian, MohdSahaidKalil, Wan Mohtar Wan Yusoff, and Aidil Abdul Hamid. (2018). Enhanced butanol production by optimization of medium parameters using *Clostridium acetobutylicum* YM1. Saudi Journal of Biological Sciences, 25(7): 1308–21.

Amiri, Hamid, Reza Azarbaijani, LalehParsa Yeganeh, AbolhassanShahzadehFazeli, Meisam Tabatabaei, Ghasem Hosseini Salekdeh, and Keikhosro Karimi. (2016). *Nesterenkonia* sp. strain F, a halophilic bacterium producing acetone, butanol and ethanol under aerobic conditions. Scientific Reports, 6(1): 1–10.

Cai, Di, Tao Zhang, Jia Zheng, Zhen Chang, Zheng Wang, Pei-yong Qin, and Tian-wei Tan. (2013). Biobutanol from sweet sorghum bagasse hydrolysate by a hybrid pervaporation process. Bioresource Technology, 145: 97–102.

Cai, Di, Yong Wang, Changjing Chen, Peiyong Qin, Qi Miao, Changwei Zhang, Ping Li, and Tianwei Tan. (2016). Acetone–butanol–ethanol from sweet sorghum juice by an immobilized fermentation-gas stripping integration process. Bioresource Technology, 211: 704–10.

Callegari, Arianna, Silvia Bolognesi, Daniele Cecconet, and Andrea G. Capodaglio. (2020). Production technologies, current role, and future prospects of biofuels feedstocks: a state-of-the-art review. Critical Reviews in Environmental Science and Technology, 50(4): 384–436.

Cheng, Chi, Teng Bao, and Shang-Tian Yang. (2019). Engineering clostridium for improved solvent production: recent progress and perspective. Applied Microbiology and Biotechnology, 103(14): 5549–66.

Dewoolkar, Karan D., and Prakash D. Vaidya. (2017). New hybrid materials for improved hydrogen production by the sorption-enhanced steam reforming of butanol. Energy Technology, 5(8): 1300–1310.

Farmanbordar, Sara, Keikhosro Karimi, and Hamid Amiri. (2018). Municipal solid waste as a suitable substrate for butanol production as an advanced biofuel. Energy Conversion and Management, 157: 396–408.

Gao, Kai, Simone Boiano, Antonio Marzocchella, and Lars Rehmann. (2014). Cellulosic butanol production from alkali-pretreated switchgrass (Panicum Virgatum) and phragmites (Phragmites Australis). Bioresource Technology, 174: 176–81.

García, V., Päkkilä, J., Ojamo, H., Muurinen, E., and Keiski, R.L. (2011). Challenges in biobutanol production: how to improve the efficiency? Renewable and Sustainable Energy Reviews, 15(2): 964–980.

Generoso, Wesley Cardoso, Virginia Schadeweg, MislavOreb, and Eckhard Boles. (2015). Metabolic engineering of saccharomyces cerevisiae for production of butanol isomers. Current Opinion in Biotechnology, 33: 1–7.

Gürgen, Samet, BedirÜnver, and İsmail Altın. (2018). Prediction of cyclic variability in a diesel engine fueled with n-butanol and diesel fuel blends using artificial neural network. Renewable Energy, 117: 538–44.

Han, Xiaoye, Zhenyi Yang, Meiping Wang, Jimi Tjong, and Ming Zheng. (2017). Clean combustion of n-butanol as a next generation biofuel for diesel engines. Applied Energy, 198: 347–59.

He, Ai-Yong, Chun-Yan Yin, Hao Xu, Xiang-Ping Kong, Jia-Wei Xue, Jing Zhu, Min Jiang, and Hao Wu. (2016). Enhanced butanol production in a microbial electrolysis cell by *Clostridium beijerinckii* IB4." Bioprocess and Biosystems Engineering, 39(2): 245–54.

Hijosa-Valsero et al. (2018). A global approach to obtain biobutanol from corn stover. Renewable Energy, 148: 223–33.

Hijosa-Valsero, María, Jerson Garita-Cambronero, Ana I. Paniagua-García, and Rebeca Díez-Antolínez. (2018). Biobutanol production from coffee silverskin. Microbial Cell Factories, 17(1): 1–9.

Hijosa-Valsero, M., Garita-Cambronero, J., Paniagua-García, A.I., and Díez-Antolínez, R. (2020). A global approach to obtain biobutanol from corn stover. Renewable Energy, 148: 223–233.

Hou, Xiaoru, Nikolaj From, IriniAngelidaki, Wouter J. JHuijgen, and Anne-Belinda Bjerre. (2017). Butanol fermentation of the brown seaweed laminaria digitata by *Clostridium beijerinckii* DSM-6422. Bioresource Technology, 238: 16–21.

Huang, Ching-Ning, Wolfgang Liebl, and Armin Ehrenreich. (2018). Restriction-deficient mutants and marker-less genomic modification for metabolic engineering of the solvent producer *Clostridium saccharobutylicum*. Biotechnology for Biofuels, 11(1): 1–13.

Huzir, Nurhamieza Md., Md. Maniruzzaman A. Aziz, S.B. Ismail, Bawadi Abdullah, Nik Azmi Nik Mahmood, N.A. Umor, and Syed AnuarFaua'ad Syed Muhammad. (2018). Agro-industrial waste to biobutanol production: eco-friendly biofuels for next generation. Renewable and Sustainable Energy Reviews, 94: 476–85.

Ibrahim, Mohamad Faizal, Norhayati Ramli, Ezyana Kamal Bahrin, and Suraini Abd-Aziz. (2017). Cellulosic biobutanol by clostridia: challenges and improvements. Renewable and Sustainable Energy Reviews, 79: 1241–54.

Iliev, Simeon. (2021). A comparison of ethanol, methanol, and butanol blending with gasoline and its effect on engine performance and emissions using engine simulation. Processes, 9(8): 1322.

Jang, Yu-Sin, Joungmin Lee, Alok Malaviya, Do Young Seung, Jung Hee Cho, and Sang Yup Lee. (2012). Butanol production from renewable biomass: rediscovery of metabolic pathways and metabolic engineering. Biotechnology Journal, 7(2): 186–98.

Joshi, S.M., Waghmare, J.S., Sonawane, K.D., and Waghmare, S.R. (2015). Bio-ethanol and bio-butanol production from orange peel waste. Biofuels, 6(1-2): 55–61.

Karimi, Keikhosro, Meisam Tabatabaei, Ilona SárváriHorváth and Rajeev Kumar. (2015). Recent trends in acetone, butanol, and ethanol (ABE) production. Biofuel Research Journal, 2(4): 301–8.

KazemiShariat Panahi, Hamed, Mona Dehhaghi, James E. Kinder, and Thaddeus Chukwuemeka Ezeji. (2019). A review on green liquid fuels for the transportation sector: a prospect of microbial solutions to climate change. Biofuel Research Journal, 6(3): 995–1024.

Kulkarni, Naveen V., William W. Brennessel, and William D. Jones. (2018). Catalytic upgrading of ethanol to n-butanol via manganese-mediated guerbet reaction. ACS Catalysis, 8(2): 997–1002.

Kumar, Satish, Jae Hyun Cho, Jaedeuk Park, and I.l. Moon. (2013). Advances in diesel–alcohol blends and their effects on the performance and emissions of diesel engines. Renewable and Sustainable Energy Reviews, 22: 46–72.

Liu, Yuzhong, Pablo Cruz-Morales, Amin Zargar, Michael S. Belcher, Bo Pang, Elias Englund, Qingyun Dan, Kevin Yin, and Jay D Keasling. (2021). Biofuels for a sustainable future. Cell.

Luo, Wei, Zhangmin Zhao, Hepeng Pan, Lankun Zhao, Chuangao Xu, and Xiaobin Yu. (2018). Feasibility of butanol production from wheat starch wastewater by *Clostridium acetobutylicum*. Energy, 154: 240–48.

Lv, Yang, Yujia Jiang, Wenfang Peng, Yan Fang, Weiliang Dong, Jie Zhou, Wenming Zhang, Fengxue Xin, and Min Jiang. (2021). Genetic manipulation of non-solvent-producing microbial species for effective butanol production." Biofuels, Bioproducts and Biorefining 15(1): 119–30.

Mariano, Adriano Pinto, Marina O.S. Dias, Tassia L. Junqueira, Marcelo P. Cunha, Antonio Bonomi, and Rubens Maciel Filho. (2013). Butanol production in a first-generation brazilian sugarcane biorefinery: technical aspects and economics of greenfield projects. Bioresource Technology, 135: 316–23.

Martins, Florinda, Carlos Felgueiras, MiroslavaSmitkova, and Nídia Caetano. (2019). Analysis of fossil fuel energy consumption and environmental impacts in European Countries. Energies, 12(6): 964.

Milano, Jassinnee, HwaiChyuan Ong, H. HMasjuki, W.T. Chong, Man Kee Lam, Ping Kwan Loh, and ViknesVellayan. (2016). Microalgae biofuels as an alternative to fossil fuel for power generation. Renewable and Sustainable Energy Reviews, 58: 180–97.

Mishra, Neeraj, and Akhilesh Dubey. (2017). Biobutanol: an alternative biofuel. Advances in Biofeedstocks and Biofuels, 155.

Moon, Hyeon Gi, Yu-Sin Jang, Changhee Cho, Joungmin Lee, Robert Binkley, and Sang Yup Lee. (2016). One hundred years of clostridial butanol fermentation. FEMS Microbiology Letters, 363(3).

Nanda, Sonil, Ajay K. Dalai, and Janusz A. Kozinski. (2017). Butanol from renewable biomass: highlights of downstream processing and recovery techniques. In Sustainable Utilization of Natural Resources, 187–211. CRC Press.

Nanda, Sonil, DasantilaGolemi-Kotra, John C. McDermott, Ajay K. Dalai, Iskender Gökalp, and Janusz A. Kozinski. (2017). Fermentative production of butanol: perspectives on synthetic biology. New Biotechnology, 37: 210–21.

Neethu, A., and Murugan, A. (2018). Bioconversion of sago effluent and oil cakes for bio-butanol production using environmental isolates. Biofuels.

Ng, Cheng Ying Chloe, Katsuyuki Takahashi, and Zhibin Liu. (2016). Isolation, characterization, and optimization of an aerobic butanol-producing bacterium from Singapore. Biotechnology and Applied Biochemistry, 63(1): 86–91.

Obergruber, Michal, VladimírHönig, Petr Procházka, Viera Kučerová, Martin Kotek, JiříBouček, and Jakub Mařík. (2021). Physicochemical properties of biobutanol as an advanced biofuel. Materials, 14(4): 914.

Oliva-Rodríguez, Alejandra G., Julián Quintero, Miguel A. Medina-Morales, Thelma K. Morales-Martínez, José A. Rodríguez-De la Garza, Mayela Moreno-Dávila, Germán Aroca, and Leopoldo J. Rios González. (2019). *Clostridium* strain selection for co-culture with *Bacillus subtilis* for butanol production from agave hydrolysates. Bioresource Technology, 275: 410–15.

Onay, Melih. (2018). Investigation of biobutanol efficiency of *Chlorella* sp. cultivated in municipal wastewater. Journal of Geoscience and Environment Protection, 6(10): 40–50.

Patakova, Petra, Daniel Maxa, MojmirRychtera, Michaela Linhova, Petr Fribert, ZlataMuzikova, Jakub Lipovsky, Leona Paulova, Milan Pospisil, and Gustav Sebor. (2011). Perspectives of biobutanol production and use. Biofuel's Engineering Process Technology, 11: 243–61.

Perrone, OlavoMicali, Francesco Lobefaro, Michele Aresta, Francesco Nocito, Mauricio Boscolo, and Angela Dibenedetto. (2018). Butanol synthesis from ethanol over CuMgAl mixed oxides modified with palladium (II) and Indium (III). Fuel Processing Technology, 177: 353–57.

Procentese, Alessandra, Francesca Raganati, Giuseppe Olivieri, Maria Elena Russo, and Antonio Marzocchella. (2017). Pre-treatment and enzymatic hydrolysis of lettuce residues as feedstock for bio-butanol production. Biomass and Bioenergy, 96: 172–79.

Procentese, Alessandra, Francesca Raganati, Giuseppe Olivieri, Maria Elena Russo, Marco de la Feld, and Antonio Marzocchella. (2017). Renewable feedstocks for biobutanol production by fermentation. New Biotechnology, 39: 135–40.

Qureshi, Nasib. (2009). Solvent production. In Encyclopedia of Microbiology, 512–28. Elsevier Ltd.

Qureshi, Nasib, Cotta, M.A., and Saha, B.C. (2014). Bioconversion of barley straw and corn stover to butanol (a Biofuel) in integrated fermentation and simultaneous product recovery bioreactors. Food and Bioproducts Processing, 92(3): 298–308.

Reddy, L. Veeranjaneya, A. Shree Veda, and Wee, Y.-J. (2020). Utilization of banana crop residue as an agricultural bioresource for the production of acetone-butanol-ethanol by *Clostridium beijerinckii* YVU1. Letters in Applied Microbiology, 70 (1): 36–41.

Ruan, Roger, Yaning Zhang, Paul Chen, Shiyu Liu, Liangliang Fan, Nan Zhou, Kuan Ding, Peng Peng, Min Addy, and Yanling Cheng. (2019). Biofuels: Introduction. In Biofuels: Alternative Feedstocks and Conversion Processes for the Production of Liquid and Gaseous Biofuels, 3–43. Elsevier.

Saekhow, B., Chookamlang, S., Leksawasdi, N. and Sanguanchaipaiwong, V. (2020). Enzymatic hydrolysis of cassava stems for butanol production of isolated *Clostridium* sp. Energy Reports, 6: 196–201.

Sanguanchaipaiwong, Vorapat, and NoppolLeksawasdi. (2018). Butanol production by *Clostridium beijerinckii* from pineapple waste juice. Energy Procedia, 153: 231–36.

Santacesaria, E., Carotenuto, G., Tesser, R., and di Serio, M. (2012). Ethanol dehydrogenation to ethyl acetate by using copper and copper chromite catalysts. Chemical Engineering Journal, 179: 209–20.

Santos, Bruno A.S., Suéllen, P.H. Azambuja, Patrícia F. Ávila, Maria Teresa B. Pacheco, and Rosana Goldbeck. (2020). n-butanol production by saccharomyces cerevisiae from protein-rich agro-industrial by-products. Brazilian Journal of Microbiology, 51(4): 1655–64.

Satlewal, Alok, Ruchi Agrawal, SamarthyaBhagia, Parthapratim Das, and Arthur J. Ragauskas. (2018). Rice straw as a feedstock for biofuels: availability, recalcitrance, and chemical properties. Biofuels, Bioproducts and Biorefining, 12(1): 83–107.

Sauer, Michael. (2016). Industrial production of acetone and butanol by fermentation—100 Years Later. FEMS Microbiology Letters, 363(13): fnw134.

Shi, Shuobo, Tong Si, Zihe Liu, Hongfang Zhang, Ee Lui Ang, and Huimin Zhao. (2016). Metabolic engineering of a synergistic pathway for n-butanol production in *Saccharomyces cerevisiae*. Scientific Reports 6(1): 1–10.

Shukor, Hafiza, Peyman Abdeshahian, Najeeb Kaid Nasser Al-Shorgani, Aidil Abdul Hamid, Norliza A. Rahman, and MohdSahaidKalil. (2016). Enhanced mannan-derived fermentable sugars of palm kernel cake by mannanase-catalyzed hydrolysis for production of biobutanol. Bioresource Technology, 218: 257–64.

Survase, Shrikant A., Adriaan van Heiningen, and Tom Granström. (2012). Continuous bio-catalytic conversion of sugar mixture to acetone–butanol–ethanol by immobilized *Clostridium acetobutylicum* DSM 792. Applied Microbiology and Biotechnology, 93(6): 2309–16.

Tri, Chu Luong, and Ichiro Kamei. (2020). Butanol production from cellulosic material by anaerobic co-culture of white-rot fungus Phlebia and bacterium clostridium in consolidated bioprocessing. Bioresource Technology, 305: 123065.

Uyttebroek, Maarten, Wouter van Hecke, and KarolienVanbroekhoven. (2015). Sustainability metrics of 1-Butanol. Catalysis Today, 239: 7–10.

Valdez-Vazquez, Idania, Marisol Pérez-Rangel, Adán Tapia, Germán Buitrón, Carlos Molina, Gustavo Hernández, and Lorena Amaya-Delgado. (2015). Hydrogen and butanol production from native wheat straw by synthetic microbial consortia integrated by species of enterococcus and clostridium. Fuel, 159: 214–22.

Varol, Y., Öner, C., Öztop, H.F. and Altun, Ş. (2014). Comparison of methanol, ethanol, or n-butanol blending with unleaded gasoline on exhaust emissions of an SI engine. Energy Sources, Part A: Recovery, Utilization, and Environmental Effects, 36(9): 938–48.

Veza, Ibham, Mohd Farid Muhamad Said, and Zulkarnain Abdul Latiff. (2021). Recent advances in butanol production by acetone-butanol-ethanol (ABE) fermentation. Biomass and Bioenergy, 144: 105919.

Wei, Haiqiao, Dengquan Feng, Mingzhang Pan, JiaYing Pan, XiaoKang Rao, and Dongzhi Gao. (2016). Experimental investigation on the knocking combustion characteristics of n-butanol gasoline blends in a DISI engine. Applied Energy, 175: 346–55.

Wen, Zhiqiang, Rodrigo Ledesma-Amaro, Jianping Lin, Yu Jiang, and Sheng Yang. (2019). Improved N-butanol production from *Clostridium cellulovorans* by integrated metabolic and evolutionary engineering. Applied and Environmental Microbiology, 85(7): e02560-18.

Woo, Ji Eun, and Yu-Sin Jang. (2019). Metabolic engineering of microorganisms for the production of ethanol and butanol from oxides of carbon. Applied Microbiology and Biotechnology, 103(20): 8283–92.

Wu, Xianyuan, Geqian Fang, Yuqin Tong, Dahao Jiang, Zhe Liang, WenhuaLeng, Liu Liu, Pengxiang Tu, Hongjing Wang, and Jun Ni. (2018). Catalytic upgrading of ethanol to n-butanol: progress in catalyst development. ChemSusChem, 11(1): 71–85.

Xiao, Min, Lan Wang, Youduo Wu, Chi Cheng, Lijie Chen, Hongzhang Chen, and Chuang Xue. (2019). Hybrid dilute sulfuric acid and aqueous ammonia pretreatment for improving butanol production from corn stover with reduced wastewater generation. Bioresource Technology, 278: 460–63.

Xue, Chuang, Guang-Qing Du, Jian-Xin Sun, Li-Jie Chen, Shuai-Shi Gao, Ming-Liang Yu, Shang-Tian Yang, and Feng-Wu Bai. (2014). Characterization of gas stripping and its integration with acetone–butanol–ethanol fermentation for high-efficient butanol production and recovery. Biochemical Engineering Journal, 83: 55–61.

Xue, Chuang, Xiaotong Zhang, Jufang Wang, Min Xiao, Lijie Chen, and Fengwu Bai. (2017). The advanced strategy for enhancing biobutanol production and high-efficient product recovery with reduced wastewater generation. Biotechnology for Biofuels, 10(1): 1–11.

Yang, Po-Ming, Yuan-Chung Lin, Kuang C Lin, Syu-RueiJhang, Shang-Cyuan Chen, Chia-Chi Wang, and Ying-Chi Lin. (2015). Comparison of carbonyl compound emissions from a diesel engine generator fueled with blends of n-butanol, biodiesel and diesel. Energy, 90: 266–73.

Yu, Xiumin, Zezhou Guo, Ling He, Wei Dong, Ping Sun, Weibo Shi, Yaodong Du, and Fengshuo He. (2018). Effect of gasoline/n-butanol blends on gaseous and particle emissions from an SI direct injection engine. Fuel, 229: 1–10.

Zhang, Jie, and Baolei Jia. (2018). Enhanced butanol production using *Clostridium beijerinckii* SE-2 from the waste of corn processing. Biomass and Bioenergy, 115: 260–66.

Zhen, Xudong, Yang Wang, and Daming Liu. (2020). Bio-butanol as a new generation of clean alternative fuel for SI (Spark Ignition) and CI (Compression Ignition) engines. Renewable Energy, 147: 2494–2521.

5

Progress in Butanol Generation and Associated Challenges

Bikash Kumar[1,2] and *Pradeep Verma*[1,*]

1. Introduction

Butanol is a four-carbon consisting of straight-chain or branched alcohol. There are different isomers of butanol that are based on the location of the carbon chain structure and -OH bond. These isomers are n-butanol, 2-butanol, iso-butanol, and tert-butanol with motor octane numbers 78, 32, 94, and 89, respectively. In general, butanol is used as an organic solvent in antibiotics, cosmetics, detergents, drugs, hormones, hydraulic fluids, and vitamins. Butanol can also act as a chemical intermediate for several high-value compounds (methacrylate and butyl acrylate) and is used in the extraction of pharmaceutical products. Iso-butanol and tert-butanol have high octane numbers and are being used as gasoline additives as octane boosters (Liu et al. 2013a).

1.1 Butanol: A promising fuel

Butanol can be used as a fuel additive but in recent years, it has attracted the attention of the scientific community as a potential liquid fuel due to its efficient properties such as high heat value, hydrophobicity, and

[1] Bioprocess and Bioenergy Laboratory, Department of Microbiology, Central University of Rajasthan, NH-8, Bandarsindri, Ajmer 305817, Rajasthan, India.
[2] Department of Biosciences and Bioengineering, Indian Institute of Technology, Guwahati, Surjyamukhi Road, North Guwahati, 781039, Assam, India.
* Corresponding author: vermaprad@yahoo.com, pradeepverma@curaj.ac.in

viscosity accompanied by low corrosiveness and volatility (Freeman et al. 1988, Dean 2005). Apart from this, a comparative evaluation of different properties of homologous potential transportation fuels are provided in Table 1.

Biofuels such as ethanol and butanol are produced in limited amounts as compared to the current requirement for them. Thus, to overcome the challenge of availability as well as the need to look for an alternative greener transportation fuel, blending biofuel with the gasoline has been suggested for the last 30 years and is currently in practice. This blending is also included in government policies for meeting the growing energy needs and to create a greener environment. The blending of bioethanol with gasoline is in practice but has several disadvantages such as:

a. Low (1/6th) heating value of ethanol as compared to gasoline, thus the requirement of retrofitted engines to minimize fuel consumption.
b. Generation of acetic acid during ethanol burning leads to the corrosion of engine and other components of the vehicle.
c. Requirement of preservatives when the ethanol proportion is increased above 15%.
d. Ethanol is hygroscopic, thus a high-water proportion is required for the liquid phase separation.
e. The difficulty associated with preservation and allocation, storage, and transition as compared to gasoline.

Replacing butanol with ethanol in blending has several advantages:

a. Has a high energy content and burning efficiency and thus can be useful in long-distance traveling.
b. The energy content and air to fuel ratio is closer to gasoline.
c. It is less explosive, less volatile, and has a lower vapor pressure with higher flash points, making it safer to handle in transportation and storage.
d. No requirement of retrofitting to use butanol blended gasoline.

These advantages can potentially help in overcoming the limitations presented by ethanol blending (Dürre 2007, Pfromm et al. 2010, Liu et al. 2013a).

1.2 History of butanol being preferred as a transportation fuel

In 1861, Pasteur reported for the first time biobutanol synthesis at the laboratory scale (Dürre 1998), and industrial-level biobutanol production via fermentation was demonstrated in 1912–1914 (Jones and Woods 1986). The butanol was produced primarily for the application as a solvent in

Table 1. A comparative evaluation of different properties of homologous fuels (adapted from Burhani et al. 2019, Freeman et al. 1988, Liu et al. 2013a, Sarangi and Nanda 2018).

Fuel	Methanol	Ethanol	Butanol	Gasoline
Formula	CH_3OH	CH_3CH_2OH	C_4H_9OH	HC_4H_{12}
Structure				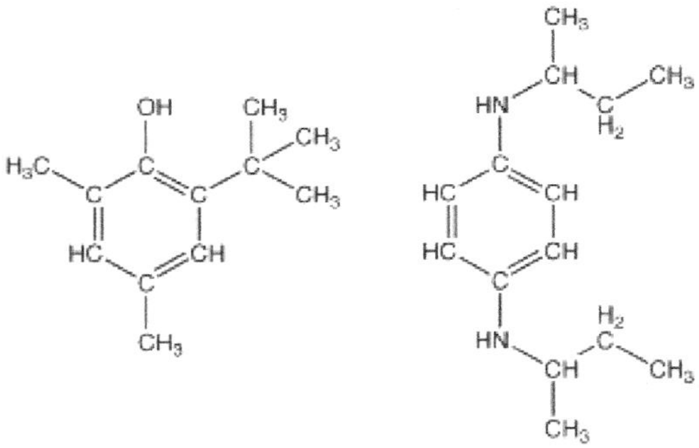
Boiling point (°C)	65	78	118	32–210
Auto-ignition temperature (°C)	435	365	343	280
Air Fuel Ratio	6.5:1	9:1	11.2:1	14.6:1
Cetane Number	3	8	25	0-10
Combustion energy (MJ/dm³)	16	19.6	29.2	32
Density at 20°C (g/m³)	0.797	0.789	0.81	0.7
Energy density (MJ/Kg)	22.7	26.9	33.1	32
Evaporation heat (MJ/Kg)	1.2	0.92	0.43	0.36

Table 1 contd. ...

...Table 1 contd.

Fuel	Methanol	Ethanol	Butanol	Gasoline
Flammability limits	6–36.5	4.3–19	1.4–11.2	0.6–0.8
Heat of vaporization (MJ/Kg)	1.20	0.92	0.43	0.36
Higher heating value (MJ/Kg)	37.18	29.8	37.3	46.5
Lower heating value (MJ/Kg)	22.7	26.9	34.3	43.4
Octane Number	111	108	96	80–99
Oxygen content (% weight)	-	34.8	21.6	-
Viscosity at 25°C (mPa.s)	0.544	1.074	2.573	0.6
Saturation pressure (kPa) at 38°C	31.69	13.8	2.27	31.01
Water solubility (mL/100 mL)	-	Miscible	9.1	< 0.01

antibiotics, cosmetics, detergent formulations, drugs, hormones, hydraulic fluids, and vitamins, and as a precursor or intermediate compound for the production of butyl acrylate and methacrylate (Liu et al. 2013a). Butanol is also used as an extraction agent in pharmaceutical manufacturing (Qureshi 2009). Prior to 2005, the application of butanol was limited to chemical synthesis but David Ramey demonstrated the potential of butanol as a renewable fuel when he drove his car on 100% butanol from Ohio to California. After that, several giant companies such as DuPont and BP have shown interest in modernizing butanol production plants. In the 1990s, a pilot scale operation of continuous fermentation with new technologies was demonstrated in Austria, which proved to be economically feasible by utilization of agricultural waste like potatoes.

2. Production of butanol

Butanol can be produced by chemical as well as biological processes. Chemical technologies such as oxo synthesis and aldol condensation are most popular for biobutanol production (Park 1996, Zverlov et al. 2006). Among biological processes, anaerobic fermentation using strains of the *Clostridia* genus is the most popular method for large-scale butanol fermentation (Karimi et al. 2015).

2.1 Chemical process for butanol production

The two major chemical-based synthesis systems for butanol production are aldol condensation and oxo synthesis.

2.1.1 Aldol Condensation

The chemical process for the production of butanol via aldol condensation involves the condensation and dehydration from two molecules of acetic aldehyde to generate crotonaldehyde. This intermediate product is then subjected to hydrogenation at 180°C and 0.2 MPa to form n-butanol.

$$CH_3CH{=}CHCHO{+}2H_2 \text{ --------->} \ CH_3CH_2CH_2CH_2OH$$

2.1.2 Oxo synthesis

Oxo synthesis method involves the reaction of propylene with carbon monoxide and hydrogen in the presence of cobalt or rhodium as the catalyst.

$$CH_3\text{-}CH\text{-}CH_2 + CO + H_2 \text{----->} \ CH_3CH_2CH_2CHO + (CH_3)_2CHCHO$$

(In presence of Co/Rh) (1.1)

In reaction 1.1., when cobalt is used as a catalyst, the reaction occurs at 130–160°C at 10–20 MPa and the products n-butyraldehyde and iso-butyraldehyde are in the ratio of 1:3. Whereas when rhodium is used as a catalyst, it requires lower temperature (80–120°C) and low pressure (at 0.7–3 MPa) with a high product ratio (8–16) of n-butyraldehyde and iso-butyraldehyde. The n- and isobutyraldehyde mixture is then hydrogenated to the corresponding n-butanol and isobutyl alcohols (Liu et al. 2013a).

$$CH_3CH_2CH_2CHO + H_2 \text{ --------->} CH_3CH_2CH_2CH_2OH \text{ (a)}$$

$$(CH_3)_2CHCHO + H_2 \text{ ---------->} (CH_3)_2CHCH_2OH \text{ (b)}$$

The hydrogenation processes are performed either in the presence of nickel or copper catalyst in the gaseous phase or nickel in the liquid phase. During oxo synthesis, some by-products can be transferred into butanol at a high temperature and high pressure that will further increase the purity of the product. The oxo synthesis method is the most preferred method for the industrial production of n-butanol due to several advantages such as easily available materials, comparable moderate reaction conditions, and a high ratio of n-butanol to isobutyl alcohol (Zverlov et al. 2006).

2.2 Biological process for butanol production

The biological method for butanol production is mediated by a microbe-assisted fermentation process. *Clostridia* are the most common genus that is key for butanol production under anaerobic conditions. The major fermentation products are acetone, butanol, and ethanol and this is called ABE fermentation (Karimi et al. 2015). There are several advantages associated with a biological system such as it can utilize a wide range of biological substrates, mild fermentation condition, high product selectivity, high security, lesser by-products, and easier recovery (Nimcevic and Gapes 2000, Liu et al. 2013a).

3. ABE Fermentation for biobutanol production

As discussed earlier, the most popular mechanism for biobutanol production is ABE fermentation, which is even employed at an industrial scale. This section will give a perspective on available microbes and biomass suitable for biobutanol production using ABE fermentation.

3.1 Microorganisms suitable for butanol production

Several microorganisms are capable of application in the biological ABE fermentation process. Among them, the *Clostridium* genus includes a variety of organisms capable of butanol production. *Clostridium acetobutylicum, C. aurantibutyricum, C. beijerinckii, C. saccharoacetobutylicum,* and *C. sporogenes* are major butanol-producing bacteria from the *Clostridia* genus (Kharkwal et al. 2009, Ni & Sun 2009, Patakova et al. 2013).

At both the laboratory and industrial scale, *C. acetobutylicum,* and *C. beijerinckii* has demonstrated high efficiency (Mo et al. 2015) and several modified forms of these two strains has also been applied in ABE fermentation (Ni and Sun 2009, Komonkiat and Cheirsilp 2013, Patakova et al. 2013, Li et al. 2014a,b). *Clostridium acetobutylicum* (Weizmann's organism) is a rod-shaped, gram-positive, obligately anaerobic, spore-forming bacteria that has been utilized for the first time at an industrial scale for ABE fermentation (Maddox 1989, Kharkwal et al. 2009, Ni and Sun 2009, Moo-Young 2019).

Naturally, butanol production using *Clostridia* strains is limited by product inhibition, that is, it cannot tolerate butanol concentration beyond 13–20 g/L (Garcia et al. 2011). Thus, co-culturing techniques have been used to overcome this limitation where one anaerobic or one aerobic strain (Tran et al. 2010) or two different anaerobic strains (Li et al. 2013) have been used for enhanced butanol production and minimizing the strenuous anaerobic environment (Kushwaha et al. 2019).

The selection of microorganisms for the ABE fermentation depends on several factors such as type of substrate, the requirement of additional nutrients, required production rate, and resistance to contaminations via bacteriophages (Kumar and Gayen 2011). Several biotechnological advances such as mutagenesis and evolutionary engineering work have been demonstrated by several groups of scientists. The mutation of *C. acetobutylicum* was demonstrated to have enhanced butanol production up to 20 g/L, which was much higher than commercial ABE fermentation, resulting in 12 g/L butanol (Xue et al. 2012, Jiang et al. 2014b). The success of mutagenesis and evolutionary engineering is subjective and greatly depends on chance. Studies have suggested that the artificial stimulation of bio-evolution, that is, repetitive evolutionary training or domestication of the *Clostridiaacetobutylicum* strain could enhance butanol tolerance up to 4% (Liu et al. 2013b).

4. Substrates/biomass for butanol

Clostridia genus plays a vital role in biobutanol production via ABE fermentation. The *Clostridia* spp. is capable of utilizing a wide range of sugars such as monosaccharides (galactose, glucose, and xylose), disaccharides (lactose, maltose, and sucrose), and polysaccharides (starch and dextrin). However, the utilization ability may vary for different sugars; for example, fructose, glucose, mannose, lactose, sucrose, starch, and dextrins are completely utilized whereas arabinose, galactose, insulin, raffinose are fermented partially. The sugars such as melibiose, rhamnose, and trehalose are not fermented by *Clostridia* sp. (Kumar and Gayen 2011). The ability of *Clostridium* sp. to completely utilize starch without the need of any intermediate hydrolysis step is one of its salient properties (Li et al. 2014a, Thang and Kobayashi 2014). Studies show that microbes can utilize the different generations of feedstock for generating fermentation for the production of butanol. The first-generation feedstocks such as rice, cassava, cereal grains, and sugarcane can be suitable for butanol production (Bušić et al. 2018). However, their direct competition with food availability limits their applications in biobutanol production (Kumar et al. 2020). Thus, food industry wastes such as cooked food, hotel waste, vegetable waste, fruit peels, etc., can be used for biobutanol production (Isah and Ozbay 2020).

The second-generation feedstocks, primarily lignocellulosic biomass such as rice straw, wheat bran, sugarcane bagasse, corn stover, etc., are a rich source of cellulosic and hemicellulosic components. Studies suggest Cl*ostridium* can utilize cellulose and hemicellulose due to their ability to produce cellulolytic enzymes (Berezina et al. 2009). Berezina et al. (2009) demonstrated the high hemicellulolytic activity in *C. saccharobutylicum* strain thus resulting in butanol yield utilizing lignohemicellulosic biomass.

The four different *Clostridia* strains that is, *C. acetobutylicum, C. beijerinckii, C. saccharoperbutylacetonicum,* and *C. saccharobutylicum,* are commonly used for butanol production. Out of these four, *C. acetobutylicum* is phylogenetically distinct and is the original starch-fermenting strain. All four *Clostridia* strains can be utilized in the glucose-containing medium. *C. beijerinckii* gave the lowest solvent yield of 28% when subjected to a 4% glucose TYA medium while the other strains showed a solvent yield of more than 30% in each case (Shaheen et al. 2000). *C. acetobutylicum* gave the best solvent yield of 19 g/L with a standard supplement maize medium whereas other strains such as *C. beijerinckii, C. saccharoperbutylacetonicum,* and *C. saccharobutylicum* gave a solvent yield of 16, 14, and 11 g/L, respectively. *C. acetobutylicum* cannot utilize molasses well when compared to other *Clostridia* strains. *C. beijerinckii* and *C. saccharobutylicum* are the best molasses-fermenting strains (Shaheen et al. 2000). *C. saccharoperbutylacetonicum* can utilizes sugar, molasses, and maize. *C. beijerinckii* can withstand acetic acid and formic acid concentration as compared to *C. acetobutylicum,* thus it is advantageous to use it while utilizing lignocellulosic hydrolysate pretreated with acetic and formic acids (Cho et al. 2012).

Some *C. beijerinckii* strains can utilize isopropanol instead of acetone (George et al. 1983). Also, some microbes such as

Butyribacteriummethylotrophicum, *C. autoethanogenum*, *C. ljungdahlii*, and *C. carboxidiworans* capable of utilizing carbon monoxide (CO) and molecular hydrogen (H_2), including acetogens. The genomic study of *C. carboxidivorans* strain P7(T) suggested that it possesses a Wood-Ljungdahl pathway gene cluster that is responsible for the utilization of CO, and hydrogen and conversion to acetyl-CoA (Bruant et al. 2010).

The major limitation associated with the application of lignocellulosic biomass are the variations in the cellulosic and hemicellulosic content of biomass. Thus, variations in the choice of pretreatment method for the preprocessing of biomass prior to fermentation is required. The other limitations with utilizing lignocellulosic biomass as feedstock for butanol production are geographical and seasonal variations, high lignin content causing inhibitor formation during pretreatment, requirements of arable land, and constant water supply for their growth (Kushwaha et al. 2019). The limitations of first-generation and second-generation feedstock has motivated scientist to look for alternatives. The algal biomass having higher carbohydrate content and lower lignin presence makes it a sustainable and promising alternative for the production of butanol (Chen et al. 2013, Suutari et al. 2015). Also, algae can utilize wastewater as a substrate, as it can assimilate carbon dioxide and remove inorganic nutrients (Oswald 2003, Kushwaha et al. 2019), thus, serving the dual purpose of biomass generation and bioremediation. Table 2 summarizes the list of several feedstock used for biobutanol production. Ibrahim et al. (2018) summarized the different stages of the conversion of different feedstock to biobutanol (Figure 1).

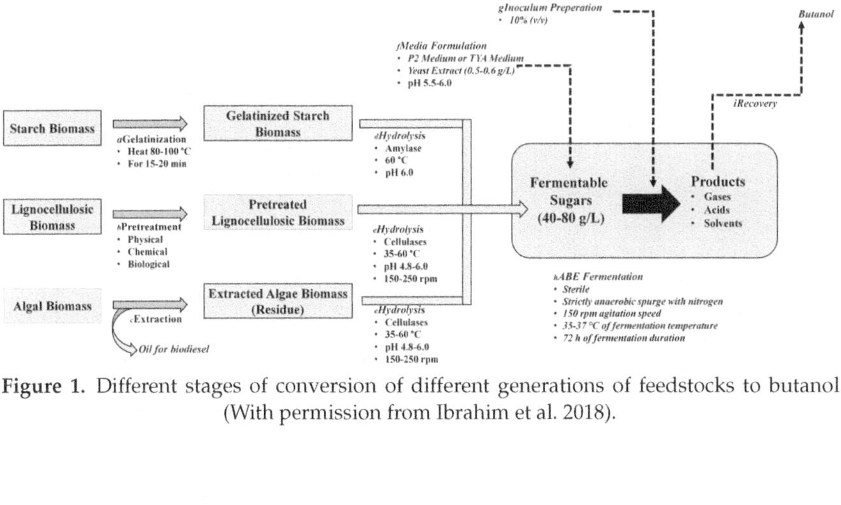

Figure 1. Different stages of conversion of different generations of feedstocks to butanol (With permission from Ibrahim et al. 2018).

Table 2. Different generations of feedstock for biobutanol production.

Feedstock	Critical observations	References
First-generation feedstock		
Cane molasses	Reducing sugar content up to 50–55%	(Jiang et al. 2009, Ni et al. 2012, der Merwe et al. 2013)
Cheese whey	Better substrate for butanol production as compared to other lactose substrates	(Foda et al. 2010, Becerra et al. 2015)
Food industry waste	Provides the additional advantage of waste management	(Stoeberl et al. 2011, Huang et al. 2015)
Cooked rice (food waste)	Techno-economic evaluation of butanol production via two-step fermentation. Also, process optimization and reduction in operational cost for the aerobic ABE fermentation and microbial saccharification	(Ozturk et al. 2021)
Pineapple peel (fruit waste)	ABE fermentation yield of 5.23 g/L using *C. acetobutylicum* B 527. Removal of inhibitors using activated carbon	(Khedkar et al. 2017)
Sugarcane	Overexpression of scrA, scrB, scrK, and adhE2 gene of *C. tyrobutyricum* for enhanced butanol production	(Zhang et al. 2017)
Second-generation feedstocks		
Rice straw	Batch fermentation of alkali pretreated rice straw using *Clostridium acetobutylicum* ATCC 824 resulted in a biobutanol titer of 9.10 g/L	(Tsai et al. 2020)
Rice straw	Acid (4% sulfuric acid) pretreated rice straw was subjected to fermentation via non-acetone-producing *Clostridium sporogenes* BE01. Acid pretreatment resulted in inhibitors formation and caused a decrease in butanol yield. Detoxification of inhibitors via using amberlite resins results in enhanced butanol yield, that is, 5.52 g/L as compared to 3.43 g/L with inhibitors	(Gottumukkala et al. 2013)

Second-generation feedstocks		
Sugarcane bagasse	Batch fermentation of alkali pretreated rice straw using *Clostridium acetobutylicum* ATCC 824 resulted in a biobutanol titre of 8.40 g/L	(Tsai et al. 2020)
Oil palm empty fruit bunch (OPEFB)	Enhanced butanol production via simultaneous saccharification and fermentation. Delayed yeast extract feeding at the 39th hour of fermentation resulted in a 42% increase in butanol yield. *In situ* gas stripping resulted in a 72% enhancement in butanol recovery	(Salleh et al. 2019)
Corn stover	Corn stover is an efficient feedstock for butanol production. Feedstock processing and optimizing the fermentation process design is key to enhanced production	(Baral et al. 2016)
Third-generation feedstock		
Neochloris oleoabundans HK-129	Medium condition (nitrogen starvation, iron) physical parameter (light intensity) impacts the triacylglyceride/carbohydrate production and fatty acid profile of microalgae. This impacts the suitability of microalgal biomass for butanol production	(Sun et al. 2014)
Microalgae-based biodiesel residues	Utilization of lipid-extracted algal residue for butanol fermentation is an economically feasible option	(Cheng et al. 2015)
Chlorella vulgaris JSC-6	Pretreated of microalgae via 1% NaOH and 3% H_2SO_4 resulting in 13.1 g/l of butanol. There is no need for a detoxification process.	(Wang et al. 2016)
C. vulgaris UTEX 2714	Microalgal carbohydrates generated via ionic liquid pretreatment of lipid extracted microalgal biomass when subjected to fermentation resulted in a high butanol concentration of 8.05 g/L	(Gao et al. 2016)
Microalgae	Unhydrolyzed microalgal biomass in ABE fermentation using immobilized *C. acetobutylicum* resulted in 4.32 g/L biobutanol	(Tsai et al. 2020)

5. Biobutanol production

5.1 Mechanism of biobutanol production

Biobutanol production can broadly be divided into two different steps, that is, extracellular breakdown of polymeric carbohydrates using enzymes secreted by *Clostridia* strain, and intracellular two-stage ABE fermentation (Pugazhendhi et al. 2019) (Figure 2).

In the extracellular breakdown, the bacterial enzymes secreted by *Clostridia* breakdown the carbohydrate polymers present in the biomass into monomers (Ezeji et al. 2004a, Wang et al. 2017). The α-&ß amylase, glucoamylase, α-glucosidase, and pullulanases are required for the breakdown of starchy components. Similarly, cellulases and ß-glucosidase hydrolyze cellulose, resulting in the generation of glucose (Pugazhendhi et al. 2019). Hemicellulose is broken down into xylose and arabinose via the action of hemicellulases. Further, the subsequent action of transketolase and transaldolases result in the generation of fructose 6-phosphate and glyceraldehyde-3-phosphate through the Embden-Meyerhof-Parnas (EMP) pathway (Madhavan et al. 2012, Paritosh et al. 2017). Subsequently, utilizing the glucose phosphotransferase system (PTS), the glucose penetrates through the cell membrane where the intracellular

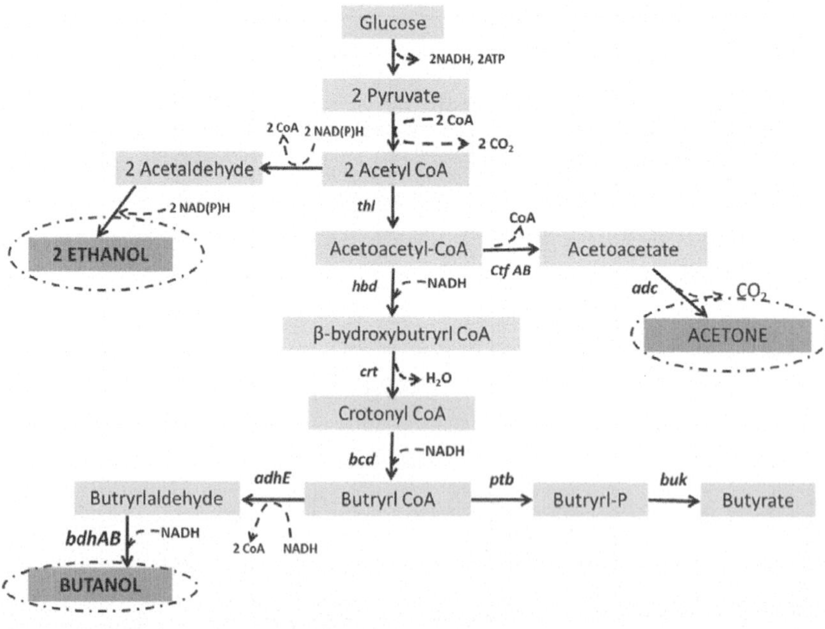

Figure 2. Metabolic pathways in *Clostridia* showing acidogenic and solventogenesis (With permission from Pugazhendhi et al. 2019).

ABE fermentation takes place (Xiao et al. 2011). The waste biomass consists of polysaccharides, proteins, and lipids; as discussed earlier, the polysaccharides are converted to monosaccharides and oligosaccharides whereas the residual lipids and proteins are converted to fatty acids and glycerol (Paritosh et al. 2017).

Two-stage ABE fermentation takes place intracellularly, resulting in acetate, butanol, and ethanol. As explained earlier, the monosaccharides penetrate the cell utilizing the membrane-bound transport systems where it is metabolized by glycolysis (Pugazhendhi et al. 2019). There are two discrete phases for ABE fermentation namely, acidogenesis, and solventogenesis (Bankar et al. 2015, Luo et al. 2017). During acidogenesis, acetic and butyric acids are produced along with the co-evolution of hydrogen (Patakova et al. 2013, Al-Shorgani et al. 2018). Acidogenesis results in a decrease in the pH due to the accumulation of organic acids. This initiates solventogenesis resulting in the generation of acetone, butanol, and ethanol accompanied by constant cell growth and reduction in hydrogen release. The initial growth phase involves the partial uptake of acids (Luo et al. 2017). Toward the end, *C. acetobutylicum* undergoes aging, loses bioactivity, and forms spores by autolysis. Overall, ABE fermentation is regulated via carbon and electron transfer and the fermentation yield greatly depends on the choice of strains and culture conditions (Pugazhendhi et al. 2019).

5.2 Fermentation system for butanol production

ABE fermentation can be conducted by researchers in different fermentation modes, that is, batch, fed-batch, and continuous, two-stage continuous mode, and alternative fermentation under anaerobic conditions (Pugazhendhi et al. 2019). The batch bioreactor operates by fermenting all media in one go whereas the fed-batch system involves the addition of media when the fermentation is under process. The continuous mode, as the name suggests, is a continuous process that involves the continuous addition of medium and collection/removal of fermentation broth/product removal. The continuous mode can be further subdivided into single-stage continuous fermentation and two-stage continuous fermentation (Pugazhendhi et al. 2019).

5.2.1 Batch fermentation

The simplest mode of fermentation for industrial production of butanol is the batch mode which typically involves substrate levels of 40–80g/L. Studies showed that efficiency decreased with an increase in substrate concentration over 80 g/L (Shaheen et al. 2000). Isar and Rangaswamy (2012) demonstrated that under optimized physiological and nutritional parameters, solvent tolerant *C. beijerinckii* ATCC 10132 resulted in 20g/L n-butanol in 72h. *C. acetobutylicum* JB200, a butanol-tolerant species capable of large-scale production of butanol when subjected to the repeated batch mode process. *C. acetobutylicum* JB200 cells were immobilized in a fibrous bed bioreactor and used for 16 consecutive batches for 800 h, resulting in a butanol yield of 16–20g/L (Jiang et al. 2014b). Biobutanol production using algal biomass growing in wastewater has been subjected to biobutanol production in batch mode. Ellis et al. (2012) demonstrated ABE production using wastewater algae with and without enzymatic hydrolysis. Maximum ABE production of 9.74 g/l was obtained after enzymatic hydrolysis, while a comparatively low solvent yield of 0.73 g/l was observed for non-hydrolyzed algal biomass. Batch mode attracts attention due to the relatively low rate of substrate utilization, and low risk of contamination and strain mutation (Kushwaha et al. 2019). On an industrial scale, batch fermentation can result in a butanol yield of 80,000 tons in 25,000 m^3 with 0.5 g/L/h of *Clostridial* inoculum. In a comparison of an average fermentation plant, batch fermentation results in a lower butanol yield as compared to continuous fermentation (Kujawska et al. 2015).

Thus, the major limitations associated with batch fermentation are the low yields of butanol that subsequently affect the recovery process efficiency which requires high energy, and involves high cost as compared to the overall yield (Kujawska et al. 2015). The low biobutanol yield in the batch system is due to the low glucose concentration supplied for fermentation (less than 100 g/L) to prevent substrate inhibition. Further, several inhibitors present in pretreated biomass hydrolysate also inhibit the growth of bacterial cells during the early stage, that negatively affects the batch fermentation with high substrate loading (Ibrahim et al. 2018). Some of the other drawbacks of batch fermentation are the loss of substrate and residual medium (Pugazhendhi et al. 2019).

5.2.2 Fed-batch fermentation

In order to overcome the limitation of batch fermentation, that is, substrate inhibition or catabolite repression, the fed-batch mode is adapted (Pugazhendhi et al. 2019). In the fed-batch reactor, to avoid substrate inhibition, a relatively low substrate concentration is loaded. Once the substrate is utilized, a small volume of the concentrated substrate is loaded at a low working volume, thus keeping the substrate concentration in a limit below the toxic level (Ibrahim et al. 2018, Pugazhendhi et al. 2019). High substrate concentration can help in reduction in the hydraulic load and wastewater generated during the process. In fed-batch system, uniformity in the cell population is maintained by monitoring cell density at regular time intervals. The rate of nutrient supply and culture withdrawal is decided by evaluating the turbidity of the culture (Pugazhendhi et al. 2019). Further, in the remaining culture, the substrate is removed at the end of the reaction. Tashiro et al. (2004) demonstrated fed-batch fermentation using glucose and butyric acid as a substrate, resulting in a high butanol yield of 15 g/L with low ethanol formation due to a combination of substrates used in fermentation. Utilization of lignocellulosic substrate such as cassava bagasse hydrolysate with *C. acetobutylicum* in batch fermentation accompanied with continuous gas stripping resulted in a biobutanol yield of 76.4 g/L of biobutanol (Lu et al. 2012). Similarly, Pang et al. (2016), demonstrated the application of alkali pretreated sugarcane bagasse as a substrate for fed-batch fermentation by *Clostridium acetobutylicum* GX01 resulting in a butanol yield of 14.07 g/L.

Fed-batch fermentation helps in minimizing the loss of substrate with aiding the enhancement of the final titer of the target product (Dolejš et al. 2014). Although fed-batch fermentation results in an improvement in the biobutanol titer, the biobutanol yield is relatively low as compared to the batch fermentation. This is because of high substrate loading and solvent toxicity/product inhibition due to large solvent accumulation.

Therefore, most fed-batch processes are designed with the integration of appropriate downstream systems to recover the solvent (Song et al. 2010). The suitable systems that can be used for solvent recovery are gas-stripping, liquid-liquid extraction, perstraction, pervaporation, etc. (Qureshi and Maddox 1995, Qureshi and Blaschek 2001). This recovery could help in the reduction of the solvent-mediated inhibitory effect on cells. Qureshi et al. (2001) demonstrated the integration of the pervaporation system (silicalite-silicone composite membrane) with fed-batch fermentation; this resulted in the enhanced solvent recovery of 154.97 g/L ABE in 870 h).

Some of the other limitations of the fed-batch method include enormous energy utilization in butanol recovery, low effluent streams, and longer durations for butanol production. Thus, there is a need to develop several scale-up techniques to improve the yield of ABE fermentation (Mariano et al. 2015, Qureshi 2015).

5.2.3 Continuous fermentation

The continuous process is designed to overcome limitations of the batch and fed-batch system, that is, substrate and product inhibition, respectively. Continuous fermentation involves the transfer of half of the medium to the culture vessel. In the fermentation vessel, the bacterial inoculum is recycled which helps in a higher fermentation yield. In the fermentation vessel consisting of recycled inoculum, fresh media is added for continuous production. This systemic technique of continuous nutrient feeding combined with product removal using an appropriate recovery system is known as the continuous fermentation mode. The continuous fermentation enables a reduction in sterilization, butanol inhibition, and re-inoculation of microorganisms. This result in higher yield, low utilization of substrate, and fermentation continuing for a longer duration of time (Pugazhendhi et al. 2019).

Pierrot et al. 1986 has demonstrated a continuous system involving cell ultrafiltration and recycling system for recovery and recycling resulting in a solvent yield of 6.5 g/L/h with an ABE ratio of 1:8:6. On comparing the batch, fed-batch, and continuous mode of fermentation, the continuous system resulted in improved yield and productivity accompanied with high stability and low deviation for *C. acetobutylicum* in continuous fermentation (Li et al. 2011, Lipovsky et al. 2016). Continuous fermentation is suitable both at laboratory and large-scale conditions as it is easy to handle, time-efficient, and economical (Ranjan and Moholkar 2012). An automated continuous mode of butanol production is easy to operate but has several disadvantages such as low utilization efficiency of cells, low product quantity and quality due to dilution of the medium, and limitations during harvesting (Klutz et al. 2015). These limitations can be handled via cell immobilization and integration of appropriate recovery techniques (Pugazhendhi et al. 2019). Some of the advantages of utilizing cell immobilizations are the easy separation of cells from the product, thus avoiding product inhibition. Also, immobilization help in minimizing nutrient depletion, efficient recycling, high cell concentration, and productivity (Liu et al. 2013a). Lipovsky et al. (2016) reported a packed-bed continuous culture of *Clostridium pasteurianum* NRRL B-598 that run for 700 h, almost 35 cycles resulting in large solvent productivity of 0.73 g/l/h at a dilution rate of 0.121 h^{-1}.

Integration of suitable recovery techniques helps in the improvement of continuous fermentation efficiency. Cai et al. (2017) demonstrated a two-stage pervaporation process for efficient solvent recovery of 782.5 g/L with high (451.98 g/L) butanol concentration. Therefore, to amplify the ratio of butanol in ABE fermentation several technical and biological modifications can be suggested.

5.2.3.1 Two-stage continuous fermentation

Several groups suggested a two-stage continuous fermentation for the microbes not having simultaneous growth and synthetic activity. The first stage involves the acidogenic turbidostatic stage, which involves maximum cell growth at a high growth rate. The growth medium is then transferred to the next stage where the synthetic phase of the microbes occurs (Pugazhendhi et al. 2019) (Figure 3). Bahl et al. (1982) demonstrated two-stage continuous fermentation using glucose as the substrate under

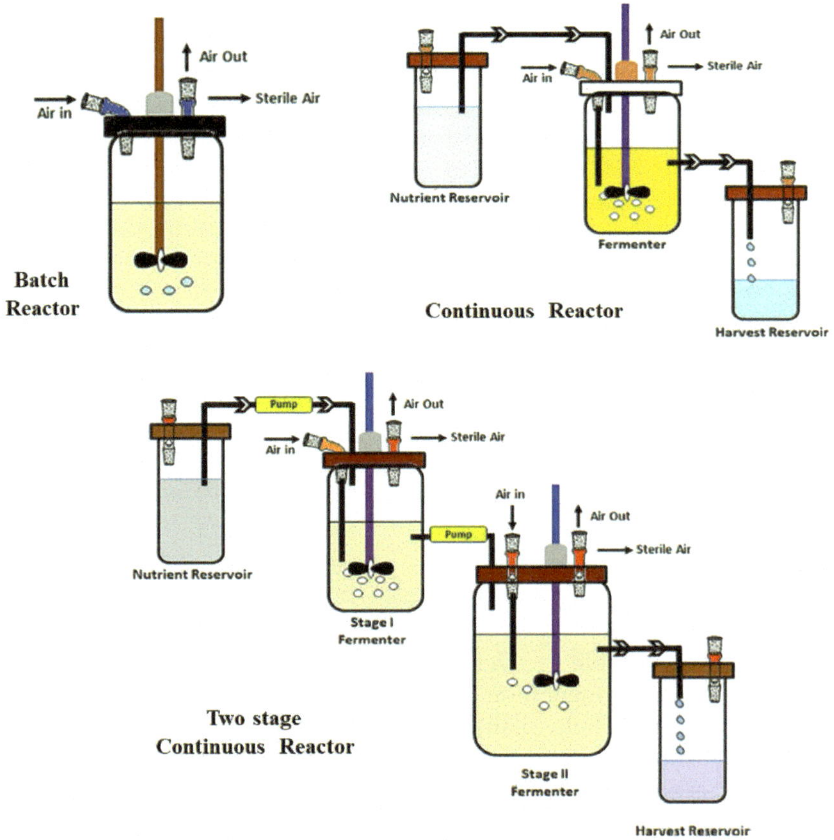

Figure 3. Mode of ABE fermentations (Adapted with permission from Pugazhendhi et al. 2019).

phosphate limited chemostat. Around 87% of substrate is converted to solvent (ABE). Around 12.5% of substrate resulted in acetate and butyrate.

5.2.4 Co-culture of microbial strains

To avoid a highly anaerobic and toxic environment, the co-culturing of two anaerobic or mixed culture of anaerobic and aerobic strains are suggested. This will help in overcoming the limitations and resulting in enhanced butanol yield.

Tran et al. (2010) demonstrated the cultivation of anaerobic *Clostridium butylicum* TISTR 1032 and aerobic *Bacillus subtilis* WD161 with cassava starch as a substrate resulting in 7.4 g/l ABE. This co-culturing will help in accompanied enhanced butanol yield (6.5-fold) without the need of strenuous anaerobic conditions. Co-culturing of two different anaerobic *Clostridial* strains is also suggested where one strain produces one intermediate for the butanol generation pathway and the other strain utilizes these intermediate precursor molecules for the generation of butanol. A similar study was performed by Li et al. (2013) where they demonstrated the utilization of continuous immobilized-cell co-culture of *Clostridium tyrobutyricum* ATCC 25755 and *C. beijerinckii* ATCC 55025 utilizing cassava starch as a substrate for the enhanced solvent yield of 13.39 g/L ABE. Here, *Clostridium tyrobutyricum* demonstrated high butyric acid activity which was later utilized by *C. beijerinckii* ATCC 55025 for butanol production (Jiang et al. 2009, 2014a).

Four strains isolated from the hydrogen-forming sludge of sewage were co-cultured and resulted in a high solvent yield (Liu et al. 2013a). The possible mechanism behind the enhanced yield is that different strains possess various advantages such as a larger carbon substrate utilization ability, high substrate and product tolerance, higher butanol yield, etc. Thus, understanding the mechanism of each strain is required along with changes in the metabolic pathways when the strains are co-cultured.

6. Downstream processing

Utilization of suitable recovery/downstream processing is key to the success of ABE fermentation. The continuous removal or phase-wise removal of products helps in the minimization of product-based inhibition. Apart from this, a suitable recovery process is required for higher yield and better economics. Some of the downstream processing strategies are discussed below.

6.1 Adsorption

Adsorption is an energy-efficient technology used for butanol recovery where the butanol is adsorbed by an adsorbent, then via utilization of a displacer or rising the temperature the butanol is desorbed from adsorbent (Kushwaha et al. 2019). Different adsorbent materials such as bonopore, charcoal, activated charcoal, polyvinylpyridine, resins, and silicalite are utilized for butanol recovery (Kushwaha et al. 2019). Silicalite is one of the prominent silica-based compounds with a zeolite-like structure consisting of hydrophobic properties. It is capable of selectively adsorbing small organic compounds (C1–C5 alcohols) from dilute aqueous solutions (Zheng et al. 2009). Silicalite shows low (97 mg/g) butanol adsorption but is capable of complete desorption with a very low energy requirement (1948 kcal/kg) as compared to other adsorbents. Other adsorbents such as activated charcoal and bone charcoal showed an adsorption efficiency of 255 mg/g, and 206 mg/g, respectively (Qureshi et al. 2005, Kushwaha et al. 2019). Milestone and Bibby (1981) demonstrated fermentation with *in situ* adsorbents, which resulted in 54.6 g/L of butanol with active carbon (Norit ROW 0.8) showing the best adsorbent ability. Resins such as polystyrene-codivinyl benzene resin are hydrophobic and can be efficiently employed as adsorbents (Xue et al. 2016). Lin et al. (2012) demonstrated the application of three different adsorbents, that is, XD-41, H-511, and KA-I with variations in their polarities for butanol recovery. Among tested resins, KA-I resin showed a maximum recovery of 99.7% with fast and high adsorption ability, low cost, and efficient desorption ability. The resins can be efficiently regenerated as well. Some biological adsorbents made up of starch and lignocelluloses have been developed to efficiently remove water from the alcohol-water vapor mixture (Chang et al. 2006, Huang et al. 2008). Although, these bio-adsorbents have a low separation capacity but are capable of low-temperature regeneration. Adsorption is a simple technique, but it is plagued with limitations such as low yield and the fact that it cannot be scaled up to the industrial level, concerns about compatibility, and chances of contamination (Martin-Calvo et al. 2018). Therefore, extensive research is required to choose the most feasible adsorbents with good desorption and regeneration ability.

6.2 Gas stripping

Gas stripping is utilized for butanol recovery from the ABE fermentation bioreactor. The gas stripping method involves the introduction of N_2, H_2, or CO_2 gases bubbles through the fermentation broth for selectively capturing the solvents. Following this, solvents are condensed and are collected in a receiver. The gases are then recovered and recycled back for the next cycle (Kujawska et al. 2015). The efficiency of the gas stripping methods depends on gas flow rate, agitation speed, media composition, and foam formation owing to the large gaseous influx (Ezeji et al. 2005). Ennis et al. (1986) suggested that improvising the gas stripping parameters may lead to the increased efficiency of the process. The application of high superficial gas velocity can result in improvement of solvent recovery because of a reduction in the liquid side mass transfer coefficient (Liao et al. 2014). The intermittent gas stripping in the fermentation medium with *Clostridium acetobutylicum* JB200 resulted in an improved n-butanol yield up to 60% (w/v)(Xue et al. 2012).

Ezeji demonstrated the application of gas stripping on the fed-batch fermentation. Results suggest that 233 g/L solvents was produced by the utilization of 500 g glucose with the productivity and yield of 1.16 g/(Lh) and 0.47 g/g, respectively. Similarly, when gas stripping is combined with continuous fermentation, the solvent yield of 460 g/l was obtained via consumption of 1163 g glucose (Ezeji et al. 2004a, Ezeji et al. 2004b). The integration of the batch fibrous bed bioreactor with a two-stage gas stripping method resulted in improved butanol recovery, that is, 175.6 g/L in the condensate of the first stage, and 612.3 g/L and 101.3 g/L of organic and aqueous phase, respectively, of the phase separation stage. Studies suggest that integration of fermentation with recovery system helps in overcoming the limitations caused by the substrate and product inhibition. Huang et al. (2014) demonstrated that *in situ* product recovery with fed-batch fermentation resulted in a four-fold increase in recovery but requires high energy (5220 kcal/kg). Further, Setlhaku et al. (2013) demonstrated novel two-stage ABE fermentation with the first stage operated in continuous mode whereas the second stage involved fed-batch mode integrated with a gas stripping unit. This helped in improving productivity and selectivity. However, prior to the application of this approach, parameters such as operational energy requirement and butanol selectivity need to be thoroughly evaluated (Kushwaha et al. 2019).

6.3 Pervaporation

Pervaporation is a widely used membrane-based process that helps in the separation of desired products from low concentration mixtures using a selective membrane. This technique utilizes the difference in partial vapor pressure of mixture components for their separation (Kushwaha et al. 2019, Pugazhendhi et al. 2019, Abdehagh et al. 2014). This technique also utilizes the molecular interactions pattern of the membrane with the feed components (Van Hecke et al. 2012). The fermentation broth is kept in contact with the membrane, and the volatile liquids or solvents diffused through this membrane are recovered by condensation of vapor under the influence of vacuum to the side of permeate (Liu et al. 2013a). The choice of a suitable membrane for pervaporation is crucial for pervaporation efficiency. Hydrophobic polymer membranes such as silicon rubber sheets, and polydimethylsiloxane membrane and organophilic membranes such as polyetrafluoroethylene, polypropylene, poly(ether block amide) (PEBA), polydimethylsiloxane (PDMS), and poly[1-(trimethylsilyl)-1-propyne]have been used as pervaporation membranes (Liu et al. 2013a, 2014).

Studies suggest that PDMS membranes demonstrate stable separation and cost efficiency, and are thus best suited for the recovery of butanol. The PEBA membrane is thin and possesses permselective properties that can help in enhanced recovery due to reduced resistance (Liu et al. 2005). Yen et al. (2012b) demonstrated a comparative analysis of PEBA and PDMS efficiency and suggested that the PEBA membrane permits higher permeation flux of 9.975 $g/m^2 h$, as compared to PDMS (3.911 $g/m^2 h$). Further, the integration of batch and fed-batch fermentation with PEBA membrane utilizing pervaporation technology resulted in 43% and 39%

higher butanol productivity as compared to that achieved with the simple process.

Mixed matrix membranes (MMMs) are also used as pervaporation membranes. These MMMs are made up of a polymeric matrix with inorganic fillers that provide strength to the membrane matrix. They possess high permeability and selectivity and are free from the limitations demonstrated by conventional organic/inorganic membranes (Liu et al. 2014). The integration of PEBA with zeolitic imidazolate resulted in the generation of MMMs with an improved butanol separation factor (18.8 at 37°C) and permitted a high flux of 520.2 g/m^2h. Some of the other inorganic fillers are hydrophobic silica, zeolite, MFI zeolite, and metal-organic frameworks (Fouad and Feng 2009, Claes et al. 2012, Tan et al. 2013).

Abdehagh et al. (2014) performed a comparative study on the adsorption and pervaporation approach and suggested that these two methods are the most energy-efficient butanol recovery methods. However, as a search of efficient desorption methods for the adsorbent method is still under research, therefore pervaporation can be the preferred method at a large scale due to the high separation factor and permeate flux. Also, the application of carbon nanotubes (CNTs) can enhance recovery manyfold. The preparation of a composite membrane consisting of PEBA and CNTs resulted in improvement in the mechanical strength of the PEBA and also resulted in enhancement of flux by 61% as compared to the stand-alone PEBA system (Yen et al. 2012a).

As discussed earlier, the coupled fermentation recovery system showed enhanced efficiency. Thus the application of synthetic silicalite-poly(dimethylsiloxane) nanocomposite membrane-mediated pervaporation coupled with fermentation resulted in high flux and a separation factor of 5–11.2 kg/m^2h and 25–41.6, respectively (Liu et al. 2011). The major limitation of the pervaporation system is the high cost and low pressure (Xue et al. 2017). Li et al. (2014c) demonstrated the integration of fermentation with pervaporation in conjunction with ultrafiltration. This helped in a decrease in butanol toxicity and the application of ultrafiltration helped in nurturing the microorganisms in the reactor. Silicalite-silicone membrane and zeolite-mixed PDMS membranes integrated with the fermentation system showed a recovery of 105.4 g/L and 34.5 g/L butanol, respectively (Qureshi et al. 2001, Xue et al. 2015).

7. Conclusion

Due to outstanding characteristics, biobutanol has gained renewed interest as an unconventional biofuel. Several scientists and researchers have been studying different aspects of biobutanol production, which leads to significant advances in biobutanol production. Yet, many challenges still need to be overcome and currently, research is in progress toward developing technologies for biobutanol production. Lignocellulosic biomass (LCB) is among the good substrates suitable for sustainable biobutanol production at a large scale. Cellulose and hemicellulose content of LCB can efficiently be converted to biobutanol after appropriate pretreatment and hydrolysis. The efficient conversion of LCB to hydrolysate with a suitable concentration of sugars, inhibitors, and stimulators is the key challenge for economical biobutanol production. The strain improvement is another critical aspect for cost-effective biobutanol production. Among them spore formation, butanol tolerance limit, comparatively slow growth rate, and contamination are the main disadvantages of the solvent-producing bacteria.

References

Abdehagh, N., Tezel, F.H., and Thibault, J. (2014). Separation techniques in butanol production: Challenges and developments. Biomass and Bioenergy, 60: 222–246.

Abdehagh, N., Dai, B., Thibault, J., and Handan Tezel, F. (2017). Biobutanol separation from ABE model solutions and fermentation broths using a combined adsorption--gas stripping process. J. Chem. Technol. & Biotechnol., 92: 245–251.

Al-Shorgani, N.K.N., Kalil, M.S., Yusoff, W.M.W., and Hamid, A.A. (2018). Impact of pH and butyric acid on butanol production during batch fermentation using a new local isolate of Clostridium acetobutylicum YM1. Saudi J. Biol. Sci., 25: 339–348.

Bahl, H., Andersch, W., and Gottschalk, G. (1982). Continuous production of acetone and butanol by Clostridium acetobutylicum in a two-stage phosphate limited chemostat. Eur. J. Appl. Microbiol. Biotechnol., 15: 201–205.

Bankar, S.B., Jurgens, G., Survase, S.A. et al. (2015). Genetic engineering of Clostridium acetobutylicum to enhance isopropanol-butanol-ethanol production with an integrated DNA-technology approach. Renew Energy, 83: 1076–1083.

Baral, N.R., Slutzky, L., Shah, A. et al. (2016). Acetone-butanol-ethanol fermentation of corn stover: current production methods, economic viability and commercial use. FEMS Microbiol. Lett., 363:fnw033.

Becerra, M., Cerdán, M.E., and González-Siso, M.I. (2015). Biobutanol from cheese whey. Microb. Cell Fact, 14: 1–15.

Berezina, O.V., Brandt, A., Yarotsky, S. et al. (2009). Isolation of a new butanol-producing Clostridium strain: high level of hemicellulosic activity and structure of solventogenesis genes of a new *Clostridium saccharobutylicum* isolate. Syst. Appl. Microbiol., 32: 449–459.

Bruant, G., Lévesque, M.-J., Peter, C. et al. (2010). Genomic analysis of carbon monoxide utilization and butanol production by *Clostridium carboxidivorans* strain P7T. PLoS One, 5: e13033.

Burhani, D., Triwahyuni, E., and Setiawan, R. (2019). Second Generation Biobutanol: An Update. Reaktor, 19: 101–110.

Bušić, A., Mar\djetko, N., Kundas, S. et al. (2018). Bioethanol production from renewable raw materials and its separation and purification: A review. Food Technol. Biotechnol., 56: 289–311.

Cai, D., Hu, S., Miao, Q. et al. (2017). Two-stage pervaporation process for effective *in situ* removal acetone-butanol-ethanol from fermentation broth. Bioresour. Technol., 224: 380–388.

Chang, H., Yuan, X.-G., Tian, H., and Zeng, A.-W. (2006). Experimental investigation and modeling of adsorption of water and ethanol on cornmeal in an ethanol-water binary vapor system. Chem. Eng. & Technol. Ind. Chem. Equipment-Process Eng., 29: 454–461.

Chen, C.-Y., Zhao, X.-Q., Yen, H.-W. et al. (2013). Microalgae-based carbohydrates for biofuel production. Biochem. Eng. J., 78: 1–10.

Cheng, H.-H., Whang, L.-M., Chan, K.-C. et al. (2015). Biological butanol production from microalgae-based biodiesel residues by *Clostridium acetobutylicum*. Bioresour. Technol., 184: 379–385.

Cho, D.H., Shin, S.-J., and Kim, Y.H. (2012). Effects of acetic and formic acid on ABE production by *Clostridium acetobutylicum* and *Clostridium beijerinckii*. Biotechnol. Bioprocess Eng., 17: 270–275.

Claes, S., Vandezande, P., Mullens, S. et al. (2012). Preparation and benchmarking of thin film supported PTMSP-silica pervaporation membranes. J. Memb. Sci., 389: 265–271.

da Silva Trindade, W.R., and dos Santos, R.G. (2017). Review on the characteristics of butanol, its production and use as fuel in internal combustion engines. Renew Sustain Energy Rev., 69: 642–651.

Dean, J.A. (2005). Lange's handbook of chemistry. McGraw-Hill Education.

der Merwe, A.B., Cheng, H., Görgens, J.F., and Knoetze, J.H. (2013). Comparison of energy efficiency and economics of process designs for biobutanol production from sugarcane molasses. Fuel, 105: 451–458.

Dolejš, I., Kras\vnan, V., Stloukal, R. et al. (2014). Butanol production by immobilised Clostridium acetobutylicum in repeated batch, fed-batch, and continuous modes of fermentation. Bioresour. Technol., 169: 723–730.

Dürre, P. (2007). Biobutanol: an attractive biofuel. Biotechnol. J. Healthc. Nutr. Technol., 2: 1525–1534.

Dürre, P. (1998). New insights and novel developments in clostridial acetone/butanol/isopropanol fermentation. Appl. Microbiol. Biotechnol., 49: 639–648.

Ellis, J.T., Hengge, N.N., Sims, R.C., and Miller, C.D. (2012). Acetone, butanol, and ethanol production from wastewater algae. Bioresour. Technol., 111: 491–495.

Ennis, B.M., Marshall, C.T., Maddox, I.S., and Paterson, A.H.J. (1986). Continuous product recovery by *in-situ* gas stripping/condensation during solvent production from whey permeate using *Clostridium acetobutylicum*. Biotechnol. Lett., 8: 725–730.

Ezeji, T.C., Qureshi, N., and Blaschek, H.P. (2004a). Acetone butanol ethanol (ABE) production from concentrated substrate: reduction in substrate inhibition by fed-batch technique and product inhibition by gas stripping. Appl. Microbiol. Biotechnol., 63: 653–658.

Ezeji, T.C., Qureshi, N., and Blaschek, H.P. (2004b). Butanol fermentation research: upstream and downstream manipulations. Chem. Rec., 4: 305–314.

Ezeji, T.C., Karcher, P.M., Qureshi, N., and Blaschek, H.P. (2005). Improving performance of a gas stripping-based recovery system to remove butanol from *Clostridium beijerinckii* fermentation. Bioprocess Biosyst. Eng., 27: 207–214.

Ezeji, T.C., Qureshi, N., and Blaschek, H.P. (2007). Bioproduction of butanol from biomass: from genes to bioreactors. Curr. Opin. Biotechnol., 18: 220–227.

Foda, M.I., Dong, H., and Li, Y. (2010). Study the suitability of cheese whey for bio-butanol production by Clostridia. J. Am. Sci., 6: 39–46.

Fouad, E.A., and Feng, X. (2009). Pervaporative separation of n-butanol from dilute aqueous solutions using silicalite-filled poly (dimethyl siloxane) membranes. J. Memb. Sci., 339: 120–125.

Freeman, J., Williams, J., Minner, S. et al. (1988). Alcohols and ethers: a technical assessment of their application as fuels and fuel components. API Publ 4261:New York: American Institute of Physics; 1988.

Gao, K., Orr, V., and Rehmann, L. (2016). Butanol fermentation from microalgae-derived carbohydrates after ionic liquid extraction. Bioresour. Technol., 206: 77–85.

Gracia, V., Päkkilä, J., Ojamo, H. et al. (2011). Challenges in biobutanol production: how to improve the efficiency? Renew Sustain Energy Rev., 15: 964–980.

George, H.A., Johnson, J.L., Moore, W.E.C. et al. (1983). Acetone, isopropanol, and butanol production by *Clostridium beijerinckii* (syn. *Clostridium butylicum*) and *Clostridium aurantibutyricum*. Appl. Environ. Microbiol., 45: 1160–1163.

Gottumukkala, L.D., Parameswaran, B., Valappil, S.K. et al. (2013). Biobutanol production from rice straw by a non acetone producing *Clostridium sporogenes* BE01. Bioresour. Technol., 145: 182–187.

Huang, H.-J., Ramaswamy, S., Tschirner, U.W., and Ramarao, B.V. (2008) A review of separation technologies in current and future biorefineries. Sep. Purif Technol., 62: 1–21.

Huang, H.-J., Ramaswamy, S., and Liu, Y. (2014). Separation and purification of biobutanol during bioconversion of biomass. Sep. Purif Technol., 132: 513–540.

Huang, H., Singh, V., and Qureshi, N. (2015). Butanol production from food waste: a novel process for producing sustainable energy and reducing environmental pollution. Biotechnol. Biofuels, 8: 1–12.

Huang, W.-C., Ramey, D.E., and Yang, S.-T. (2004). Continuous production of butanol by *Clostridium acetobutylicum* immobilized in a fibrous bed bioreactor. *In*: Proceedings of the Twenty-Fifth Symposium on Biotechnology for Fuels and Chemicals Held May 4–7, 2003, in Breckenridge, CO. pp. 887–898.

Ibrahim, M.F., Kim, S.W., and Abd-Aziz, S. (2018). Advanced bioprocessing strategies for biobutanol production from biomass. Renew Sustain Energy Rev., 91: 1192–1204.

Isah, S., and Ozbay, G. (2020). Valorization of food loss and wastes: Feedstocks for biofuels and valuable chemicals. Front Sustain Food Syst., 4: 82.

Isar, J., and Rangaswamy, V. (2012). Improved n-butanol production by solvent tolerant *Clostridium beijerinckii*. Biomass and Bioenergy, 37: 9–15.

Jiang, L., Wang, J., Liang, S. et al. (2009). Butyric acid fermentation in a fibrous bed bioreactor with immobilized *Clostridium tyrobutyricum* from cane molasses. Bioresour. Technol., 100: 3403–3409.

Jiang, M., Chen, J., He, A. et al. (2014a). Enhanced acetone/butanol/ethanol production by *Clostridium beijerinckii* IB4 using pH control strategy. Process Biochem., 49: 1238–1244.

Jiang, W., Zhao, J., Wang, Z., and Yang, S.-T. (2014b). Stable high-titer n-butanol production from sucrose and sugarcane juice by *Clostridium acetobutylicum* JB200 in repeated batch fermentations. Bioresour. Technol., 163: 172–179.

Jones, D.T., and Woods, D.R. (1986). Acetone-butanol fermentation revisited. Microbiol. Rev., 50: 484–524.

Karimi, K., Tabatabaei, M., Sárvári Horváth, I., and Kumar, R. (2015). Recent trends in acetone, butanol, and ethanol (ABE) production. Biofuel Res. J., 2: 301–308.

Kharkwal, S., Karimi, I.A., Chang, M.W., and Lee, D.-Y. (2009). Strain improvement and process development for biobutanol production. Recent Pat. Biotechnol., 3: 202–210.

Khedkar, M.A., Nimbalkar, P.R., Gaikwad, S.G. et al. (2017). Sustainable biobutanol production from pineapple waste by using *Clostridium acetobutylicum* B 527: drying kinetics study. Bioresour. Technol., 225: 359–366.

Klutz, S., Magnus, J., Lobedann, M. et al. (2015). Developing the biofacility of the future based on continuous processing and single-use technology. J. Biotechnol., 213: 120–130.

Komonkiat, I., and Cheirsilp, B. (2013). Felled oil palm trunk as a renewable source for biobutanol production by *Clostridium* spp. Bioresour. Technol., 146: 200–207.

Kujawska, A., Kujawski, J., Bryjak, M., and Kujawski, W. (2015). ABE fermentation products recovery methods—a review. Renew Sustain Energy Rev., 48: 648–661.

Kumar, B., Bhardwaj, N., Agrawal, K. et al. (2020). Current perspective on pretreatment technologies using lignocellulosic biomass: An emerging biorefinery concept. Fuel Process Technol., 199: 106244.

Kumar, M., and Gayen, K. (2011). Developments in biobutanol production: new insights. Appl. Energy, 88: 1999–2012.

Kushwaha, D., Srivastava, N., Mishra, I. et al. (2019). Recent trends in biobutanol production. Rev. Chem. Eng., 35: 475–504.

Li, H., Luo, W., Wang, Q., and Yu, X. (2014a). Direct fermentation of gelatinized cassava starch to acetone, butanol, and ethanol using *Clostridium acetobutylicum* mutant obtained by atmospheric and room temperature plasma. Appl. Biochem. Biotechnol., 172: 3330–3341.

Li, J., Baral, N.R., and Jha, A.K. (2014b). Acetone--butanol--ethanol fermentation of corn stover by *Clostridium* species: present status and future perspectives. World J. Microbiol. Biotechnol., 30: 1145–1157.

Li, J., Chen, X., Qi, B. et al. (2014c). Efficient production of acetone--butanol--ethanol (ABE) from cassava by a fermentation—pervaporation coupled process. Bioresour. Technol., 169: 251–257.

Li, L., Ai, H., Zhang, S. et al. (2013). Enhanced butanol production by coculture of *Clostridium beijerinckii* and *Clostridium tyrobutyricum*. Bioresour. Technol., 143: 397–404.

Li, S.-Y., Srivastava, R., Suib, S.L. et al. (2011). Performance of batch, fed-batch, and continuous A--B--E fermentation with pH-control. Bioresour. Technol., 102: 4241–4250.

Liao, Y.-C., Lu, K.-M., and Li, S.-Y. (2014). Process parameters for operating 1-butanol gas stripping in a fermentor. J. Biosci. Bioeng., 118: 558–564.

Lienhardt, J., Schripsema, J., Qureshi, N., and Blaschek, H.P. (2002). Butanol production by *Clostridium beijerinckii* BA101 in an immobilized cell biofilm reactor. Appl. Biochem. Biotechnol., 98: 591–598.

Lin, X., Wu, J., Jin, X. et al. (2012). Selective separation of biobutanol from acetone--butanol--ethanol fermentation broth by means of sorption methodology based on a novel macroporous resin. Biotechnol. Prog., 28: 962–972.

Lipovsky, J., Patakova, P., Paulova, L. et al (2016). Butanol production by *Clostridium pasteurianum* NRRL B-598 in continuous culture compared to batch and fed-batch systems. Fuel Process Technol., 144: 139–144.

Liu, F., Liu, L., and Feng, X. (2005). Separation of acetone--butanol--ethanol (ABE) from dilute aqueous solutions by pervaporation. Sep. Purif Technol., 42: 273–282.

Liu, G., Wei, W., and Jin, W. (2014). Pervaporation membranes for biobutanol production. ACS Sustain Chem. & Eng., 2: 546–560.

Liu, H., Wang, G., and Zhang, J. (2013a). The promising fuel-biobutanol. Liq Gaseous Solid Biofuels-Conversion Tech., 175–198.

Liu, X.-B., Gu, Q.-Y., and Yu, X.-B. (2013b). Repetitive domestication to enhance butanol tolerance and production in *Clostridium acetobutylicum* through artificial simulation of bio-evolution. Bioresour. Technol. 130: 638–643.

Liu, X., Li. Y,, Liu, Y. et al. (2011). Capillary supported ultrathin homogeneous silicalite-poly (dimethylsiloxane) nanocomposite membrane for bio-butanol recovery. J. Memb. Sci., 369: 228–232.

Lu, C., Zhao, J., Yang, S.-T., and Wei, D. (2012). Fed-batch fermentation for n-butanol production from cassava bagasse hydrolysate in a fibrous bed bioreactor with continuous gas stripping. Bioresour. Technol., 104: 380–387.

Luo, H., Zhang, J., Wang, H. et al. (2017). Effectively enhancing acetone concentration and acetone/butanol ratio in ABE fermentation by a glucose/acetate co-substrate system incorporating with glucose limitation and *C. acetobutylicum/S. cerevisiae* co-culturing. Biochem. Eng. J., 118: 132–142.

Maddox, I.S. (1989). The acetone-butanol-ethanol fermentation: recent progress in technology. Biotechnol. Genet Eng. Rev., 7: 189–220.

Madhavan, A., Srivastava, A., Kondo, A., and Bisaria, V.S. (2012) Bioconversion of lignocellulose-derived sugars to ethanol by engineered *Saccharomyces cerevisiae*. Crit. Rev. Biotechnol., 32: 22–48.

Madihah, M.S., Ariff, A.B., Khalil, M.S. et al. (2001). Anaerobic fermentation of gelatinized sago starch-derived sugars to acetone—1-butanol—Ethanol solvent by *Clostridium acetobutylicum*. Folia Microbiol. (Praha), 46: 197–204.

Mariano, A.P., Ezeji, T.C., and Qureshi, N. (2015). Butanol production by fermentation: efficient bioreactors. *In*: Commercializing Biobased Products. pp. 48–70.

Martin-Calvo, A., der Perre, S., Claessens, B. et al. (2018). Unravelling the influence of carbon dioxide on the adsorptive recovery of butanol from fermentation broth using ITQ-29 and ZIF-8. Phys. Chem. Chem. Phys., 20: 9957–9964.

Milestone, N.B., and Bibby, D.M. (1981). Concentration of alcohols by adsorption on silicalite. J. Chem. Technol. Biotechnol., 31: 732–736.

Mitchell, W.J. (1997). Physiology of carbohydrate to solvent conversion by clostridia. Adv. Microb. Physiol., 39: 31–130.

Mo, X., Pei, J., Guo, Y. et al. (2015). Genome sequence of *Clostridium acetobutylicum* GXAS18-1, a novel biobutanol production strain. Genome Announc., 3: e00033--15.

Moo-Young, M. (2019). Celluar systems. *In*: Comprehensive Biotechnology. Elsevier, pp. 11–232.

Ni, Y., and Sun, Z. (2009). Recent progress on industrial fermentative production of acetone—butanol--ethanol by *Clostridium acetobutylicum* in China. Appl. Microbiol. Biotechnol., 83: 415–423.

Ni, Y., Wang, Y., and Sun, Z. (2012). Butanol production from cane molasses by *Clostridium saccharobutylicum* DSM 13864: batch and semicontinuous fermentation. Appl. Biochem. Biotechnol., 166: 1896–1907.

Nimcevic, D., and Gapes, J.R. (2000). The acetone-butanol fermentation in pilot plant and pre-industrial scale. J. Mol. Microbiol. Biotechnol., 2: 15–20.

Oswald, W.J. (2003). My sixty years in applied algology. J. Appl. Phycol., 15: 99–106.

Ozturk, A.B., Arasoglu, T., Gulen, J. et al. (2021). Techno-economic analysis of a two-step fermentation process for bio-butanol production from cooked rice. Sustain Energy & Fuels, 3705–3718.

Pang, Z.-W., Lu, W., Zhang, H. et al. (2016). Butanol production employing fed-batch fermentation by *Clostridium acetobutylicum* GX01 using alkali-pretreated sugarcane bagasse hydrolysed by enzymes from *Thermoascus aurantiacus* QS 7-2-4. Bioresour. Technol., 212: 82–91.

Paritosh, K., Kushwaha, S.K., Yadav, M. et al. (2017). Food waste to energy: an overview of sustainable approaches for food waste management and nutrient recycling. Biomed. Res. Int.

Park, C.-H. (1996). Pervaporative butanol fermentation using a new bacterial strain. Biotechnol. Bioprocess Eng., 1: 1.

Patakova, P., Linhova, M., Rychtera, M. et al. (2013). Novel and neglected issues of acetone--butanol--ethanol (ABE) fermentation by clostridia: Clostridium metabolic diversity, tools for process mapping and continuous fermentation systems. Biotechnol. Adv., 31: 58–67.

Pfromm, P.H., Amanor-Boadu, V., Nelson, R. et al. (2010). Bio-butanol vs. bio-ethanol: a technical and economic assessment for corn and switchgrass fermented by yeast or *Clostridium acetobutylicum*. Biomass and Bioenergy, 34: 515–524.

Pierrot, P., Fick, M., and Engasser, J.M. (1986). Continuous acetone-butanol fermentation with high productivity by cell ultrafiltration and recycling. Biotechnol. Lett., 8: 253–256.

Pugazhendhi, A., Mathimani, T., Varjani, S. et al. (2019). Biobutanol as a promising liquid fuel for the future-recent updates and perspectives. Fuel, 253: 637–646.

Qureshi, N., Maddox, I.S., and Friedl, A. (1992). Application of continuous substrate feeding to the ABE fermentation: relief of product inhibition using extraction, perstraction, stripping, and pervaporation. Biotechnol. Prog., 8: 382–390.

Qureshi, N., and Maddox, I.S. (1995). Continuous production of acetone-butanol-ethanol using immobilized cells of *Clostridium acetobutylicum* and integration with product removal by liquid-liquid extraction. J. Ferment Bioeng., 80: 185–189.

Qureshi, N., Meagher, M.M., and Hutkins, R.W. (1999) Recovery of butanol from model solutions and fermentation broth using a silicalite/silicone membrane. J. Memb. Sci., 158: 115–125.

Qureshi, N., and Blaschek, H.P. (2000). Economics of butanol fermentation using hyper-butanol producing *Clostridium beijerinckii* BA101. Food Bioprod Process, 78: 139–144.

Qureshi, N., and Blaschek, H.P. (2001). Recovery of butanol from fermentation broth by gas stripping. Renew Energy, 22: 557–564.

Qureshi, N., Meagher, M.M., Huang, J., and Hutkins, R.W. (2001). Acetone butanol ethanol (ABE) recovery by pervaporation using silicalite--silicone composite membrane from fed-batch reactor of *Clostridium acetobutylicum*. J. Memb. Sci., 187: 93–102.

Qureshi, N., Hughes, S., Maddox, I.S., and Cotta, M.A. (2005). Energy-efficient recovery of butanol from model solutions and fermentation broth by adsorption. Bioprocess Biosyst. Eng., 27: 215–222.

Qureshi, N. (2009). Solvent production. *In*: Schaechter, M. (ed.). Encyclopedia of Microbiology (Third Edition), Third Edition. Academic Press, Oxford, pp. 512–528.

Qureshi, N. (2015). Butanol production by fermentation: efficient bioreactors. Commer. Biobased Prod Oppor Challenges, Benefits, Risks 48.

Ranjan, A., and Moholkar, V.S. (2012). Biobutanol: science, engineering, and economics. Int. J. Energy Res., 36: 277–323.

Salleh, M.S.M., Ibrahim, M.F., Roslan, A.M., and Abd-Aziz, S. (2019). Improved biobutanol production in 2-L simultaneous saccharification and fermentation with delayed yeast extract feeding and *in-situ* recovery. Sci. Rep., 9: 1–9.

Sarangi, P.K., and Nanda, S. (2018). Recent developments and challenges of acetone-butanol-ethanol fermentation. pp. 111–123. *In*: Recent Advancements in Biofuels and Bioenergy Utilization. Springer.

Setlhaku, M., Heitmann, S., Górak, A., and Wichmann, R. (2013). Investigation of gas stripping and pervaporation for improved feasibility of two-stage butanol production process. Bioresour. Technol., 136: 102–108.

Shaheen, R., Shirley, M., Jones, D.T. others. (2000). Comparative fermentation studies of industrial strains belonging to four species of solvent-producing clostridia. J. Mol. Microbiol. Biotechnol., 2: 115–124.

Song, H., Eom, M.-H., Lee, S. et al (2010). Modeling of batch experimental kinetics and application to fed-batch fermentation of *Clostridium tyrobutyricum* for enhanced butyric acid production. Biochem. Eng. J., 53: 71–76.

Stoeberl, M., Werkmeister, R., Faulstich, M., and Russ, W. (2011). Biobutanol from food wastes--fermentative production, use as biofuel an the influence on the emissions. Procedia Food Sci., 1: 1867–1874.

Sun, X., Cao, Y., Xu, H. et al. (2014). Effect of nitrogen-starvation, light intensity and iron on triacylglyceride/carbohydrate production and fatty acid profile of Neochloris oleoabundans HK-129 by a two-stage process. Bioresour. Technol., 155: 204–212.

Suutari, M., Leskinen, E., Fagerstedt, K. et al. (2015). Macroalgae in biofuel production. Phycol. Res., 63: 1–18.

Tan, H., Wu, Y., and Li, T. (2013). Pervaporation of n-butanol aqueous solution through ZSM-5-PEBA composite membranes. J. Appl. Polym. Sci., 129: 105–112.

Tashiro, Y., Takeda, K., Kobayashi, G. et al. (2004). High butanol production by *Clostridium saccharoperbutylacetonicum* N1-4 in fed-batch culture with pH-stat continuous butyric acid and glucose feeding method. J. Biosci. Bioeng., 98: 263–268.

Thang, V.H., and Kobayashi, G. (2014). A novel process for direct production of acetone--butanol--ethanol from native starches using granular starch hydrolyzing enzyme by *Clostridium saccharoperbutylacetonicum* N1-4. Appl Biochem. Biotechnol., 172: 1818–1831.

Tran, H.T.M., Cheirsilp, B., Hodgson, B., and Umsakul, K. (2010). Potential use of *Bacillus subtilis* in a co-culture with *Clostridium butylicum* for acetone--butanol--ethanol production from cassava starch. Biochem. Eng. J., 48: 260–267.

Tsai, T.-Y., Lo, Y.-C., Dong, C.-D. et al. (2020). Biobutanol production from lignocellulosic biomass using immobilized *Clostridium acetobutylicum*. Appl Energy 277: 115531.

Van Hecke, W., Vandezande, P., Claes, S. et al. (2012). Integrated bioprocess for long-term continuous cultivation of *Clostridium acetobutylicum* coupled to pervaporation with PDMS composite membranes. Bioresour. Technol. 111: 368–377.

Wang, Y., Guo, W., Cheng, C.-L. et al. (2016). Enhancing bio-butanol production from biomass of *Chlorella vulgaris* JSC-6 with sequential alkali pretreatment and acid hydrolysis. Bioresour. Technol., 200: 557–564.

Wang, Y., Ho, S.-H., Yen, H.-W. et al. (2017). Current advances on fermentative biobutanol production using third generation feedstock. Biotechnol. Adv., 35: 1049–1059.

Wiehn, M., Staggs, K., Wang, Y., and Nielsen, D.R. (2014). *In situ* butanol recovery from *Clostridium acetobutylicum* fermentations by expanded bed adsorption. Biotechnol. Prog., 30: 68–78.

Xiao, H., Gu, Y., Ning, Y. et al. (2011). Confirmation and elimination of xylose metabolism bottlenecks in glucose phosphoenolpyruvate-dependent phosphotransferase system-deficient *Clostridium acetobutylicum* for simultaneous utilization of glucose, xylose, and arabinose. Appl. Environ. Microbiol., 77: 7886–7895.

Xue, C., Zhao, J., Lu, C. et al. (2012). High-titer n-butanol production by *Clostridium acetobutylicum* JB200 in fed-batch fermentation with intermittent gas stripping. Biotechnol. Bioeng., 109: 2746–2756.

Xue, C., Yang, D., Du, G. et al. (2015). Evaluation of hydrophobic micro-zeolite-mixed matrix membrane and integrated with acetone--butanol--ethanol fermentation for enhanced butanol production. Biotechnol. Biofuels, 8: 1–9.

Xue, C., Liu, F., Xu, M. et al. (2016). Butanol production in acetone-butanol-ethanol fermentation with *in situ* product recovery by adsorption. Bioresour. Technol., 219: 158–168.

Xue, C., Zhao, J., Chen, L. et al. (2017). Recent advances and state-of-the-art strategies in strain and process engineering for biobutanol production by *Clostridium acetobutylicum*. Biotechnol. Adv., 35: 310–322.

Yen, H.-W., Chen, Z.-H., and Yang, I.-K. (2012a). Use of the composite membrane of poly (ether-block-amide) and carbon nanotubes (CNTs) in a pervaporation system incorporated with fermentation for butanol production by *Clostridium acetobutylicum*. Bioresour. Technol., 109: 105–109.

Yen, H.-W., Lin, S.-F., and Yang, I.-K. (2012b). Use of poly (ether-block-amide) in pervaporation coupling with a fermentor to enhance butanol production in the cultivation of *Clostridium acetobutylicum*. J. Biosci. Bioeng., 113: 372–377.

Zhang, J., Yu, L., Lin, M. et al. (2017). n-Butanol production from sucrose and sugarcane juice by engineered *Clostridium tyrobutyricum* overexpressing sucrose catabolism genes and adhE2. Bioresour. Technol., 233: 51–57.

Zheng, Y.-N., Li, L.-Z., Xian, M. et al. (2009). Problems with the microbial production of butanol. J. Ind. Microbiol. Biotechnol., 36: 1127–1138.

Zverlov, V.V., Berezina, O., Velikodvorskaya, G.A., and Schwarz, W.H. (2006). Bacterial acetone and butanol production by industrial fermentation in the Soviet Union: use of hydrolyzed agricultural waste for biorefinery. Appl. Microbiol. Biotechnol., 71: 587–597.

6

Mechanisms and Applications of Biofuel:

Acetone-Butanol-Ethanol Fermentation

*Ketaki Nalawade, Vrushali Kadam, Shuvashish Behera,**
Kakasaheb Konde and *Sanjay Patil**

1. Introduction

The rapid diminishing of fossil fuel reserves and growing environmental concerns resulting from fuel emissions have made researchers explore alternative biofuel sources. The production of biofuels such as biobutanol and bioethanol using natural sources have appeared as propitious transportation fuels due to its sustainability and ecological benefits, reducing the reliance on crude oil. Nowadays, biodiesel and bioethanol are utilized as substitute fuels for gasoline and diesel as it decreases noxious emissions, that is, CO, HC and haze pollution from the exhaust (Yusoff et al. 2015). Bioethanol as fuel needs significant modifications made to conventional engines, and it is more corrosive than gasoline. Biobutanol and gasoline utilize similar feedstock for its production. Moreover, biobutanol is a viable biofuel for IC engines due to its physicochemical properties that enhance engine performances. The microbial biobutanol production is more multi-faceted than bioethanol, however biobutanol offers more merits than gasoline and bioethanol (Rathour et al. 2018, Karthick and Nanthagopal 2021, Timung et al. 2021).

Department of Alcohol Technology and Biofuels, Vasantdada Sugar Institute, Manjari (Bk.), Pune-412307, India.
* Corresponding authors: behera.shuvashish@gmail.com; sv.patil@vsisugar.org.in

The innovations in biological processes and numerous novel fermentation techniques are used to convert abundantly accessible waste biomass into biobutanol.

Biobutanol has been mostly produced by acetone butanol ethanol (ABE) fermentation using various feedstocks such as whey, sugar beets, barley, potato and corn starch, which is called as the first-generation biofuel (Patil et al. 2019). Nonetheless, the conflict in utilizing food as a feedstock for fuel conversion and its cost becomes the utmost communal impact on biobutanol production due to its competition for land requisite and allied resources (Wang et al. 2017). To overcome this problem, researchers started utilizing municipal, agricultural, and forest waste materials for ABE production (Behera and Kumar 2019). Furthermore, researchers focused on algal biomass for biobutanol production as an economically feasible process due to its increased solvent-producing capability and higher energy content. The algal biomass eliminates rigorous labor work and production costs. Microalgae utilizes wastewater as nutrient sources for the bioethanol production (Wang et al. 2017).

Clostridium strains such as *Clostridium saccharoperbutylacetonicum, Clostridium acetobutylicum, Clostridium saccharobutylicum* and *Clostridium beijerinckii* are the most commonly used strains for butanol production (Jimenez-Bonilla and Wang 2018). The significant problem related to the bacterial-mediated ABE fermentation is its self-inhibition owing to butanol toxicity (Behera et al. 2018). Solvent toxicity and nutrient depletion during fermentation processes resulted in the cessation of the fermentation and ultimately led to a lower production rate with poor solvent yield. Moreover, lower solvent concentration resulted in a relatively high recovery cost and purification techniques (Maiti et al. 2016). Many researchers implemented various strategies such as isolation and identification of an ideal strain for improved biobutanol productivity, design of bioprocessing approaches to increase the its productivity and yield rate, and an effectual recovery technique to outstrip its toxicity (Patil et al. 2019, Pugazhendhi et al. 2019).

Scientists have identified solventogenic *Clostridia* tolerant strains for biobutanol fermentation and controlled its spore formation. As the problems with *Clostridia* strains became more evident, researchers started utilizing a rapidly growing *E. coli*. However, bacteria-mediated butanol production has several disadvantages such as the potential for phage infection, the complexity of the two-phase fermentation, sporulation during solventogenesis, high levels of by-products and very low titre, fermentation spoilage, product degradation, and product toxicity. Moreover, metabolically engineered *S. cerevisiae*, a strain tolerant to elevated levels of n-butanol has been used for butanol fermentation (Swidah et al. 2015, Jimenez-Bonilla and Wang 2018, Rathour et al. 2018,

Pugazhendhi et al. 2019). On the other hand, the information about the mechanism and strategies of ABE fermentation, technological advancement for economic process development and its application is scattered. Therefore, the current chapter deals explicitly with the mechanism and limitations of biobutanol in ABE fermentation, strategies to overcome these limitations and applications of biobutanol.

2. ABE fermentation

Aldol condensation and oxo-synthesis are commonly used chemical processes for butanol production (Liu et al. 2013). However, butanol production can be done using biological process through bacterial fermentation called as ABE fermentation through which butanol, acetone, and ethanol are obtained as the main product (Abd-Alla et al. 2015, Liberato et al. 2019). *Clostridium* species are widely used bacterial species for the fermentation under strict anaerobic conditions for the production of biobutanol (Gottumukkala et al. 2017). ABE fermentation by *Clostridia* species follows a complex intracellular pathway, where glucose is converted to pyruvate by glycolysis, or the Embden-Meyerhoff pathway (EMP) where pyruvate cleaved to acetyl-CoA by using pyruvate ferredoxin oxido-reductase. Products from the intracellular pathway falls into three different categories: the first category includes solvents like acetone, butanol, and ethanol; second category includes organic acids such as lactic acid, acetic acid, and butyric acid; the third category includes gases like hydrogen and carbon dioxide (Zheng et al. 2009, Xue et al. 2013). The ABE fermentation mainly comprises of acidogenesis phase (acid production) and solventogenesis phase (solvent production) during the exponential and stationary growth (Patakova et al. 2018, Li et al. 2020). During the acidogenesis phase, exponential cell growth occurs, which decreases the pH resulting in accumulation of acetate and butyrate (Li et al. 2020). Endospore formation starts during the process of solventogenesis and then the cell enters in to the stationary phase. Phosphotransacetylase and acetate kinase catalyzes the conversion of acetate to acetyl-CoA. Butyrate synthesis occurs in two steps, that is, thiolase catalyzes the two molecules of acetyl-CoA into one molecular acetoacetyl-CoA, which is further subjected to three enzymatic (crotonase, hydroxybutyryl-CoA dehydrogenase, and butyryl-CoA dehydrogenase) activities for the formation of butyryl-CoA. Then, butyryl-CoA is converted into butyrate, which is catalyzed by phosphotransbutylase and butyrate kinase. During the fermentation, there is necessity for pH drops from growth pH to fermentation pH, which causes a change from the acidogenesis phase to the solventogenesis phase through which accumulated organic acid (acetate and butyrate) is converted to solvents like acetone, butanol, and ethanol. Butyrate and acetate again convert into the butyryl-CoA and acetyl-CoA, respectively in presence of catalytic activity of enzyme CoA transferase. The formation of alcohols requires the same key enzymes, such as NAD(P)H-dependent aldehyde/alcohol dehydrogenases where butanol dehydrogenase is encoded by the bdh gene, which helps in the formation of butanol. Acetone formation is a two-step process, where first acetoacetate is formed from acetoacetyl-CoA under CoA transferase, which finally converts into acetone and molecular CO_2 with the help of decarboxylase (Liu et al. 2013).

Clostridia genus is well-known natural and quality butanol producer and some of the strains like *C. beijerinckii* P260, *C. acetobutylicum* P262 (which is also known as *C. saccharobutylicum*), *C. beijerinckii* 8052, *C. acetobutylicum* B18, *C. acetobutylicum* NRRL B643, *C. beijerinckii* BA101, *C. acetobutylicum* ATCC 824, *C. saccharobutylicum* P262, and *C. beijerinckii* LMD27.6 are used at the industrial level for biobutanol production (Qureshi et al. 2016). The ratio of ABE production is 3:6:1, in which butanol is the highest one (Qureshi et al. 2010). Microorganism selection is an important task in ABE fermentation, which depends on various factors such as desired productivity, initial substrate, supplementary nutrients, resistance to bacteriophages, etc. (Kumar and Gayen 2011). Several researchers are also working in the area of metabolic engineering and molecular biology for improvisation of butanol production by evolutionary engineering, genomic studies, transcriptional analysis, and mutagenesis (Kumar and Gayen 2011, Cooksley et al. 2012, Li et al. 2013, Karimi et al. 2015). Ngoc-Phuong-Thao et al. (2016) studied *Clostridium acetobutylicum* metabolism for enhancing the production of butanol with high yield and selectivity, where they came across the Cap0037 protein, which acts as the novel regulator affecting bothacidogenesis and solventogenesis. Recently, non-clostridial organisms such as genetically engineered *E. coli* and *Saccharomyces cerevisiae* were utilized in butanol production (Atsumi et al. 2008, Steen et al. 2008, Branduardi et al. 2013). Li et al. (2014) also showed an optimistic approach towards butanol production with its tolerance through isolation and acclimatization process.

3. Mechanism of biobutanol production

ABE fermentation is a cost-effective process that can be successful with consideration of some of the parameters like selection of the cost-effective as well as environment friendly substrate, which does not

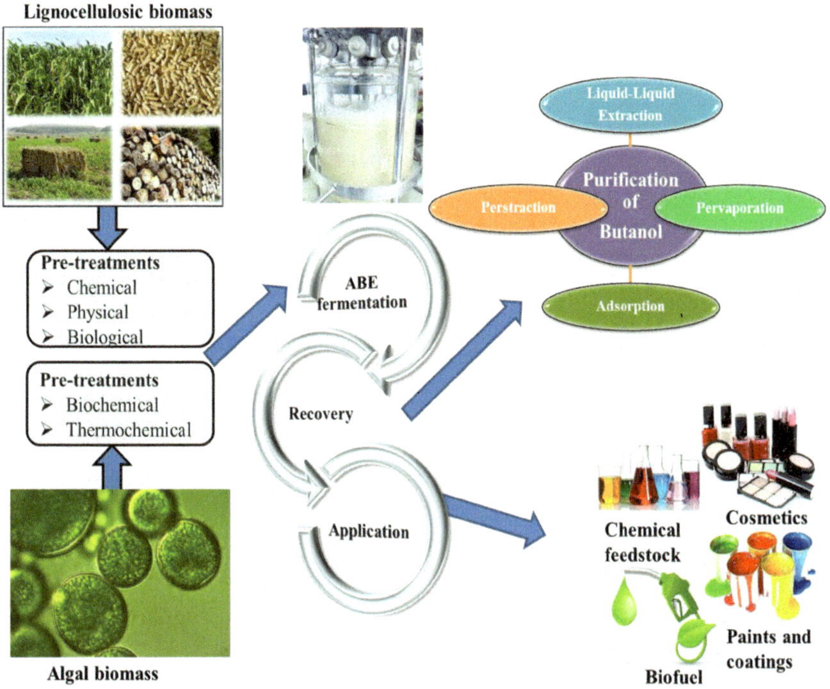

Figure 1. Process of lignocellulose and algal biomass based biobutanol production.

compete with the food chain products; proper strain selection maximizes the solvent ratios and yield in the fermentation with high- end product tolerance; the selection of best method for the solvent separation from the fermentation broth (Gottumukkala et al. 2017, Xue and Chi 2019, Veza et al. 2021). Biobutanol produced through the first generation resulted in the competition with the food demand, which led to the utilization of lignocellulosic and algal biomasses called as second- and third-generation biofuels (Wang et al. 2017, Kiss et al. 2019, Behera and Kumar 2019). The overall process for lignocellulose and algal biomass for biobutanol production is shown in Figure 1.

3.1 Lignocellulosic biomass route

ABE-producing *Clostridia* has a broad range of substrate utilization ability, which can utilize different types of carbon sources like sucrose, lactose, glucose, starch, xylan, glycerol, xylose, inulin, and arabinose (Jones et al. 1986, Mitchell et al. 1997, Andrade et al. 2003). ABE fermentation by using different strains and different biomass substrates has been reported by several researchers. Different lignocellulosic biomass such as sugarcane bagasse, wheat straw, paddy straw, whey, hardwood, sago starch, excess sludge, domestic organic waste, palm oil waste, agricultural waste, corn fiber, crop straw, animal manure, etc., can be used for fermentation, which may help in the reduction of production cost (Okolie et al. 2021, Timung et al. 2021). Substrate selection is also dependent on its profusion, annual supply, processing cost, the sugar monomers obtained, and the generation of inhibitory substances in the process (Maiti et al. 2016). Feedstock with a higher percentage of compounds like lignin and biopolymers are not considered as promising feedstock for biobutanol production as they cause a crucial inhibitory effect due to the liberation of the excess of soluble lignin as well as the phenolic compounds (Takkellapati et al. 2018, Birgen et al. 2019).

Lignocellulosic biomass is a complex polymer structure that needs to be pretreated and saccharified for the release of fermentable sugars (Carvalheiro et al. 2008). Butanol-producing bacteria utilises both hexoses and pentoses as the sources of carbon. Raganati et al. (2012) presented the conversion of both hexose and pentoses into ABE using *Clostridium acetobutylicum* and how the utilization of sugars occurs in a sequential manner such as glucose-mannose-arabinose-xylose. Different pretreatment techniques such as mechanical, physical, chemical, and biological are used to enhance the release of fermentable sugars from agro-residues. The types of pretreatment methods vary with the different types of biomasses and also affects in the biobutanol production as mentioned in Table 1. Various physical pretreatment techniques have been studied like grinding, hydrothermal, steam explosion, microwave irradiation, and ultrasonication (Birgen et al. 2019). Chemical techniques include diluted or concentrated acid hydrolysis, catalysis-based hydrolysis, alkaline delignification, and AFEX (Birgen et al. 2019). Alkaline pretreatment results in overall lower sugar yield but it generates less inhibitory compounds. Biological routes with mild operating conditions have been reported in two ways: one by *in situ* enzymatic hydrolysis and other *ex situ* by using different enzymes like cellulase, amylase, and xylanase (Maiti et al. 2016). ABE-fermenting *Clostridia*, which are cellulolytic in nature, have been screened for the butanol, acetone, and ethanol production from avicel (Virunanon et al. 2008).

The hydrolysis process generates various compounds like polyphenolic aromatic compounds (vanilline, ferulic acid, syrindic aldehyde), aliphatic acids (formic acid, acetic acid, levulinic acid), and furan derivatives, which are inhibitory to the bacteria. However, acid hydrolysis enables the conversion of fermentable sugar into the furfural and hydroxyl methyl furfural (HMF). The removal of the inhibitory compound is

Table 1. Different lignocellulosic biomasses with their pretreatment for biobutanol production.

Sr. no.	Substrate	Microrganism	Pretreatment	Yield (g/g)	Butanol (g/L)	References
1	Rice straw	*Clostridium* spp.	Alkali	0.49	2.93	Cheng et al. 2012
2	Sugarcane bagasse	*Clostridium* spp.	Alkali	0.37	1.95	Cheng et al. 2012
3	Sweet potato vines	*Clostridium acetobutylicum*	Alkali	0.18	6.4	He et al. 2017
4	Palm kernel cake	*Clostridium saccharoperbutylacetonicum* N1-4	Dilute acid	-	3.59	Shukor et al. 2014
5	Lettuce leaves	*Clostridium acetobutylicum* DSMZ 792	Alkali	-	1.1	Procentese et al. 2017
6	Corn cob	*Clostridium saccharobutylicum* DSM 13864	Alkali	0.35	12.27	Gao and Rehman 2014
7	Sweet Sorghum bagasse	*Clostridium beijerinckii* P260	Liquid hot water	0.38–4.6	8.45	Qureshi et al. 2016
8	Switch grass	*Clostridium saccharobutylicum* DSM 13864	Alkali	0.4	13.0	Gao et al. 2014
9	Sugarcane bagasse	*Clostridium beijerinckii* NCIMB 8052	Sequential-combinational	-	6.4	Su et al. 2015
10	Sweet sorghum bagasse	*Clostridium acetobutylicum* ABE1201	Dilute acid	-	12.4	Cai et al. 2013
11	Willow Biomass	*Clostridium beijerinckii*	Acid	-	4.5	Han et al. 2013
12	Corn fibre	*Clostridium beijerinckii mutant* RT66	Dilute acid	0.35	9.3	Guo et al. 2013

the most important step for the successful ABE Fermentation (Birgen et al. 2019). Application of the detoxifying agent like $Ca(OH)_2$ increases the production of the ABE solvent. In the case of corn cob hydrolysate fermentation, by treating it with the $Ca(OH)_2$ detoxification process, the production of ABE was enhanced from 3.8 g/L to 16 g/L (Zhang et al. 2012). Also, supplementation of glycerol converts the furfural into furfural alcohol, while the lime treatment oxidizes inhibitory compounds into carbon dioxide in the production of biobutanol from lignocellulosic feedstock (Ujor et al. 2014). Cai et al. (2013) used pervaporation membrane for both the process of detoxification of sweet sorghum bagasse (SSB) hydrolysate and for solvent separation from the fermentation broth (in which they achieved 94.5% furfural reduction). They also performed laccase detoxification through which 87.5% phenolics were removed. For the utilization of SSB pretreated with dilute acetic acid, followed by detoxification process and in combination with hybrid pervaporation for ABE fermentation, they could achieve 201.9 g/L of butanol and 76.2 g/L of acetone. Su et al. (2015) performed the sequential combinatorial lignocellulose pretreatment procedure (SCLPP) for ABE production by using sugarcane bagasse as a substrate and *Clostridium beijerinckii* NCIMB 8052 as the fermenting organism. SCLPP involves a series of processes such as microwave decomposition, enzymatic hydrolysis, ammonia immersion, decomposition of microorganisms, and hot water treatment (which is used to maximize sugar yield and reduce the inhibitory effect on ABE fermentation), leading to a total ABE concentration of 11.9 g/L with a concentration of 6.4 g/L of butanol without any detoxification process.

3.2 Algal biomass route

Biofuel production by using agricultural and lignocellulosic residues has created different concerns, which has generated research on third-generation biofuel (the most effective and suitable alternative for the economical greener cost-effective approach) (Aratboni et al. 2019, Khan et al. 2018). Algal biomass has several advantages over lignocellulosic biomass such as the availability and production of algal biomass throughout the year, utilization of CO_2 gas, higher growth potential, less water requirement for its cultivation, use of wastewater as the source for its cultivation, and it does not require any pesticides or herbicides for its cultivation (Behera et al. 2015, Aratboni et al. 2019). Algae (which divided into two types; micro and macro) are present everywhere in this ecosystem, which is one of the oldest types of life form and can survive under broad environmental conditions (Onay 2020). *Chlorococcum infusionum* and *Chlorella vulgaris* are unicellular microalgae, whereas aquatic plants, that

Table 2. Different algal biomasses with their pretreatment for biobutanol production.

Sr. no.	Substrate	Microorganism	Pretreatment	Yield (g/g)	Butanol (g/L)	References
1	Macroalgae (*Rhizoclonium* spp.)	*Clostridium beijerinckii* TISTR 1461	Acid	0.135	-	Salaeh et al. 2019
2	Green macroalgae (*Enteromorpha intestinalis*)	*Clostridium acetobutylicum* KCTC 1790	Acid	0.36	-	Nguyen et al. 2019
3	Brown seaweed (*Laminaria digitata*)	*Clostridium beijerinckii* DSM-6422	Thermal	0.420	7.16	Hou et al. 2017
4	Green Macroalgae (*Ulva lactuca*)	*Clostridium beijerinckii* ATTC 35702	Acid	0.26	4.0	Potts et al. 2012
5	Green Macroalgae (*Ulva lactuca*)	*Clostridium beijerinckii* NCIMB 8052	Hot water	0.25	-	Hetty et al. 2013
6	Microalgae (*Nannochloropsis gaditana*)	*Clostridium acetobutylicum*	Acid	0.145	2.9	Onay 2020

is, macroalgae which are large in size and multicellular (Timung et al. 2021). Microalgae are capable of immense production of proteins (20 to 50%), lipids (20 to 60%), and carbohydrates (33 to 64%) within a short duration of time period, which can be further converted into the biofuels as well as value- added products (Halder et al. 2019).

The cultivation of microalgae is dependent on the available sunlight, which limits the commercial-scale production of microalgae in regions with high solar radiation. Sugar content in some of the microalgae is high; that is, 30–40% in the *Chlorella* (Kamiński et al. 2011). Algae can be cultivated in open ponds and closed photo-bioreactors (Medipally et al. 2015). Inorganic nutrients such as phosphorous and nitrogen, potassium, and sulphur are essential for the proliferation of cells in algae production (Onay 2020, Brennan et al. 2010). ABE fermentation with the help of algal biomass is effectively studied using gram-positive, spore-forming anaerobic bacterium such as *Clostridium saccharoperbutylacetonicum* and *Saccharolytic Clostridium* spp. (Obando et al. 2011). The production of biofuel using algal biomass follows three steps, that is, cultivation of algae, followed by harvesting and dewatering of algae, and the third step, which is the disruption and fractionation of cell wall to release sugars to further convert them into biofuel or value-added products (Halder et al. 2019).

Similar to lignocellulosic biomass, the polysaccharides present in the algal biomass cannot be utilized directly during the fermentation (Zhang et al. 2021). The effective utilization of microalgal biomass requires biochemical and thermochemical conversion processes (Nigam et al. 2011, Khan et al. 2018). The biochemical conversion route includes photobiological hydrogen production, alcoholic fermentation, and anaerobic digestion and the thermochemical conversion process includes pyrolysis, gasification, and a direct combustion process for which the conversion rate is very low, as it requires a long time for processing (Halder et al. 2019). The pretreatment of microalgae can be done by various methods like hydrolysis, nourishment, and sterilisation (Le et al. 2019). Wang et al. (2016) studied the technical feasibility of butanol production from algal biomasses *Chlorella vulgaris* JSC-6, which was treated with alkali followed by acid pretreatment, and they were able to produce 13.1 g/L of butanol.

Another study showed that acid treatment using HCL was mostly effective among the other acids like H_2SO_4 and H_3PO_4 for the release of carbohydrate content from microalgae *Nannochloropsis gaditana* (Onay 2020). Castro et al. (2015) found that acid treatment is effective at 1 M concentration for 120 min at 80 to 90°C for the microalgal biomass conversion, where they were able to achieve 3.74 g/L of biobutanol production. Ellis et al. (2012) carried out ABE production from algae biomass using the fermenting microbe *Clostridium saccharoperbutylacetonicum* N1-4 and could achieve a total ABE of 2.74 g/L from pretreated algae, which was enhanced to 7.27 g/L with supplementation of glucose, and 9.74 g/L through supplementation of enzymes. However, pretreated algal biomass needs to be detoxified to increase butanol concentration during the fermentation process where amino acids and peptides are mainly considered as the inhibitory compound (Gao et al. 2016). In this study, detoxified starches from pretreated biomass were able to produce 8.05 g/L of butanol, which was 6.33 g/l without any detoxification process. In another study, *Ulva lactuca*, a green seaweed (macro algae) was first treated with hot water and then its polysaccharides were further hydrolyzed by a cellulase cocktail and used for ABE fermentation, where a 7.5 g/L of butanol concentration was achieved by using *Clostridium beijerinckii* NCIMB 8052 (Bikker et al. 2016). Further, Cheng and his co-worker stated that butanol production is possible from microalgae-based biodiesel residues through ABE fermentation using *Clostridium acetobutylicum*, which produced 3.86 g/L of butanol with a 0.13 g/g yield (Cheng et al. 2015). Wang et al (2016) studied the technical viability of butanol production from algal biomasses *Chlorella vulgaris* JSC-6, which was treated with alkali followed by acid pretreatment and they were able to produce 13.1 g/L of butanol concentration.

4. Limitations and Challenges of ABE fermentation

Biobutanol produced from ABE fermentation cannot match economically with the petrochemical route and an efficient economical biobutanol production process is essential to overcome major problems like high substrate costs, yield of butanol, butanol recovery cost, and low productivity that occur during batch fermentation (Gottumukkala et al. 2017, Birgen et al. 2019, Kushwaha et al. 2022). Some of the limitations such as the strict anaerobic nature of *Clostridium* species, strain degeneration and bacteriophage infection are observed in the butanol production process. Starch and sugar-based substrates used in the ABE fermentation like rye, millet, wheat, maize, and molasses are high-cost substrates, which enhance the process's economics (Dong et al. 2011). During ABE fermentation, butanol is the most toxic and it inhibits the metabolic activity of cells as compared to ethanol (Behera et al. 2018). Industrial-scale fermentation faces a major issue with butanol toxicity, which depends on its polarity and hydrophobicity. Accumulation of high butanol concentration in the plasma membrane causes structural damage as well as the inhibition and destruction of the physiological function and membrane bound by ATPase activity (Dong et al. 2011). Feedback inhibition occurs in ABE fermentation in which its own product butanol causes the toxicity to the cells, which leads to low productivity, yield, and titer, which further increases the separation cost (Xue and Chi 2019).

5. Strategies to overcome limitations of butanol production

To overcome this situation related to strict anaerobic nature, an aerotolerant species need to develop through the metabolic and genetic engineering approach for biobutanol production (Xue and Chi 2019). Major problems like strain degeneration and bacteriophage infection occur during ABE fermentation by using *C. acetobutylicum*, which can be overcome by rigorous sterilization and proper strain selection (Gottumukkala et al. 2017). Different low-cost alternative substrates, which are renewable in nature like cellulose, hemicellulose, and lignocellulose can be used in the ABE fermentation process.

Butanol production can also be achieved from algal biomasses, which shows high yield. Cell density and productivity can be increased by two ways; immobilization of cells and cell recycle. Cell immobilization is the process which can be done by cell entrapment, that is, the adsorption method (Xue and Chi 2019). Immobilization techniques have been used by several researchers to improve and exploit the efficiency of butanol-producing bacterial cells, which provide several advantages as compared to the free cell system (Qureshi et al. 2000, Lienhardt et al. 2002, Huang et al. 2004). It separates the cells from the product decreasing inhibition, which leads to higher cell concentration resulting in high productivity. During this process, nutrient depletion and the inhibition of the product decreases (Tashiro et al. 2010). Cell recycling can be achieved by centrifugation or the membrane process. However, researchers are facing major issues like membrane fouling (Du et al. 2012, Xue and Chi 2019). Some genetic engineering techniques used to increase butanol production are random mutagenesis, targeted mutagenesis, genome mining, and genome annotation (Patakova et al. 2018). An advanced integrated approach of continuous solvent separation from the fermentation broth prevents the solvent inhibition and enhances the productivity (Sukumaran et al. 2011). Problems related to product recovery can be solved by genetic engineering, where strains can be modified in such a manner that they can tolerate high sugar as well as high product concentration. Simultaneous fermentation and a product- removal approach can be used in the reactor to minimize the toxicity level of the product (Qureshi et al. 2008).

Different fermentation techniques like batch, fed-batch, and continuous medium integrated with the butanol recovery was found to be an effective phenomenon, which maintains low butanol levels in the fermentation broth and leads towards an increased production due to less toxicity; this integrated approach reduces the cost of separation (Timung et al. 2021). Ezeji et al. (2004) evaluated the integrated approach for ABE fermentation in order to overcome product and substrate inhibition. The integrated approach includes fed batch fermentation using *Clostridium beijerinckii* BA101 with a product recovery system like gas stripping using carbon dioxide and hydrogen as the carrier gas. Solvent productivity was improved by 400% with 232.8 g of total solvents 201 h of fed batch fermentation.

6. Mechanism of biobutanol purification

Butanol recovery, which contributes in process efficiency, is the most energy-intensive step in biobutanol fermentation. Conventionally it is recovered by distillation, which is a more expensive and high-energy-consuming process with low selectivity (Ezeji et al. 2013, Kujawska et al. 2015). Considering the bottlenecks, researchers have developed *in situ* recovery methods such as supercritical extraction, aqueous two-phase separation, liquid-liquid extraction, adsorption, ionic liquids, pervaporation, gas stripping, and perstraction. Each method has its own merits and demerits, like antifouling properties, energy contribution, and easy scaling up and operation (Jimenez-Bonilla and Wang 2018). Silicalite, polymeric resins, zeolite, polyvinylpyridine, and activated carbon are the commercial adsorbents typically used in the butanol recovery process. Application of such adsorbents for butanol recovery has been limited as it needs a higher concentration of butanol in the fermentation broth, along with its lower butanol adsorption capacity, desorption efficiency, and selectivity (Abdehagh et al. 2014). The adsorption methods are majorly applicable for laboratory-scale butanol recovery due to the limited capacity of adsorbents, which restrict their applicability on the industrial or semi-technical scale (Karthick and Nanthagopal 2021). Butanol was selectively recovered from fermentation by a liquid-liquid extraction technique using a water-insoluble organic extractant. Moreover, organic extractants with a higher distribution coefficient are toxic to microorganisms while nontoxic extractants exhibits lower distribution coefficients (Zheng et al. 2009). Oleyl alcohol is the most commonly used extractant for butanol recovery as it exhibits relatively higher extraction efficiency and reduced toxicity towards butanol-producing microorganisms (Abdehagh et al. 2014). However, the extraction of butanol in fermentation broth resulted in phase separation problems and subsequently, contamination of broth with added chemicals (Kujawska et al. 2015).

In situ butanol recovery during ABE fermentation can be achieved using the gas stripping method, which involves bubbling of fermentation gases or nitrogen through the fermentation broth, followed by condensation. The bubbling of gas in the fermenter resulted in the capture of ABE followed by condensation and collection. The recycling of gas is carried out to capture produced ABE till the end of fermentation process (Rathour et al. 2018, Pugazhendhi et al. 2019). Xue et al. (2012) reported the butanol recovery of 113.3 g/L from 474.9 g/L glucose, using gas stripping with fed-batch fermentation of *Clostridium acetobutylicum* JB200 within 326 h. However, they observed the production of 19.1 g/L butanol from 86.4 g/L glucose within 78 h without gas stripping. Similarly, Sharif Rohani et al. (2015) performed non-integrated and integrated butanol production. The product inhibition effect was parly alleviated by the integration of gas stripping, pervaporation, and vacuum separation techniques. Thirteen g/L of butanol concentration was obtained in non-integrated fermentation. However, *in situ* recovery methods resulted in the increased butanol concentration of 30 g/L. Furthermore, increased sugar conversion was observed in integrated fermentation as compared to non-integrated fermentation. In another study, fed-batch fermentation of corn stover bagasse hydrolysate using immobilized *Clostridium acetobutylicum* resulted in a butanol concentration of 18.6 g/L with an ABE yield of 83.2 g/kg, obtained from 97.6 g/L of total fermentable sugars (75.3 g/L of glucose and 22.3 g/L of xylose) (Cai et al. 2017). Thus, the gas stripping method offers several merits, like easy scaling up, simple operating, reduced energy costs, and capital investment for butanol recovery. On the other hand, the gas stripping method usually eliminates enormous quantities of water from the butanol creek and needs a higher energy input owing to its inferior selectivity than the other product recovery method.

Nowadays, two-stage *in situ* product recovery techniques are used for recovery of ABE from fermentation media by taking into account the individual benefits of various methods. In some studies, stripper is applied to strip off the solvents and then the stripped effluent is further recycled back into the fermentation broth. Diez-Antolinez et al. (2018) reported 350–400 g/L and 477 g/L of butanol concentration in the condensate and organic phase, respectively, in a two-stage gas stripping process combined with the cheese whey fermentation using *Clostridium beijerinckii* CECT 508. This two-stage gas stripping reduces energy utilization in water removal during purification of butanol. Abdehagh et al. (2016) reported butanol recovery of 261 mg/g from a binary solution containing 5.8 mg/L butanol in the vapour phase using activated carbon F-400. Moreover, ABE model solutions and the fermentation broth containing 5.1 mg/L and 2.3 mg/L butanol in the vapor phase exhibit the adsorption capacities of 212 and 220 mg/g, respectively, further indicating the effectiveness of combined adsorption—gas striping method for recovery ofbiobutanol.

The two-stage pervaporation technique using the polydimethylsiloxane/polyvinylidene fluoride (PDMS/PVDF) membrane exhibits total ABE of 782.5 g/L and 451.98 g/L of n-butanol, which was ~ 41-fold higher than that of the fermentation broth. The first stage permeate of the *in situ* pervaporation was utilized as a feed in the second stage of pervaporation, utilizing the same PDMS/PVDF membrane (Cai et al. 2017). The pervaporation technique involves a selective diffusion of target molecules through the membrane by applying a vacuum. Membrane properties determine the efficiency of the separation process. PDMS, referred to as silicone rubber, is good membrane material due to its high hydrophobic properties with excellent chemical and thermal stability (Chen et al. 2013). A close-circulating vapor stripping-vapor permeation process exhibits the highest butanol separation factor of 142.7 using the PDMS membrane with 339.3 g/L ABE and 212.7 g/L butanol, with the decreased energy utilization of 19.6 kJ/g-butanol (Zhu et al. 2018). Further, Cabezas et al. (2019) reported that PDMS-coated ionic liquid membranes exhibit higher selectivity for butanol than a single PDMS membrane. A permeate obtained with the trihexyl(tetradecyl) phosphonium bis(trifluoromethanesulfonyl)imidebased membrane contains 36% acetone, 54.2% butanol, 4.8% ethanol, and 5% water. The major disadvantage of pervaporation is membrane fouling due to the complex nature of fermentation broth (Qureshi and Blaschek 1999). Another study used fluorinated PDMS membrane with the anti-biofouling property to recover butanol from the fermentation broth. The ABE recovery was enhanced by 51% in fed-batch fermentation as compared to batch fermentation using the fluorinated PDMS membrane (Zhu et al. 2020). The introduction of the fluoroalkyl group reduces the surface energy, resulting in excellent hydrophobicity and lipophobicity and readily alleviating microbial adhesion onto the membrane resulted in an excellent anti-biofouling property.

Perstraction is a combination of liquid-liquid extraction and pervaporation, where a membrane separates the extractant and the fermentation broth. The butanol diffuses through the membrane, which was followed by extraction with the retention of the other components. The polytetrafluoroethylene (PTFE) membrane and 1-dodecanol resulted in improved butanol production from 16.0 to 20.1 g/L using *C. saccharoperbutylacetonicum* N1-4 (Tanaka et al. 2012). Qureshi et al. (1992) reported 8.89 g/L of butanol production in the first cycle, which reached to 10.29 g/L in the second operation cycle. The significant advantage of this process is the indirect contact of the organic phase and the aqueous phase. Extractants having a relatively high distribution coefficient and selectivity are usually employed for the perstraction process. Perstraction offers several advantages such as reduced product inhibition, enhanced cell growth, increased productivity, and energy-efficiency. However, the membrane barrier curbs the rate of extraction, resulting in lower productivity, which is the apparent disadvantage of the perstraction process (Huang et al. 2014).

7. Application of biobutanol

Butanol is a valuable chemical feedstock used in cosmetics, textile, paint and coatings, pharmaceutical, and food and plastic industries as represented in Table 3. Concerns about fossil fuel depletion and greenhouse gas emissions, biobutanol is a renewable and environmentally friendly biofuel. It is readily mixable with diesel fuel and gasoline, having an elevated calorific value with lesser vapor pressure. Moreover, it is utilized as an extractant in the manufacturing of vitamins, hormones, and antibiotics (Qureshi and Blaschek 2000, Liu et al. 2013). As reported by Welt Bernard (1950), the pain was relieved far more rapidly by n-butanol than with common analgesics.

Table 3. Applications of butanol and its derivatives.

Butanol Derivatives	Applications
1-butanol	Solvents, gasoline additive,plasticizers, chemical intermediate, cosmetics.
2-butanol	Perfumes, artificial flavours, solvent, industrial cleaners, chemical intermediate.
Iso-butanol	Industrial cleaners,additive for paint, solvent, ink ingredient gasoline additive.
Tert-butanol	Gasoline additive for octane booster,industrial cleaners, solvent, denaturant for ethanol.

7.1 Transportation biofuel

Engine performance characteristics, exhaust emissions, and combustion quality depend on the fuel's physical and chemical properties. High viscosity, low volatility, high hydrophobicity, anti-corrosive nature, similar A/F ratio, low moisture, and higher energy content makes biobutanol an alternative to non-renewable petroleum fuel for current and future generations (Liu et al. 2013). n-Butanol can be readily blended with gasoline as it exhibits almost analogous chemical and physical

Table 4. Physical and chemical properties of diesel, ethanol, gasoline, and n-butanol (Kolesinska et al. 2019).

Fuel	Energy Density MJ/L)	A/F	Energy Content/ Btu/US gallon	Octane Number	Water Solubility (%)
Diesel	35.5	14.7	130,000	-	Negligible
Ethanol	19.6	8.94	84,000	96	100
Gasoline	32	14.6	114,000	81-89	Negligible
Butanol	29.2	11.12	105,000	78	7

properties (Table 4). In contrast, iso-butanol, 2-butanol, and 3-butanol were used as additives in engine applications (Qureshi and Ezeji 2008). Thirty percent v/v biobutanol could be added to fuel without any further engine modification. Moreover, the butanol fuel contains oxygen atoms, exhibiting a smaller A/F ratio than gasoline, resulting in reduced HC and CO emissions. Butanol and its mixtures could be utilized directly in the transportation tanks and re-fuelling infrastructure (Patakova et al. 2011).

7.2 Chemical

Apart from its most important application as biofuel, it is used as a solvent for fats, vegetable oils, waxes, lacquers, shellac, dyes, natural and synthetic resins, gums, hydraulic fluids, varnish, coatings, alkaloids, paints and detergents, etc. Also, used as a plasticizer, hydraulic fluids, dyeing assistant, and dehydrating agent. It is used as a swelling and solubilizer agent in the textile industry and spinning baths, respectively. n-Butanol is used to synthesize many chemicals, such as butyl acrylate, dibutyl phthalate, butyl acetate, amine resins, and a wide variety of butyl esters (Kirschner 2006). Moreover, it is widely used as a solvent in the production of perfumes, oils, pharmaceuticals, and plastic industry. Further, it is used as a food-grade extractant and enhancer in the nitrocellulose lacquersformation. Further, it is used as an intermediate in the manufacturing of perfumes, rubber cement, artificial leather, and photographic films (Elder 1987).

7.3 Cosmetics

It is used in eye makeup, personal hygiene, shaving cosmetic products, and solvent in nail care cosmetic products. Further, used as a clarifying substance in the production of shampoo (Elder 1987).

7.4 Food application

n-Butanol is generally recognized as safe by the FDA as a flavoring agent (Elder 1987). n-Butanol can be used as an indirect food additive. It is an adjuvant in the production of polymeric coatings for polyolefin films, cellophane, and polysulfide polymerpolyepoxy resins intended to use as a food contact surface of articles. Moreover, it is used to produce adhesives for the components used in manufacturing, transporting, packaging, and holding food. n-Butanol is used as defoaming agent in the production of coatings for paper and paperboard, which can be used in food packaging (Elder 1987, Zhang et al. 2016).

8. Conclusion and future prospects

Higher substrate cost, low butanol titre, and high product recovery cost are some of the prime problems in conventional ABE fermentation. In the recent decade, advances in traditional techniques have overcome glitches in ABE fermentation. The advancement in the field of biotechnology and genetic engineering can improve the yield as well as the quality of the product. The production of n-butanol by the enzymatic action of genetically modified solventogenic microorganisms is an economically feasible process due to its increased solvent-producing capability. Further, consistency in research for *in situ* butanol recovery through membrane-based perstraction and pervaporation, gas stripping, adsorption, and liquid-liquid extraction techniques can lead to enhanced n-butanol production. Also, butanol production from cheap feedstock in combination with effective pretreatment, developed in-house enzymes, and integrated solvent recovery will help to solve critical issues of the present scenario for the evolution of a better cost-effective process.

Acknowledgement

The authors sincerely acknowledge the support and encouragement provided by Mr. Shivajirao Deshmukh, Director General, VSI, for encouragement, valuable guidance, and permission to publish this paper.

References

Abd-Alla, Mohamed Hemida, Abdel-Naser Ahmed Zohri, and Abdel-Wahab Elsadek El-Enany et al. (2015). Acetone–butanol–ethanol production from substandard and surplus dates by Egyptian native *Clostridium* strains. Anaerobe, 32: 77–86.

Abdehagh, Niloofar, Bo Dai, and Jules Thibault et al. (2017). Biobutanol separation from ABE model solutions and fermentation broths using a combined adsorption–gas stripping process. Journal of Chemical Technology & Biotechnology, 92(1): 245–251.

Abdehagh, Niloofar, F., Handan Tezel, and Jules Thibault. (2014). Separation techniques in butanol production: Challenges and developments. Biomass and Bioenergy, 60: 222–246.

Andrade, José Carlos, and Isabel Vasconcelos. (2003). Continuous cultures of *Clostridium acetobutylicum*: culture stability and low-grade glycerol utilisation. Biotechnology letters, 25(2): 121–125.

Aratboni, Hossein Alishah, Nahid Rafiei, and Raul Garcia-Granados et al. (2019). Biomass and lipid induction strategies in microalgae for biofuel production and other applications. Microbial Cell Factories, 18(1): 1–17.

Atsumi, Shota, Anthony F. Cann, and Michael R. Connor et al. (2008). Metabolic engineering of *Escherichia coli* for 1-butanol production. *Meta*bolic Engineering, 10(6): 305–311.

Behera, Shuvashish, Richa Singh, and Richa Arora et al. (2015). Scope of algae as third generation biofuels. Frontiers in Bioengineering and Biotechnology, 2: 90.

Behera, Shuvashish, Nilesh Kumar Sharma, and Sachin Kumar. 2018. Prospects of solvent tolerance in butanol fermenting bacteria. pp. 249–264. *In*: Biorefining of Biomass to Biofuels. Springer, Cham.

Behera, Shuvashish, and Sachin Kumar. (2019). Potential and prospects of biobutanol production from agricultural residues. Liquid Biofuel Production, (2019): 285–318.

Bikker, Paul, Marinus M. van Krimpen, and Piet van Wikselaar et al. (2016). Biorefinery of the green seaweed *Ulva lactuca* to produce animal feed, chemicals and biofuels. Journal of Applied Phycology, 28(6): 3511–3525.

Birgen, Cansu, Peter Dürre, Heinz A. Preisig et al. (2019). Butanol production from lignocellulosic biomass: revisiting fermentation performance indicators with exploratory data analysis. *Biotechnology for Biofuels* 12(1): 1–15.

Branduardi, Paola, Valeria Longo, Nadia Maria Berterame et al. (2013). A novel pathway to produce butanol and isobutanol in *Saccharomyces cerevisiae*. Biotechnology for Biofuels 6(1): 1-12.

Brennan, Liam, and Philip Owende. (2010). Biofuels from microalgae—a review of technologies for production, processing, and extractions of biofuels and co-products. Renewable and Sustainable Energy Reviews, 14(2): 557–577.

Cabezas, R., Suazo, K., Merlet, G. et al. (2019). Performance of butanol separation from ABE mixtures by pervaporation using silicone-coated ionic liquid gel membranes. Royal Society of Chemistry Advances, 9(15): 8546–8556.

Cai, Di, Changjing Chen, Changwei Zhang et al. (2017). Fed-batch fermentation with intermittent gas stripping using immobilized *Clostridium acetobutylicum* for biobutanol production from corn stover bagasse hydrolysate. Biochemical Engineering Journal, 125: 18–22.

Cai, Di, Tao Zhang, Jia Zheng et al. (2013). Biobutanol from sweet sorghum bagasse hydrolysate by a hybrid pervaporation process. Bioresource Technology, 145: 97–102.

Carvalheiro, Florbela, Luís C Duarte, and Francisco, M. Gírio. (2008). Hemicellulose biorefineries: a review on biomass pretreatments. *Journal of Scientific & Industrial Research,* (2008): 849–864.

Castro, Yessica A., Joshua T. Ellis, Charles D. Miller et al. (2015). Optimization of wastewater microalgae saccharification using dilute acid hydrolysis for acetone, butanol, and ethanol fermentation. Applied Energy, 140: 14–19.

Chen, Chunyan, Zeyi Xiao, Xiaoyu Tang et al. (2013). Acetone–butanol–ethanol fermentation in a continuous and closed-circulating fermentation system with PDMS membrane bioreactor. Bioresource Technology, 128: 246–251.

Cheng, Chieh-Lun, Pei-Yi Che, Bor-Yann Chen et al. (2012). Biobutanol production from agricultural waste by an acclimated mixed bacterial microflora. Applied Energy, 100: 3–9.

Cheng, Hai-Hsuan, Liang-Ming Whang, and Kun-Chi Chan et al. (2015). Biological butanol production from microalgae-based biodiesel residues by *Clostridium acetobutylicum*. Bioresource Technology, 184: 379–385.

Cooksley, Clare M., Ying Zhang, Hengzheng Wang et al. (2012). Targeted mutagenesis of the *Clostridium acetobutylicum* acetone–butanol–ethanol fermentation pathway. Metabolic Engineering, 14(6): 630–641.

Díez-Antolínez, Rebeca, M. Hijosa-Valsero, A.I. Paniagua-Garcia et al. (2018). *In situ* two-stage gas stripping for the recovery of butanol from acetone-butanol-ethanol (ABE) fermentation broths. Chemical Engineering Transactions, 64: 37–42.

Dong, H., Zhang, Y., Zhu, Y. et al. (2011). Biofuels and bioenergy: acetone and butanol. pp. 71–85. *In*: Comprehensive Biotechnology: Industrial Biotechnology and Commodity Products, Elsevier.

Du, Jianjun, Amy McGraw, Nicole Lorenz et al. (2012). Continuous fermentation of *Clostridium tyrobutyricum* with partial cell recycle as a long-term strategy for butyric acid production. Energies, 5(8): 2835–2848.

Elder, R.L. (1987). Final report on the safety assessment of n-Butyl Alcohol." Journal. of the Aamerican. College of Toxicology, 6: 403–424.

Ellis, Joshua T., Neal N. Hengge, Ronald C. Sims et al. (2012). Acetone, butanol, and ethanol production from wastewater algae. Bioresource Technology, 111: 491–495.

Ezeji, Thaddeus Chukwuemeka, Nasib Qureshi, and Blaschek, H.P. (2004). Acetone butanol ethanol (ABE) production from concentrated substrate: reduction in substrate inhibition by fed-batch technique and product inhibition by gas stripping. Applied Microbiology and Biotechnology, 63(6): 653–658.

Ezeji, Thaddeus Chukwuemeka, Nasib Qureshi, and Hans Peter Blaschek. (2013). Microbial production of a biofuel (acetone–butanol–ethanol) in a continuous bioreactor: impact of bleed and simultaneous product removal. Bioprocess and Biosystems Engineering, 36(1): 109–116.

Gao, Kai, and Lars Rehmann. (2014). ABE fermentation from enzymatic hydrolysate of NaOH-pretreated corncobs. Biomass and Bioenergy, 66: 110–115.

Gao, Kai, Simone Boiano, Antonio Marzocchella et al. (2014). Cellulosic butanol production from alkali-pretreated switchgrass (*Panicum virgatum*) and *phragmites* (*Phragmites australis*). Bioresource Technology, 174: 176–181.

Gao, Kai, Valerie Orr, and Lars Rehmann. (2016). Butanol fermentation from microalgae-derived carbohydrates after ionic liquid extraction. Bioresource Technology, 206: 77–85.

García, Verónica, Johanna Päkkilä, Heikki Ojamo et al. (2011). Challenges in biobutanol production: how to improve the efficiency? Renewable and Sustainable Energy Reviews, 15(2): 964–980.

Gottumukkala, Lalitha Devi, Kate Haigh, and Johann Görgens. (2017). Trends and advances in conversion of lignocellulosic biomass to biobutanol: microbes, bioprocesses and industrial viability. Renewable and Sustainable Energy Reviews, 76: 963–973.

Grant, B. (2009). Biofuels made from algae are the next big thing on alternative energy horizon. Scientist (2009): 37–41.

Guo, Ting, Ai-yong He, Teng-fei Du et al. (2013). Butanol production from hemicellulosic hydrolysate of corn fiber by a *Clostridium beijerinckii* mutant with high inhibitor-tolerance. Bioresource Technology, 135: 379–385.

Han, S.-H., Cho, D.H., Yong Hwan Kim et al. (2013). Biobutanol production from 2-year-old willow biomass by acid hydrolysis and acetone–butanol–ethanol fermentation. Energy, 61: 13–17.

Halder, Pobitra, and Azad, A.K. (2019). Recent trends and challenges of algal biofuel conversion technologies. pp. 167–179. *In*: Advanced Biofuels. Woodhead Publishing.

He, Chi-Ruei, Che-Lun Huang, Yung-Chang Lai et al. (2017). The utilization of sweet potato vines as carbon sources for fermenting bio-butanol. Journal of the Taiwan Institute of Chemical Engineers, 79: 7–13.

Hemming, David, ed. (2011). Plant sciences reviews 2010. Cabi.

Hou, Xiaoru, Nikolaj From, Irini Angelidaki et al. (2017). Butanol fermentation of the brown seaweed *Laminaria digitata* by *Clostridium beijerinckii* DSM-6422. Bioresource Technology, 238: 16–21.

Hu, Qiang, Milton Sommerfeld, Eric Jarvis et al. (2008). Microalgal triacylglycerols as feedstocks for biofuel production: perspectives and advances. The Plant Journal, 54(4): 621–639.

Huang, Bingxin, Qian Liu, Jürgen Caro et al. (2014). Iso-butanol dehydration by pervaporation using zeolite LTA membranes prepared on 3-aminopropyltriethoxysilane-modified alumina tubes. Journal of Membrane Science, 455: 200–206.

Huang, Wei-Cho, David E. Ramey, and Shang-Tian Yang. Continuous production of butanol by *Clostridium acetobutylicum* immobilized in a fibrous bed bioreactor. pp. 887–898. *In*: Proceedings of the Twenty-Fifth Symposium on Biotechnology for Fuels and Chemicals Held May 4–7, 2003, in Breckenridge, CO. Humana Press, Totowa, NJ, 2004.

Huesemann, Michael H., Li-Jung Kuo, Lindsay Urquhart et al. (2012). Acetone-butanol fermentation of marine macroalgae. Bioresource Technology, 108: 305–309.

Jiménez-Bonilla, Pablo, and Yi Wang. (2018). *In situ* biobutanol recovery from *Clostridial* fermentations: a critical review. Critical Reviews in Biotechnology, 38(3): 469–482.

Jones, David T., and David R. Woods. (1986). Acetone-butanol fermentation revisited. Microbiological Reviews, 50(4): 484–524.

Kamiński, Władysław, Elwira Tomczak, and Andrzej Gorak. (2011). Biobutanol-production and purification methods. Atmosphere, 2(3).

Karimi, Keikhosro, Meisam Tabatabaei, Ilona Sárvári Horváth et al. (2015). Recent trends in acetone, butanol, and ethanol (ABE) production. Biofuel Research Journal, 2(4): 301–308.

Karthick, C., and Nanthagopal, K. (2021). A comprehensive review on ecological approaches of waste to wealth strategies for production of sustainable biobutanol and its suitability in automotive applications. Energy Conversion and Management, 239: 114219.

Khan, Muhammad Imran, Jin Hyuk Shin, and Jong Deog Kim. (2018). The promising future of microalgae: current status, challenges, and optimization of a sustainable and renewable industry for biofuels, feed, and other products. Microbial Cell Factories, 17(1): 1–21.

Kiss, Anton A, Iulian Pătraşcu, and Costin Sorin Bîldea. (2019). From substrate to biofuel in the acetone–butanol–ethanol process. pp. 59–82. *In*: Second and Third Generation of Feedstocks.

Kirschner, M. (2006). n-Butanol. Chemical Market Reporter January 30–February 5, ABI/INFORM Global. 42.

Kolesinska, Beata, Justyna Fraczyk, Michal Binczarski et al. (2019). Butanol synthesis routes for biofuel production: trends and perspectives. Materials, 12(3): 350.

Kujawska, Anna, Jan Kujawski, Marek Bryjak et al. (2015). ABE fermentation products recovery methods—a review. Renewable and Sustainable Energy Reviews, 48: 648–661.

Kumar, Manish, and Kalyan Gayen. (2011). Developments in biobutanol production: new insights. Applied Energy, 88(6): 1999–2012.

Kushwaha, Anamika, Shivani Goswami, Afreen Sultana et al. (2022). Waste biomass to biobutanol: recent trends and advancements. Waste-to-Energy Approaches Towards Zero Waste, (2022): 393–423.

Le, Truong Giang, Dang-Thuan Tran, and Thi Cam Van Do. (2019). Design considerations of microalgal culture ponds and photobioreactors for wastewater treatment and biomass cogeneration. pp. 535–567. *In*: Microalgae Biotechnology for Development of Biofuel and Wastewater Treatment.

Li, Han-guang, Fred Kwame Ofosu, Kun-tai Li et al. (2014). Acetone, butanol, and ethanol production from gelatinized cassava flour by a new isolates with high butanol tolerance. Bioresource Technology, 172: 276–282.

Li, Han-guang, Wei Luo, Qiu-ya Gu et al. (2013). Acetone, butanol, and ethanol production from cane molasses using *Clostridium beijerinckii* mutant obtained by combined low-energy ion beam implantation and N-methyl-N-nitro-N-nitrosoguanidine induction. Bioresource Technology, 137: 254–260.

Li, Shubo, Li Huang, Chengzhu Ke et al. (2020). Pathway dissection, regulation, engineering and application: lessons learned from biobutanol production by solventogenic *Clostridia*. Biotechnology for Biofuels, 13(1): 1–25.

Liberato, Vanessa, Carolina Benevenuti, Fabiana Coelho et al. (2019). *Clostridium* sp. as bio-catalyst for fuels and chemicals production in a biorefinery context. *Catalysts* 9(11): 962.

Lienhardt, Jason, Justin Schripsema, Nasib Qureshi et al. (2002). Butanol production by *Clostridium beijerinckii* BA101 in an immobilized cell biofilm reactor. Applied Biochemistry and Biotechnology, 98(1): 591–598.

Liu, Hongjuan, Genyu Wang, and Jianan Zhang. (2013). The promising fuel-biobutanol. Liquid, Gaseous and Solid Biofuels-Conversion Techniques (2013): 175–198.

Maiti, Sampa, Gorka Gallastegui, Saurabh Jyoti Sarma et al. (2016). A re-look at the biochemical strategies to enhance butanol production. Biomass and Bioenergy, 94: 187–200.

Mata, Teresa M., Antonio A. Martins, and Nidia S. Caetano. (2010). Microalgae for biodiesel production and other applications: a review. Renewable and Sustainable Energy Reviews, 14(1): 217–232.

Medipally, Srikanth Reddy, Fatimah Md. Yusoff, Sanjoy Banerjee et al. (2015). Microalgae as sustainable renewable energy feedstock for biofuel production. BioMed Research International, 2015.

Mitchell, Wilfrid J. (1997). Physiology of carbohydrate to solvent conversion by clostridia. Advances in Microbial Physiology, 39: 31–130.

Ngoc-Phuong-Thao Nguyen, Sonja Linder, Stefanie K. Flitsch et al. (2016). Cap0037, a novel global regulator of *Clostridium acetobutylicum* metabolism. Mbio, 7.5: e01218-16.

Nguyen, Trung Hau, In Yung Sunwoo, Chae Hun Ra et al. (2019). Acetone, butanol, and ethanol production from the green seaweed *Enteromorpha intestinalis* via the separate hydrolysis and fermentation. Bioprocess and Biosystems Engineering, 42(3): 415–424.

Nigam, Poonam Singh, and Anoop Singh. (2011). Production of liquid biofuels from renewable resources. Progress in Energy and Combustion Science, 37(1): 52–68.

Obando, Juan Jacobo Jaramillo, and Carlos Ariel Cardona. (2011). Analysis of the production of biobutanol in the acetobutilyc fermentation with *Clostridium Saccharoperbutylacetonicum* N1-4 ATCC13564. Revista Facultad de Ingeniería Universidad de Antioquia, 58: 36–45.

Okolie, Jude A., Alivia Mukherjee, Sonil Nanda et al. (2021). Next-generation biofuels and platform biochemicals from lignocellulosic biomass. International Journal of Energy Research.

Onay, Melih. (2020). Enhancing carbohydrate productivity from *Nannochloropsis gaditana* for bio-butanol production. Energy Reports, 6: 63–67.

Oudshoorn, Arjan, Luuk A.M. Van Der Wielen, and Adrie J.J. Straathof. (2009). Assessment of options for selective 1-butanol recovery from aqueous solution. Industrial & Engineering Chemistry Research, 48(15): 7325–7336.

Patakova, Petra, Daniel Maxa, Mojmir Rychtera et al. (2011). Perspectives of biobutanol production and use. Biofuel's Engineering Process Technology, 11: 243–261.

Patakova, Petra, Jan Kolek, Karel Sedlar et al. (2018). Comparative analysis of high butanol tolerance and production in *Clostridia*. Biotechnology Advances, 36(3): 721–738.

Patil, Ravichandra C., Pravin G. Suryawanshi, Rupam Kataki et al. (2019). Current challenges and advances in butanol production. Sustainable Bioenergy, 225-256.

Potts, Thomas, Jianjun Du, Michelle Paul et al. (2012). The production of butanol from *Jamaica bay macro* algae. Environmental Progress & Sustainable Energy, 31(1): 29–36.

Procentese, Alessandra, Francesca Raganati, Giuseppe Olivieri et al. (2017). Pre-treatment and enzymatic hydrolysis of lettuce residues as feedstock for bio-butanol production. Biomass and Bioenergy, 96: 172–179.

Pugazhendhi, Arivalagan, Thangavel Mathimani, Sunita Varjani et al. (2019). Biobutanol as a promising liquid fuel for the future-recent updates and perspectives. Fuel, 253: 637–646.

Qureshi, Nasib, and Hans P. Blaschek. (2000). Economics of butanol fermentation using hyper-butanol producing *Clostridium beijerinckii* BA101. Food and Bioproducts Processing, 78(3): 139–144.

Qureshi, Nasib, and Hans P. Blaschek. (1999). Butanol recovery from model solution/ fermentation broth by pervaporation: evaluation of membrane performance. Biomass and Bioenergy, 17(2): 175–184.

Qureshi, Nasib, and Thaddeus Chukwuemeka Ezeji. (2008). Butanol, 'a superior biofuel' production from agricultural residues (renewable biomass): recent progress in technology. Biofuels, Bioproducts and Biorefining: Innovation for a Sustainable Economy, 2(4): 319–330.

Qureshi, Nasib, Badal C. Saha, Bruce Dien et al. (2010). Production of butanol (a biofuel) from agricultural residues: Part I–Use of barley straw hydrolysate. Biomass and bioenergy, 34(4): 559–565.

Qureshi, Nasib, Badal C. Saha, Ronald E. Hector et al. (2008). Butanol production from wheat straw by simultaneous saccharification and fermentation using *Clostridium beijerinckii*: Part I—Batch fermentation. Biomass and Bioenergy, 32(2): 168–175.

Qureshi, Nasib, Siqing Liu, Stephen Hughes et al. (2016). Cellulosic butanol (ABE) biofuel production from sweet sorghum bagasse (SSB): impact of hot water pretreatment and solid loadings on fermentation employing *Clostridium beijerinckii* P260. Bioenergy Research, 9(4): 1167–1179.

Qureshi, Nasibuddin, Ian S. Maddox, and Anton Friedl. (1992). Application of continuous substrate feeding to the ABE fermentation: relief of product inhibition using extraction, perstraction, stripping, and pervaporation. Biotechnology Progress, 8(5): 382–390.

Raganati, Francesca, Sebastian Curth, Peter Götz et al. (2012). Butanol production from lignocellulosic-based hexoses and pentoses by fermentation of *Clostridium acetobutylicum*. *Chemical Engineering Transactions,* 27(6).

Rathour, Ranju Kumari, Vishal Ahuja, Ravi Kant Bhatia et al. (2018). Biobutanol: New era of biofuels. International Journal of Energy Research, 42(15): 4532–4545.

Salaeh, Suhaila, Prawit Kongjan, Somrak Panphon et al. (2019). Feasibility of ABE fermentation from *Rhizoclonium* spp. hydrolysate with low nutrient supplementation. Biomass and Bioenergy, 127: 105269.

Sharif Rohani, Aida, Poupak Mehrani, and Jules Thibault. (2015). Comparison of *in-situ* recovery methods of gas stripping, pervaporation, and vacuum separation by multi-objective optimization for producing biobutanol via fermentation process. The Canadian Journal of Chemical Engineering, 93(6): 986–997.

Shukor, Hafiza, Najeeb Kaid Nasser Al-Shorgani, Peyman Abdeshahian et al. (2014). Production of butanol by *Clostridium saccharoperbutylacetonicum* N1-4 from palm kernel cake in acetone–butanol–ethanol fermentation using an empirical model. Bioresource Technology, 170: 565-573.

Steen, Eric J., Rossana Chan, Nilu Prasad et al. (2008). Metabolic engineering of *Saccharomyces cerevisiae* for the production of n-butanol. Microbial Cell Factories, 7(1): 1–8.

Su, Haifeng, Gang Liu, Mingxiong He et al. (2015). A biorefining process: Sequential, combinational lignocellulose pretreatment procedure for improving biobutanol production from sugarcane bagasse. Bioresource Technology, 187: 149–160.

Sukumaran, Rajeev K., Lalitha Devi Gottumukkala, Kuniparambil Rajasree et al. (2011). Butanol fuel from biomass: revisiting ABE fermentation. pp. 571–586. *In*: Biofuels. Academic Press.

Takkellapati, Sudhakar, Tao Li, and Michael A. Gonzalez. (2018). An overview of biorefinery-derived platform chemicals from a cellulose and hemicellulose biorefinery. Clean Technologies and Environmental Policy, 20(7): 1615–1630.

Tanaka, Shigemitsu, Yukihiro Tashiro, Genta Kobayashi et al. (2012). Membrane-assisted extractive butanol fermentation by *Clostridium saccharoperbutylacetonicum* N1-4 with 1-dodecanol as the extractant. Bioresource Technology, 116: 448–452.

Tashiro, Yukihiro, and Kenji Sonomoto. (2010). Advances in butanol production by *Clostridia*. Current Research, Technology and Education Topics in Applied Microbiology and Microbial Biotechnology, 2: 1383–94.

Timung, Seim, Harsimranpreet Singh, and Anshika Annu. (2021). Bio-butanol as biofuels: the present and future scope. Liquid Biofuels: Fundamentals, Characterization, and Applications, 467–485.

Ujor, Victor, Chidozie Victor Agu, Venkat Gopalan et al. (2014). Glycerol supplementation of the growth medium enhances *in situ* detoxification of furfural by *Clostridium beijerinckii* during butanol fermentation. Applied Microbiology and Biotechnology, 98(14): 6511–6521.

Van der Wal, Hetty, Bram LHM Sperber, Bwee Houweling-Tan et al. Production of acetone, butanol, and ethanol from biomass of the green seaweed *Ulva lactuca*. Bioresource Technology, 128 (2013): 431–437.

Vane, Leland M. (2005). A review of pervaporation for product recovery from biomass fermentation processes. Journal of Chemical Technology & Biotechnology: International Research in Process, Environmental & Clean Technology, 80(6): 603–629.

Veza, Ibham, Mohd Farid Muhamad Said, and Zulkarnain Abdul Latiff. (2021). Recent advances in butanol production by acetone-butanol-ethanol (ABE) fermentation. Biomass and Bioenergy, 144: 105919.

Virunanon, Chompunuch, Sam Chantaroopamai, Jessada Denduangbaripant et al. (2008). Solventogenic-cellulolytic clostridia from 4-step-screening process in agricultural waste and cow intestinal tract. Anaerobe, 14(2): 109–117.

Wang, Yue, Shih-Hsin Ho, Hong-Wei Yen et al. (2017). Current advances on fermentative biobutanol production using third generation feedstock. Biotechnology Advances, 35(8): 1049–1059.

Wang, Yue, Wanqian Guo, Chieh-Lun Cheng et al. (2016). Enhancing bio-butanol production from biomass of *Chlorella vulgaris* JSC-6 with sequential alkali pretreatment and acid hydrolysis. Bioresource Technology, 200: 557–564.

Welt, Bernard. (1950). n-Butanol: Its use in control of postoperative pain in otorhinolaryngological surgery. AMA Archives of Otolaryngology, 52(4): 549–564.

Xue, Chuang, and Chi Cheng. (2019). Butanol production by *Clostridium*. Advances in Bioenergy, 4: 35–77.

Xue, Chuang, Jingbo Zhao, Congcong Lu et al. (2012). High-titer n-butanol production by *Clostridium acetobutylicum* JB200 in fed-batch fermentation with intermittent gas stripping." Biotechnology and Bioengineering, 109(11): 2746–2756.

Xue, Chuang, Xin-Qing Zhao, Chen-Guang Liu et al. (2013).Prospective and development of butanol as an advanced biofuel. Biotechnology Advances, 31(8): 1575–1584.

Yusoff, M.N.A.M., Zulkifli, N.W.M., Masum, B.M. et al. (2015). Feasibility of bioethanol and biobutanol as transportation fuel in spark-ignition engine: a review. Royal Society of Chemisstry Advances, 5(121): 100184–100211.

Zhang, Jie, Mingyu Wang, Mintian Gao et al. (2013b). Efficient acetone–butanol–ethanol production from corncob with a new pretreatment technology—wet disk milling. Bioenergy Research, 6(1): 35–43.

Zhang, J., Wang, S. and Wang, Y. (2016). Biobutanol production from renewable resources: Recent advances. Advances in Bioenergy, 1: 1–68.

Zhang, W.L., Liu, Z.Y., Liu, Z. et al. (2012). Butanol production from corncob residue using *Clostridium beijerinckii* NCIMB 8052. Letters in Applied Microbiology, 55(3): 240–246.

Zhang, Yan, and Thaddeus Chukwuemeka Ezeji. (2013a). Transcriptional analysis of *Clostridium beijerinckii* NCIMB 8052 to elucidate role of furfural stress during acetone butanol ethanol fermentation. Biotechnology for Biofuels, 6(1): 1–17.

Zheng, Yan-Ning, Liang-Zhi Li, Mo Xian et al. (2009). Problems with the microbial production of butanol. Journal of Industrial Microbiology and Biotechnology, 36(9): 1127–1138.

Zhu, Chao, Lijie Chen, Chuang Xue et al. (2018). A novel close-circulating vapor stripping-vapor permeation technique for boosting biobutanol production and recovery. Biotechnology for Biofuels, 11(1): 1–13.

Zhu, Haipeng, Xinran Li, Yang Pan et al. (2020). Fluorinated PDMS membrane with anti-biofouling property for in-situ biobutanol recovery from fermentation-pervaporation coupled process. Journal of Membrane Science, 609: 118225.

Zverlov, V.V., Berezina, O., Velikodvorskaya, G.A. et al. (2006). Bacterial acetone and butanol production by industrial fermentation in the Soviet Union: use of hydrolyzed agricultural waste for biorefinery. Applied Microbiology and Biotechnology, 71(5): 587–597.

7

Bio-butanol:
Potential Feedstocks and Production Techniques

Anita and *Narendra Kumar**

1. Introduction

The usage of renewable energy sources must be expedited due to the rapid depletion of fossil fuel supplies and their significant environmental impact. Hence, required substitutes must be found, and a conversation to renewable energy sources must be planned that have low environmental effects and are quantifiably sufficient to meet the demand and provide energy supply security. Solar energy is gaining popularity as a non-polluting, free, and renewable source of energy. However, using solar energy for power generation is more expensive than using conventional power generation methods, necessitating new strategies to solve this issue (Jamel et al. 2013). India ranks sixth in the world in terms of total energy consumption, with imports contributing to more than 70% of its primary energy needs, primarily in the form of conventional energy sources like crude oil and natural gas. On the other hand, India has increased its power generation by installing the capacity of 1.36 GW over to 112 GW since independence, and the electrification of more than 500,000 villages has been done. This achievement is outstanding but not sufficient, because approximately 44% of households do not have access to electricity

Department of Environmental Science, School of Earth and Environmental Sciences, Babasaheb Bhimrao Ambedkar University (A Central University), Lucknow.
* Corresponding author: narendrakumar_lko@yahoo.co.in

(*MNRE Annual Report* 2019–20). The Ministry of Power established a road map to achieve "power on demand". According to the 16th Electric Power Survey, an additional 100 GW is required to meet the predicted demand. To overcome the issues of the energy crisis future alternative energy sources are the only answer. An alternative fuel, which is considered a necessity for the reasons stated above, should be technically, economically, and environmentally feasible, as well as easily accessible (shahbaz et al. 2020). The energy sources like alternative biofuels, liquefied petroleum gas (LPG), compressed natural gas (CNG), hydrogen, and biodiesel, are used as alternative sources over conventional fuels in the internal combustion engines, while hybrid, electric, and fuel cell automobiles are yet to be explored whereas, some of their important features are analyzed comparatively.

Non-conventional energy is derived from an inexhaustible source. Sustainable usage of energy resources is a matter of concern that needs to be discussed at the research level. Utilizing energy sources and considering why this is the only option must be undertaken? The majority of considerations are safety, cost, stability, efficiency, and environmental impact. Many sectors around the world are still dependent on non-renewable for electrical generations. These fuels are quite effective in terms of power generation quality, but they are not cost-effective in the long run.

The non-conventional sources of energy, such as solar and wind, will contribute to electrical capacity, but most automobiles require liquid fuel and will continue to do so for the foreseeable future. However, there is a demand for the inclusion of biofuels in an attempt to sustainably power the world's millions of automobiles. Both developing and developed economies were promoting scientific study into the manufacturing and development of biofuels as a replacement for petroleum-based energy (Owusu et al. 2016).

Feedstock and process chemistry (catalytic issues) are two crucial aspects of the success of such bio-based products/fuels. Next to feedstock, catalysts play the most imperative role in any industrial process as their adoption in the processes directly affects the feasible production routes, product costs, product quality, and yield (*Technology Roadmaps Biofuels for Transport*, IEA 2011).

Biofuels have accelerated the look for an alternative source, more sustainable and renewable than gasoline (Ravindra et al. 2019). Renewable energy derived from biomass (Kumar et al. 2011) would be cost-effective and sustainable alternative methods. Biofuels (solid, liquid, and gas fuels) are renewable fuels derived from biological feedstocks and include bioethanol or biobutanol as a gasoline-equivalent (Kapasi et al. 2010). Alternative fuels can help scale down the dependency on non-renewable

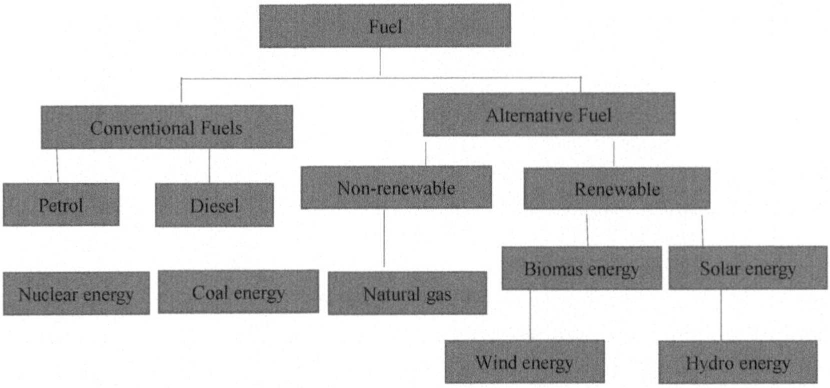

Figure 1. Categorization of energy sources.

sources fuels and enhance the socioeconomics of rural regions in order to deliver significant environmental benefits to both industrially developing and developed countries, such as reduced greenhouse gas (GHG) emissions and less air pollution (Detlef et al. 2017).

Biofuels are categorized into four generations, which are first, second, third generation, and fourth-generation biofuels. Although the first generation could bid Carbon dioxide benefits, the food feedstock made it less preferable for biofuel production causing food-fuel competition (Khan et al. 2021). The second-generation biofuel uses non-food feedstocks like waste. It was considered to be superior and preferable to first-generation biofuel (Edrei et al. 2010). Later, third-generation biofuel was introduced where algal species were used for biofuel production. The carbohydrate-rich algae were able to be converted to acetone, butanol, and ethanol (Naik 2010). Algal biomass was found to be ubiquitous and became one of the clean and renewable substrates for ABE production (Demirbas 2010). The fourth-generation biofuel involves the implementation of genetic engineering techniques to target biofuel production efficiency, metabolism of cells, feedstock utilization, and so on, especially with algal species (Lü et al. 2011). Types of fuel can be catagorized as conventional and alternative fuels, which can further be divided into different types of fuel which is given in the flow chart below.

2. Biobutanol: An alternative fuel

Due to increasing worries about the uncertainty of oil prices, depletion of oil resources, and climate change, next-generation biofuels from renewable sources have attracted the attention of researchers, industry, and governments (Dellomonaco et al. 2010). New-generation fuels from non-conventional sources have captured attention among investigators, researchers, government, and industrialists, due to major concerns about the inconstancy of oil prices, exhaustion of oil reserves, and most importantly, climate change (Dellomonaco et al. 2010). Biobutanol or biobased butanol fuel is a second-generation alcohol fuel usually produced by fermentation of biomass through the acetone-butanol-ethanol (ABE) process. The process uses *Clostridium acetobutylicum* (popularly known as the Weizmann organism after its discoverer) for degrading sugars. It was originally used for the production of acetone from starch in 1916 by Chaim Weizmann with butanol as a by-product. The process also creates several other by-products in recoverable amounts: acetic, lactic, and propionic acids, isopropanol, ethanol, and H_2.

Biobutanol, a liquid transportation biofuel, has recently attracted a lot of attention from researchers all over the world due to its better fuel qualities than ethanol. Butanol is comparatively better in blending with gasoline, high in energy content, it has a less hygroscopic nature and lower volatility than ethanol; therefore, it can be directly used in convention engines, and low corrosiveness, these are the advantageous fuel properties of butanol which are given below in Table 1. Butanol is also synthesized by chemical processes like oxo synthesis, Reppe synthesis, and crotonaldehyde dehydrogenation (Lee et al. 2008).

Biobutanol has innumerable advantages in many aspects such as reduction in air pollution, reduced greenhouse emanations, discharge of biodegradable materials, and carbon sequestration (Demirbas 2008). Butanol is a superior biofuel as it contains 33% more energy than ethanol and can be blended with gasoline. Blended butanol burns cleaner, thus reducing environmental pollution and can be used in vehicles designed for gasoline and petrol without any change in the engine (Muhammad et al. 2021). Biobutanol has a lower energy density than gasoline, resulting

Table 1. Basic properties of fuels.

Properties	Butanol	Gasoline	Ethanol	Methanol	Diesel
Boiling point (°C)	118	85	78	65	150–380
Auto-ignition temperature (°C)	343	280	365	435	350–625
Energy content/value (BTU/gal)	33.1	44.5	26.9	19.6	38.6
Air-fuel ratio	11.1	14.8	9.0	6.5	14.1
Density	0.809	0.7–0.8	0.785	0.79	0.85
Octane number	78	81-89	102	104	40-55
Kinematic Viscosity	2.6	2.5–3.2	1.8	0.5	1.0–4.0
Cetane number	25	0-10	8	3	40–50

in a comparatively high energy content among gasoline alternatives under the 1992 Energy Policy Act. Based on its structure, butanol is classified in the family of alcohol like methanol (1-carbon), ethanol (2-carbon), propanol (3-carbon), and butanol (4-carbon). It's most commonly used in the manufacture of synthetic rubber, paint, detergents, lacquer, rayon, and brake fluids, along with solvents for fat, waxes, and resins. Butanol has a higher energy content than ethanol and has more advantageous physical characteristics such as low vapor pressure, low solubility with water, and complete solubility with diesel fuel, as described (Dürre 2007).

3. Conventional techniques for biobased butanol production

One of the oldest industrial fermentations is the ABE fermentation. During World War I, there were further developments in the ABE fermentation to produce acetone for its use in smokeless munitions powder production. N-butanol was regarded as an undesired by-product of the ABE fermentation process up until the 1920s. With the discovery of n-butanol and butyl acetate, this viewpoint was revised. These were used as solvents for nitrocellulose lacquer, which is used in the automotive industry (Ranjan and moholkar 2012). In 1916, Chaim Weizmann was the first to commercialize n-butanol. To date, he is credited with inventing the historic industrial ABE fermentation, which uses a bacteria called *Clostridium acetohutylicum* to transform fermented maize starches into acetone, which can then be used to create explosives to date. *Clostridium acetohutylicum* ATCC824 remains the best investigated and utilized strain (Green 2011). Solvent fermentation continued until the 1950s. However, since 1954, when the price of petroleum fell below even that of sugar, along with increasing demands for n-butanol, fermentative synthesis has been progressively substituted by the chemical treatment in western countries. After this, the n-butanol was produced with petroleum via hydrolysis of haloalkanes or hydration of alkenes. The process was mostly based on aldol condensation of acetaldehydes, followed by the dehydration and then the hydrogenation of croton-aldehyde. The industrial discovery of oxo synthesis boosted the chemical synthesis of n-butanol (Ndaba et al. 2015).

The ABE production pathway is a complex pathway in *Clostridial* species. The important end products of ABE fermentation contain mainly solvents, organic acids, and gases.

The process of ABE fermentation through genus *Clostridium* can be divided into two distinct phases:

(1) Acidogenesis: sugars being metabolized to produce acetate and butyrate, and accumulation of these acids reduces the pH.

(2) Solventogenesis: acids are then re-utilized to produce acetone-butanol-ethanol with the increase in pH.

The shift is initiated when the n-butyric concentration increases to more than 2 g/L and the pH decreases to 5. An anaerobic microorganism has to maintain a balance of NADH and NAD+ because of its inability to utilize the electron transport system to generate ATP. The conversion of glucose to ABE is the only way to regenerate NAD+ because all other end products yield NADH. The acetoacetylCoA is converted to acetoacetate, then to acetone, and later i-propanol. This conversion takes place during solventogenesis phase of fermentation where reduction of acetate and butyrate occurs. Glycolysis is the phase of fermentation where the majority of the ATP is generated (Tashiro and Sonomoto 2010). The majority of the acetyl-CoA is converted to acids and solvents whereas some are used for lipid synthesis and growth. Acetone is considered as an undesirable co-product of the ABE process because of its poor fuel properties and corrosiveness to rubber engine parts. Iso-propanol-butanol ethanol (IBE), an alcohol biofuel mixture is considered a "green solvent".

The solvent-producing *Clostridia reutilize* acetate or butyrate catalyzed by CoA-transferase to form solvents. This uptake is a de-toxic bio-reaction to protect the cell from the inhibitory effect of the organic acids (Gheshlaghi et al. 2009). The uptake does not take place in the absence of sugars because there is no sufficient reducing power to drive the uptake mechanism (Tashiro et al. 2007). The major requirement for successful fermentation is the sugar concentration as the sufficient presence of sugar ensures the success of ABE fermentation (Karakashev et al. 2007). Thus, using n-butyric acid as a carbon source cannot be solely used for the production of n-butanol and it needs the presence of sugars in a broth. The promising results fermentation of butyrate and glucose. Tashiro et al. (2007) observed that a mixture of butyrate and glucose has shown the promising potential of n-butanol production.

The fermentation process is basically divided into batch culture, fed-batch culture, and continuous culture. Batch culture is considered as the simplest method among three modes of cultures, therefore, it is one of the most studied methods for different parameters such as pH, carbon/nitrogen (C/N) ratio, partial pressure in the fermentor headspace (hydrogen or carbon monoxide), and the addition of electron carriers. The pH is considered as the important factor for the acidogenesis or solventogenesis through *C. acetobutylicum* and *C. beijerinckii* at alkaline and acidic pH conditions, respectively. In the batch mode of culture by the *C. acetobutylicum* ATCC 824T strain, the organic acid production was > 20 g/L with less ABE production (1 g/L) at 6.0 pH while higher ABE production of 17 g/L and lower organic acid production of 3 g/L was observed at 4.5 pH. The batch culture without maintaining pH with two species *C. saccharoperbutylacetonicum* N1–4 (ATCC13564) strain and *Clostridium saccharobutylicum* NCP262T producing a significant amount of ABE (14 g/L and 6.6 g/L). Additionally, the production of ABE or acid is strongly influenced by the C/N ratio. It has been reported that the optimum C/N ratio significantly increased the yield of ABE production as compared to a low C/N ratio with *C. acetobutylicum* ATCC 824T utilizing glucose as a substrate (Hayriet al. 2021). Moreover, using sago starch and *C. saccharobutylicum* NCP 262T, C/N ratio above optimum levels significantly affected the yield and eventually ABE production.

Fed-batch cultures have significantly more benefits than batch cultures. Fed-batch cultures for ABE fermentation have been exploited to utilize such benefits for successful butanol production. A higher sugar concentration increases substrate inhibitions, leading to a longer lag phase with low productivity of butanol. Approximately, 50–60 g/L of sugar can be easily used as butanol production above 10 g/L may affect cell growth, utilization of sugar, and further production of butanol. Organic acids such as acetic acid, butyric acid, and lactic acid, beside co-substrates, are considered as the promising substrate for the production of butanol (Tashiro et al. 2004). On the other hand, high organic acid concentration induces substrate inhibitions under low pH conditions (Cho and Oshiro et al. 2010). In the broth, monitoring of the pH can be the only solution to constantly maintain the level of organic acid whereas in fed-batch culture, the high yield of butanol was achieved without monitoring the pH with substrate feeding in two rounds. It has been observed that fed culture with glucose and xylose at the constant substrate feeding by *C. acetobutylicum* ATCC 824T strain produced 12 g/L butanol. Thus, fed-batch cultures can elevate butanol production by avoiding the inhibition of the substrate.

Continuous culture is the feasible mode of butanol production as these methods can help in diluting butanol in broth and overcome the limitation of butanol inhibition, thus improving the productivity of butanol. Although in continuous culture, the butanol concentration should be lower it has higher operational stability as compared to batch and fed-batch culture. Thus, more efforts are being invested to the process, which can enhance the productivity of butanol. In continuous-chemostat culture, the balance as per the cell concentration, amount of substrate, as well as product is obtained by replacement > 3 times with fresh medium. Earlier, the effect of different factors like pH, dilution rate, and amount of substrate, nutrients, as well as electron carriers on ABE fermentation was extensively examined in continuous-chemostat cultures. Among different factors, pH plays a decisive role in butanol production. In a continuous culture with *C. acetobutylicum* and glucose, the pH level of 4.3 or 4.5 and 5.7 or 6.0 exhibited the shift of metabolic pathways towards butanol and acid production (Sukwong et al. 2019). Additionally, higher butanol productivity of 0.34 and 0.529 g/L/h were observed 5.5 and 5.6 pH with *C. beijeri* nckii and *C. saccharoperbutylacetonicum* from glucose and xylose as compared to pH 5.0 and 4.6, respectively. Utilizing *C. acetobutylicum* and *C. beijerinckii*, the addition of yeast extract, ammonium chloride, and glucose in high concentrations improves butanol productivity (Zheng et al. 2012). Apart from pH, the dilution rate is also the most studied factor for butanol production with *C. acetobutylicum*, *C. beijerinckii*, and *C. saccharoperbutylacetonicum*. Dilution rate influences the cell concentration, substrate, and ultimately, the production level. The higher dilution rate is known to stimulate acid production while the lower dilution rate during continuous culture induces solventogenesis. To overcome the limitation posed by lower and high dilution rates, a two-stage continuous culture has been believed to improve butanol production. Due to the possibility of cell washout leading to lower cell density, the dilution rate in chemostate cultures can not be kept more than 0.3/h for superior butanol production. Upon increasing the dilution rate (0.24–0.26/h), the concentration of the cell showed a drastic decline from 2.62 g/L to 0.765 g/L in the presence of the *C. saccharoperbutylacetonicum* strain. The shortcoming associated with low cell density can be easily addressed by maintaining initial high cell density in continuous culture.

Methods like cell immobilization and cell recycling help to achieve maximum cell density for the production of butanol. The dilution rate of 2 hours has been reported to improve the productivity (16.2 g/L/h), utilizing brick as a carrier by *C. beijerinckii* BA101. Similarly, cell recycling through ultrafiltration and microfiltration has also been performed at high rates of dilution (> 0.3/h). Generally, obtaining high cell density is a very time-consuming process but), but Lienhardt et al. were successfully able to obtain a cell density of 20 g/L in less time (12 h) by concentrating broth from 4 L to 0.4 L (Lienhardt et al. 2002). However, cell bleeding techniques have been observed to maintain constant cell concentration with a higher operational stability than the normal technique. A high ABE productivity (7.55 g/L/h) has been achieved using microfiltration with a 0.85/h dilution rate by *C. saccharoperbutylacetonicum* (Qureshi et al. 2000). Productivity of butanol with respect to the substrate and used oraganisms are shown in Table 2.

4. Contemporary feedstock for the production of bio-butanol

One of the most important elements determining the production of ABE fermentation's feasibility is the substrate. Various subtrates has been used for the production of butanol like corn, whey permeate, starch, molasses, etc., whereas exploration for preprogatives for higher production is increasing constantly.

Table 2. Demonstration of various treatments with respect to different types of substrate.

Substrate	Organism	Fermentation	Butanol Yield	References
Glucose	*C. acetobutylicum*	Continuous culture	175 Mm	Bahl et al. 1982
Glucose	*C. beijerinckii* NRRL B592	Continuous culture	9.1 g/L	Mustschlechener et al. 2000
Glucose	*C. heijerinckii* BA 101	Batch fermentation	32.6 g/L	Chen and Blascheck et al. 1999
Glucose	*C. beijerinckii* BA	Continuous fermentation	7.9 g/L	Qureshi et al. 2000
Glucose	*C. beijerinckii BA*	Fed-batch fermentation	232 g/L	Ezeji et al. 2004a
Glucose	*C. saccharoperbulyl-aceloiiiciim* N\-4	Batch fermentation	24.2 g/L	Thang et al. 2010
Glucose	*C. heijerinckii* DSM 6423	Continuous fermentation	4.48 g/L	Survase et al. 2011
Molasses	*C. saccharohutylicwn*	Continuous fermentation	15.27 g/L	Ni et al. 2013
Glucose	*C. acetobutylicum*	Fed-batch fermentation	14.53 g/L	Dolejs et al. 2014

4.1 Feedstock for butanol production

Despite of superior fuel properties, n-butanol has to compete in the area of production costs with ethanol in order to penetrate the larger biofuel markets. Medium composition and substrate concentrations are known to impact the productivity, yield, and final concentration of n-butanol in the process (Kheyrandish et al. 2015). The substrate cost is one of the challenges for ABE fermentation, and greatly contributes to the overall cost-effectivness of fermentation products. The reduction in substrate cost offers the best opportunity to decrease the cost of production. Thus, low-cost, generous, and sustainable feedstocks such as wastes and agricultural residues should be used to improve the cost of the process (Van der Merwe et al. 2013).

4.2 First-generation feedstock

During the early twentieth century, food industry wastes like sugarcane and cereal grains were used as the feedstocks for ABE fermentation. With the increased demand for food, the use of food waste is no longer ethical and economical (Zhang et al. 2010). There are optimum operating conditions and challenges associated with feedstocks and processes. Food waste is a possible source of carbohydrate-based biomass that might be employed as a feedstock in ABE fermentation. Furthermore, this resource is safe, has a high yearly rate of availability, and has a low cost of supply, becomes an income for the ABE fermentation process (Sarchami et al. 2015). Cheese whey has already been shown to be a preferable biobutanol feedstock to other lactose substrates (Foda et al. 2010, Becerra et al. 2015). Various researchers have reported that cane molasses (which decreases sugar content by 50–55%) is another commonly used feedstock for butanol fermentation (Jiang et al. 2009, Vander Merwe et al. 2013). Another commonly used substrate for butanol fermentation is cane molasses (which reduces sugar concentration by 50–55% (Vander Merwe et al. 2013)). In the batch mode, the efficient utilization of cane molasses produced 19.8 g/l of total solvent (13.4 g/l of butanol) in 72 hours. The semicontinuous mode was shown to have higher solvent productivity (1.05 g/l/h), greater sugar utilization, and a shorter fermentation time than the batch mode, demonstrating the method's industrial feasibility (kushwaha et al. 2019).

4.3 Second-generation feedstock

The second-generation feedstock includes lignocellulosic biomass, which is a complex combination of, hemicellulose, cellulose, and lignin along with extractives and inorganic materials. Biomass is widely available, and biobutanol production has increased significantly in several countries with a substantial agricultural area (Kumar et al. 2012). The use of non-food feedstock (agricultural wastes, wood chips, grain residues, etc.) reduces the dependency on food biomass and even lowers production costs. According to the United Nations Environment Programme (UNEP 2015), agricultural biomass is produced at a pace of 140 billion tonnes per year, which is comparable to 50 billion tonnes of oil. The UNEP has identified Costa Rica, Cambodia, and India as major producers of agricultural waste, with India alone producing 415.5 million tonnes (equivalent to 103.9 million tonnes of oil).

4.4 Third-generation feedstock

Because of the limitations of first- and second-generation feedstocks, researchers are now focusing their efforts on third-generation feedstocks, which are widely available in huge quantities and need little available land (Ullah et al. 2015). Although algae can assimilate CO_2 and eliminate inorganic nutrients from wastewater bodies, they are being considered as a potential source of clean energy (Oswald 2003). Due to its widespread availability, algal biomass would be an appropriate substrate for ABE manufacturing. Algae are thought to be the most essential substrate for future clean and renewable energy production (Demirbas 2010). Algal biomass has a higher carbohydrate and lower lignin content compared

Table 3. Various feedstocks used in the production of butanol.

Feedstock	Microorganisms	ABE production (g/l)/time (h)	References
Rice straw	*C. acetobutylicum* NCIM 2337 *C. saccharoperbutylacetonicum* N1-4 *C. acetobutylicum* NRRL B-591	13.5 butanol 7.9 butanol 10.5/72	Rajan et al. 2013a Chen et al. (2013b) Amiri et al. (2014)
Wastewater algae	*Saccharoperbutylacetonicum* ATCC 27021	0.13 g butanol/g	ernigan et al. (2013)
Sweet sorghum bagasse	*C. acetobutylicum* ABE 1201	20.9/72	Cai et al. (2013)
Sugarcane bagasse	*C. acetobutylicum* GX01 *C. acetobutylicum* XY16 *C. acetobutylicum* CH02	21.11/60 14.26/60 12.12/120	Pang et al. (2016) Li et al. (2017) Kong et al. (2016)
Bamboo	*C. beijerinckii* ATCC 55025-E604	6.45 butanol/73	Kumar et al. (2017)
Willow biomass	*C. beijerinckii* NCIMB 8052	Stem: 4.5	Han et al. (2013)

to green plants, making it a more feasible and sustainable feedstock for the production of biofuel, including biobutanol (Cheng et al. 2013, Suutari et al. 2015). *C. saccharoperbutylacetonicum* N1–4 exhibits amylolytic characteristics toward starch-based polymers, according to a study (Thang et al. 2010), which is important because many algae, such as *Scenedesmus* and *Chlorella*, contain more than 50% (dry weight) starch, cellulose, and glycogen. Several algae have cell walls made of cellulose and starch, whereas green algae have cell walls made up of cellulose and starch (Singh and Olsen 2011). This enzymatic capacity might be beneficial for using *C. saccharo-perbutylacetonicum*, as well as other saccharolytic *Clostridia* species, to produce ABE from wastewater algae efficiently. Some feedstocks with the amounts produced are given bellow in Table 3.

5. Future prospects of butanol

Currently, the rapid decline in the fuel reserve along with increasing environmental concerns is shifting the focus towards the alternative, sustainable bio-based fuel. That is, to sustain fuel supply, the transition towards sustainable, reliable, and economical alternative of biofuel is very crucial (Gao et al. 2012). Biobutanol is gaining huge impetus as the appropriate alternative and a remarkable increase in the research has been observed in production of biobutanol (Behera and Kumar 2019). It is attaining noticeable attention due to several advantages as compared to fossil based fuels. Comparing different physical and chemical properties, butanol is considered as an efficient fuel as compared to other types of biofuels (Bharathiraja et al. 2017). Due to low vapour pressure (Baral and Shah 2016), high energy content, and low emission of pollutants upon combustion, there has been an increased reliance on biobutanol (Lamani et al. 2018). In order to increase the butanol production, the selection of species, substrate, pretreatment, as well as biotechnological method can be the factors responsible for increase in butanol production (Bharathiraja et al. 2017). Biobutanol is produced from fermentation of different biomaterial such as corn (Gao et al. 2012), baggase (Lu et al. 2017), palm mill effluent (Al-Shorgani et al. 2012c), molasses (Li et al. 2013) and food waste (Arancon et al. 2013). Using waste residues of agricultural origin as the substrate for butanol production is the most suitable approach for sustainable biofuel production but at the same time can increase the challenges of the food industry (Bharathiraja et al. 2017). Biobutanol is produced on a wide scale by the fermentation process as the microbes have shown immense potential to utilize the sugar in biological material (Qureshi et al. 2008). The tradition fermentation methods using the *Clostridia* strain has been used for the biobutanol production since the twentieth century (Lütke-Eversloh and Bahl 2011). The fermentation is carried out in the presence of bacteria and the process is known as ABE fermentation (Gao et al. 2015). The plant-based feedstock for biobutanol production is often limited by inadequate photosynthetic efficiency and huge costs, which makes the butanol production more expensive than fossil fuel (Shanmugam et al. 2021). The inability to directly degrade lignocellulosic biomass and utilization of pretreatment methods such as hydrothermal treatment and acid hydrolysis can enhance the cost of biobutanol production.

The huge cost of feedstock is the major bottleneck for the biobutanol production (Al-Al-Shorgani et al. 2012a), thus, the focus has shifted towards economical and renewable feedstock (Al-Shorgani et al. 2015a). Furthermore, the inhibition during pretreatment methods has restricted the use of lignocellulosic biomass for biobutanol production. Furthermore, the low butanol yield and expensive raw material are the major limitations for decline of industrial butanol production. Hence, the continuous search for cost-effective raw material is needed to overcome the limitation posed by other biological-based material for bio-butanol production (Bharathiraja et al. 2017).

There are novel techniques of fermentation that have been developed using advanced biotechnology as well as engineering techniques to increase the conversion efficiency of waste to biobutanol (Behera and Kumar 2019). The genetically modified plants produce cellulases, hemicellulases, and reduce the use of pretrement method by modifying lignin; these are suitable approaches to address the multiple problems associated with biological material as substrate for butanol production. Exploitation of genetic engineering techniques can help to elevate the production of butanol. Further, *Escherichia coli, Pseudomonas* sp. and *Bacillus subtilis* after genetic modification can substantially enhanced the butanol yield (Bharathiraja et al. 2017).

Algal biomass is considered as the cheap and sustainable feedstock for thirdgeneration biofuel. To overcome issues associated with plant-based material, microalgae due to high photosynthetic efficiency and carbon dioxide sequestration can serve as the potential feedstock for butanol production (Shanmugam et al. 2021). The inherent capability to grow in different types of hostile conditions such as wastewater and saline water has increased the interest towards microalgae. Due to huge photosynthetic efficiency (Upadhyay et al. 2021), quick growth (Singh et al. 2020a), short life cycle, high biomass production (Singh et al. 2020b c) carbohydrate content, cheap nutrients, and small land requirement, no competition with agricultural crops (Singh et al. 2021, Singh and Singh 2021, Singh et al. 2022), low cellulose content and no lignin, microalgae can serve as the potential organism for the biobutanol production. Microalgae's ability to fix atmospheric carbon can certainly assist to decrease the intensity of global warming (Singh et al. 2021). Beside pigments, microalgae contains high content of lipid (20–50%) (Brennan and Owende 2010), proteins (60%) (Becker 2007), and carbohydrates (20–40%) (Hu 2004). Employing *Clostridial* species for fermentation of sugar extracted from algae holds the great possibility of industrial production of biobutanol. Furthermore, selecting carbohydrate-enriched algal species can be a practical approach for butanol production (Shanmugam et al. 2021).

Algal biomass does not require harsh pretreatment; however, simple treatment methods are required to convert carbohydrates into sugar (Wang et al. 2017b). The algal biomass undergoes chemical and mechanical treatment as the carbohydrate in microalgae needs to be converted into fermentable sugars. The acid and alkali pretreatment of algal biomass yields butanol production of 13.1 g/L after fermentation using *C. acetobutylicum* ATCC 824 (Wang et al. 2016). Due to minimum pretreatment, as compared to other plant-based feedstocks, microalgae can act as a suitable candidate for butanol production.

The advancement in metabolic/genome engineering can assist in introducing genes or modulating pathways responsible for the fixation of carbon dioxide and utilize the end metabolites for butanol production. Furthermore, relying on integrated technologies like butanol production with co-production of different value-added products can convert single-stage processes to multi-terminus processes and such techniques can certainly minimize the cost of butanol production. Before butanol production from microalgae biomass, several constraints such as thorough knowledge about the metabolic pathways responsible for carbohydrate production, environmental factors inducing carbohydrate accumulation, reducing capital and operational cost of algal biomass production, and feasible processes for the butanol purification need to be dealt with (Farhana et al. 2022). Thus, addressing such bottlenecks can surely make the microalgae the promising source of butanol for future generations.

6. Conclusion

The environmental protection and awareness of global emission standards are growing increasingly stringent, while traditional fuel resources are under immense pressure. One of the best strategies for meeting emissions rules and resolving energy issues is to create alternative fuels for vehicles and other applications. The use of renewable feedstock provides a cost-effective and ecologically friendly source of energy for the entire planet. Biobutanol is a promising renewable biofuel that has the potential to replace nonrenewable fossil fuels in the present and future. The fed-batch, continuous, and two-stage continuous fermentations processes have been described by various researchers, which are currently being practiced for biobutanol production. It was also shown that including bacteria in the fermentation process increased biobutanol yield and productivity. In the future, the correct combination of substrate, microorganisms, and technology will open up new opportunities in the field of fermentative biobutanol synthesis. In order to improve biobutanol production efficiency, there is still a significant gap of knowledge to be filled. Biobutanol synthesis continues to be a source of worry from an industrial standpoint. The investigation for a possible microorganism capable of enhancing or increasing biobutanol production at the commercialization level is still ongoing as several molecular methods are being developed to redesign wild-type strains with desired characteristics. As a concluding point, industrial-scale biobutanol production will enhance the possibilities for economic development and give employment benefits.

References

Ahl, H., Andersch, W., and Gottschalk, G. (1982). Continuous production of acetone and butanol by *Clostridium acetobutylicum* in a two-stage phosphate limited chemostat. Eur. J. Appl. Microbiol. Biotechnol., 15: 201–5.

Al-Shorgani, N.K.N., Kalil, M.S., Yusoff, W.M.W., Hasan, C.M.M., and Hamid, A.A. (2015b). Improvement of the butanol production selectivity and butanol to acetone ratio (B: A) by addition of electron carriers in the batch culture of a new local isolate of *Clostridium acetobutylicum* YM1. Anaerobe, 36: 65–72.

Anjan, A., and Moholkar, V.S. (2012). Biobutanol: science, engineering, and economics. Int. J. Energy Res., 36: 277–323.

Arancon, R.A.D., Lin, C.S.K., Chan, K.M., Kwan, T.H., and Luque, R. (2013). Advances on waste valorization: new horizons for a more sustainable society. Energy Sci. Eng., 1(2): 53–71.

Bahl, H., Andersch, W., and Gottschalk, G. (1982). Continuous production of aceotone and butanol by *Clostridium acetobutylicum* in a two-stage phosphate limited chemostat. Eur. J. Appl. Microbiol. Biotechnol., 15: 201–205.

Baral, N.R., and Shah, A. (2014). Microbial inhibitors: formation and effects on acetone-butanol-ethanol fermentation of lignocellulosic biomass. Appl. Microbiol. Biotechnol., 98: 9151–9172.

Becker, E. (2007). Micro-algae as a source of protein. Biotechnol. Adv., 25: 207–10.

Behera, S., and Kumar, S. (2019). Potential and prospects of biobutanol production from agricultural residues. Liquid Biofuel Production, pp. 285–318.

Bharathiraja, B., Jayamuthunagai, J., Sudharsanaa, T., Bharghavi, A., Praveenkumar, R., Chakravarthy, M., and Yuvaraj, D. (2017). Biobutanol–An impending biofuel for future: A review on upstream and downstream processing tecniques. Renewable and Sustainable Energy Reviews, 68: 788–807.

Brennan, L., and Owende, P. (2010). Biofuels from microalgae—a review of technologies for production, processing, and extractions of biofuels and co-products. Renew Sustain Energy Rev., 14: 557–77.

Changwei Zhang, ZhihaoSiBoChen, HuidongChen, WenqiangRen, ShikunCheng, and Shufeng L.I. (2021). Co-generation of acetone-butanol-ethanol and lipids by a sequential fermentation using *Clostridia acetobutylicum* and *Rhodotorulaglutinis*, spaced-out by an *ex-situ* pervaporation step. Journal of Cleaner Production, 285: 124902. ISSN 0959-6526, https://doi.org/10.1016/j.jclepro.2020.124902.

Chen, C.Y., Zhao, X.Q., Yen, H.W., Ho, S.H., Cheng, C.L., Lee, D.J., Bai, F.W., and Chang, J.S. (2013b). Microalgae-based carbohydrates for biofuel production. Biochem. Eng. J., 78: 1–10.

Cho, D.H., Shin, S.-J., and Kim, Y.-H. (2012). Effects of acetic and formic acid on ABE production by *Clostridium acetobutylicum* and *Clostridium beijerinckii*. Biotechnol. Bioprocess Eng., 17: 270–275.

Clementina Dellomonaco, Ramon Gonzalez, and Fabio Fava. (2010). The path to next-generation biofuels: Successes and Challenges in the Era of Synthetic Biology.

Dawud Ansaru, and Franziska Holz. Anticipating global energy, climate, and policy in 2055: Constructing qualitative and quantitative narratives.

Detlef, P. vanVuuren, ElkeStehfest, David, E.H.J., Gernaat, Jonathan, C., Doelman, Maartenvan den Berg, Mathijs Harmsen, Harmen Sytzede Boer, Lex, F.B., Vassilis, D., and Oreane, Y.E. (2017). Energy, land-use and greenhouse gas emissions trajectories under a green growth paradigm.

Dolejš, I., Krasňan, V., Stloukal, R., Rosenberg, M., and Rebroš, M. (2014). Butanol production by immobilised *Clostridium acetobutylicum* in repeated batch, fed-batch, and continuous modes of fermentation. Bioresour Technol., 169: 723–30.

Ellis, J.T. Neal N. Hengge, Ronald C. Sims, and Charles D. Miller. (2012). Acetone, butanol, and ethanol production from wastewater algae. Bioresource Technology, 111: 491–495. ISSN 0960-8524,https://doi.org/10.1016/j.biortech.2012.02.002.

European Biofuels Technology Platform, http://www.biofuelstp.eu/agri-residues/; 2014.

Farhana, B., Jamal, A., Huang, Z., Urynowicz, M., and Muhammad, I.A. (2022). Advancement and role of abiotic stresses in microalgae biorefinery with a focus on lipid production. Fuel, 316: 123192. ISSN 0016-2361, https://doi.org/10.1016/j.fuel.2022.123192.

Gao, K., Li, Y., Tian, S., and Yang, X. (2012). Screening and characteristics of a butanol-tolerant strain and butanol production from enzymatic hydrolysate of NaOH-pretreated corn stover. World J. Microbiol. Biotechnol., 28(10): 2963–2971.

Hamid amiri, and Keikhosro Karimi (Pretreatment and hydrolysis of lignocellulosic wastes for butanol production: Challenges and Perspectives).

Hayri Yamana, Murat Kadir Yesilyurt. (2021). The influence of n-pentanol blending with gasoline on performance, combustion, and emission behaviors of an SI engine. Engineering Science and Technology, an International Journal, 24(6): 1329–1346. ISSN 2215-0986, https://doi.org/10.1016/j.jestch.2021.03.009.

https://mnre.gov.in/annual report 2019–2020.

https://www.iea.org/reports/india-energy-outlook-2021.

Hu, Q. 2004. Environmental effects on cell composition. Wiley Online Library, pp. 83–93.

Husemann, M.H.W., and Papoutsakis, E.T. (1989). Enzymes limiting butanol and acetone formation in continuous and batch cultures of *Clostridium acetobutylicum*. Appl. Microbiol. Biotechnol., 31: 435–444.

Jamel, M.S., Abd Rahman, A., and Shamsuddin, A.H. Advances in the integration of solar thermal energy with conventional and non-conventional power plants. Centre for Renewable Energy, University Tenaga Nas.

Jin Zheng a, Yukihiro Tashiro b, Tsuyoshi Yoshida a, Ming Gao a, Qunhui Wang c. Kenji Sonomoto. (2012). Continuous Butanol Fermentation From Xylose With High Cell Density by Cell Recycling System.

Khaiwal Ravindra, Maninder Kaur-Sidhu, Suman Mor, and Siby John. (2019). Trend in household energy consumption pattern in India: A case study on the influence of socio-cultural factors for the choice of clean fuel use. Journal of Cleaner Production, 213: 1024–1034. ISSN 0959-6526, https://doi.org/10.1016/j.jclepro.2018.12.092.

Klutz, S., Magnus, J., Lobedann, M. et al. (2015). Developing the biofacility of the future based on continuous processing and single-use technology. J. Biotechnol., 213: 120–30.

Kumar, M., and Gayen, K. (2011). Developments in biobutanol production: new insights. Appl. Energy, 88: 1999–2012.

Kushwaha, Deepika, Srivastava, Neha, Mishra, Ishita, Upadhyay, Siddh Nath, and Mishra, Pradeep Kumar. (2019). Recent trends in biobutanol production. Reviews in Chemical Engineering, 35(4): 475–504. https://doi.org/10.1515/revce-2017-0041.

Lamani, V.T., Yadav, A.K., and Gottekere, K.N. (2018). Performance, emission, and combustion characteristics of twin-cylinder common rail diesel engine fuelled with butanol-diesel blends. Environ. Sci. Pollut. Res., 24(29): 23351–23362.

Li, H.G., Luo, W., Gu, Q.Y., Wang, Q., Hu, W.J., and Yu, X.B. (2013) Acetone, butanol, and ethanol production from cane molasses using *Clostridium beijerinckii* mutant obtained by combined low-energy ion beam implantation and N-methyl-N-nitro-N-nitrosoguanidine induction. Bioresour. Technol., 137: 254–260.

Li, S.-Y., Srivastava, R., Suib, S.L., Li, Y., and Parnas, R.S. (2011). Performance of batch, fed-batch, and continuous A-B-E fermentation with pH-control. Bioresour. Technol., 102(5): 4241–50.

Lienhardt, J., Schripsema, J., Qureshi, N., and Blaschek, H.P. (2002). Butanol production by *Clostridium beijerinckii* BA101 in an immobilized cell biofilm reactor: Increase in sugar utilization. Appl. Biochem. Biotechnol., 98–100L 591–598.

Lipovsky, J., Patakova, P., Paulova, L., Pokorny, T., Rychtera, M., and Melzoch, K. (2016). Butanol production by *Clostridium pasteurianum* NRRL B-598 in continuous culture compared to batch and fed-batch systems. Fuel Process Technol., 144: 139–44.

Lu, C., Yu, L., Vaghese, S., Yu, M., and Yang, S.T. (2017). Enhanced robustness in acetone-butanol-ethanol fermentation with engineered *Clostridiumbeijerinckii* overexpressing adh E2 and ctf AB. Bioresource Technol., 243: 1000–1008. ISSN 0960-8524, https://doi.org/10.1016/j.biortech.2017.07.043.

Lu, C., Zhao, J., Yang, S.-T., and Wei, D. (2012). Fed-batch fermentation for n-butanol production from cassava bagasse hydrolysate in a fibrous bed bioreactor with continuous gas stripping. Bioresour. Technol., 104: 380–7.

Lütke-Eversloh, T., and Bahl, H. (2011). Metabolic engineering of *Clostridium acetobutylicum*: recent advances to improve butanol production. CurrOpinBiotechnol., 22(5): 634–647. ISSN 0958-1669, https://doi.org/10.1016/j.copbio.2011.01.011.

Maria kosseva, and Colin Webb. (2020). Food industry waste: Assessment and Recuperation of commodities.

MervatFoda, Hongjun, Dong and Yin Li. (2010). Study the suitability of cheese whey for bio-butanol production by clostridia. Dairy Science Dept. National Research Center, Cairo, Egypt, Institute of Microbiology, Chinese Academy of Sciences, Beijing, China.

Muhammad Kamran, and Muhammad Rayyan Fazal. (2021). Fundamentals of Renewable Energy Systems.

Muhammad Shahbaz, Tareq Al-Ansari, Muhammad Aslam, Zakir Khan, Abrar Inayat, Muhammad Athar, Salman Raza Naqvi, Muahammad Ajaz Ahmed, and Gordon McKay. (2020). A state of the art review on biomass processing and conversion technologies to produce hydrogen and its recovery via membrane separation. International Journal of Hydrogen Energy, 45(30): 15166–15195,ISSN 0360-3199.

Nida Khan, Kumarasamy Sudhakar, and Rizalman Mamat. (2021). Role of Biofuels in Energy Transition, Green Economy and Carbon Neutrality.

Noomtima, P., and Cheirsilp, B. (2011). Production of butanol from palm empty fruit bunches hydrolyzate by *Clostridium acetobutylicum*. Energy Procedia, 9: 140–146.

Oshiro, M., Hanada, K., Tashiro, Y., and Sonomoto, K. (2010). Efficient conversion of lactic acid to butanol with pH-stat continuous lactic acid and glucose feeding method by *Clostridium saccharoperbutylacetonicum*. Appl. Microbiol. Biotechnol. 87: 1177–1185.

Oswald, W.J. (2003). My sixty years in applied algology. J. Appl. Phycol., 15: 99–106.

Pang, Z.-W., Lu, W., Zhang, H. et al. (2016). Butanol production employing fed-batch fermentation by *Clostridium acetobutylicum* GX01 using alkalipretreated sugarcane bagasse hydrolysed by enzymes from Thermoascusaurantiacus QS 7-2-4. Bioresour. Technol., 212: 82–91.

Phebe Asantewaa Owusu, Samuel Asumadu-Sarkodie, and Shashi Dubey. A Review of Renewable Energy Sources, Sustainability Issues, and Climate Change Mitigation.

Pierrot, P., Fick, M., and Engasser, J. (1986). Continuous acetone-butanol fermentation with high productivity by cell ultrafiltration and recycling. Biotechnol. Lett., 8: 253–6.

Procentese, A., Johnson, E., Orr, V., Garruto, A., Wood, J., Marzocchella, A., and Rehmann, L. 2015. Deep Eutectic Solvent Pretreatment and Saccharification of Corncob. Bioresour. Technol., 192: 31–36.

Qureshi, N., Schripsema, J., Lienhardt, J., and Blaschek, H.P. (2000). Continuous solvent production by *Clostridium beijerinckii* BA101 immobilized by adsorption onto brick. World J. Microbiol. Biotechnol. 16: 377–382.

Qureshi, N., Saha, B.C., Hector, R.E., Hughes, S.R., and Cotta, M.A. (2008). Butanol production from wheat straw by simultaneous saccharification and fermentation using *Clostridium beijerinckii*: part I—batch fermentation. Biomass Bioenergy, 32: 168–75.

Ranjan, A., and Moholkar, V.S. (2012), Biobutanol: science, engineering, and economics. Int. J. Energy Res., 36: 277–323. https://doi.org/10.1002/er.1948.

Sarchami, T., and Rehmann, L. (2015). Optimizing acid hydrolysis of jerusalem artichoke-derived inulin for fermentative butanol production. BioEnerResour. 8(3): 1148–1157.

Shanmugam, S., Hari, A., Kumar, D., Rajendran, K., Mathimani, T., Atabani, A.E., Brindhadevi, K., and Pugazhendhi, A. (2021). Recent developments and strategies in genome engineering and integrated fermentation approaches for biobutanol production from microalgae. Fuel, 285: 119052.

Singh, A., and Olsen, S.I. (2011). A critical review of biochemical conversion, sustainability and life cycle assessment of algal biofuels. Appl. Energy, 88: 3548–3555.

Singh, D.V., Ali, R., Kulsum, M., and Bhat, R.A. (2020a). Ecofriendly approaches for remediation of pesticides in contaminated environs. pp. 173–194. *In*: Bioremediation and Biotechnology, Vol. 3. Springer, Cham.

Singh, D.V., Bhat, R.A., Upadhyay, A.K., Singh, R., and Singh, D.P. (2020b). Microalgae in aquatic environs: A sustainable approach for remediation of heavy metals and emerging contaminants. Environmental Technology & Innovation 101340.

Singh, D.V., Upadhyay, A.K., Singh, R., and Singh, D.P. (2020c). Eco-friendly and eco technological approaches in treatment of wastewater by different algae and cyanobacteria. pp. 43–64. *In*: Algae and Sustainable Technologies. CRC Press.

Singh, D.V., and Singh, R.P. (2021). Algal consortia based metal detoxification of municipal wastewater: Implication on photosynthetic performance, lipid production, and defense responses. Science of The Total Environment, pp. 151928.

Singh, D.V., Upadhyay, A.K., Singh, R., and Singh, D.P. (2021). Microalgal competence in urban wastewater management: phycoremediation and lipid production. International Journal of Phytoremediation 1–11.

Singh, D.V., Upadhyay, A.K., Singh, R., and Singh, D.P. (2022). Implication of municipal wastewater on growth kinetics, biochemical profile, and defense system of Chlorella vulgaris and *Scenedesmus vacuolatus*. Environmental Technology & Innovation, p. 102334.

Sukwong, P.. Sunwoo, I.Y., Nguyen, T.H., Jeong, G.T., Kim, S.K. (2019). R-phycoerythrin, R-phycocyanin and ABE production from Gelidiumamansii by *Clostridium acetobutylicum*.

Suutari, M., Leskinen, E., Fagerstedt, K., Kuparinen, J., Kuuppo, P., and Blomster, J. (2015). Macroalgae in biofuel production. Phycol. Res., 63: 1–18.

Tashiro, Y., Takeda, K., Kobayashi, G., Sonomoto, K., Ishizaki, A., and Yoshino, S. (2004). High butanol production by *Clostridium saccharoperbutylacetonicum* N1–4 in fed-batch culture with pH-stat continuous butyric acid and glucose feeding method. J. Biosci. Bioeng., 98: 263–8.

Technology Roadmaps Biofuels for transport, IEA, 2011.

Thang, V.H., Kanda, K., and Kobayashi, G. (2010). Production of acetone–butanol–ethanol (ABE) in direct fermentation of cassava by Clostri*dium saccharoperbutyl-acetonicum* N1–4. Appl. Biochem. Biotech. 161: 157–170.

Umair Shahzad. The Need For Renewable Energy Sources. Department of Electrical Engineering, Riphah International University, Faisalabad, Pakistan.

Umair Shahzad. (2012). The Need For Renewable Energy Sources. Department of Electrical Engineering, Riphah International University, ISSN: 2306-708X.

Upadhyay, A.K., Singh, R., Singh, D.V., Singh, L., and Singh, D.P. (2021). Microalgal consortia technology: A novel and sustainable approach of resource reutilization, waste management and lipid production. Environmental Technology & Innovation, 101600.

Wang, J., Gao, M., Wang, Q., Zhang, W., and Shirai, Y. (2016). Pilot-scale open fermentation of food waste to produce lactic acid without inoculum addition. RSC Adv., 6(106): 104354–104358.

Zlata Mužíková, Pavel Šimáček, Milan Pospíšil, and Gustav Šebor. (2014). Density, Viscosity and Water Phase Stability of 1-Butanol-Gasoline Blends. Journal of Fuels 2014.

8

Biomaterial As Feedstocks for Butanol Biofuel:

Lignocellulosic Biomass

Kirti Bhatnagar,[1] *Neha Jaiswal,*[2] *Anju Patel,*[2,*]
Pankaj Kumar Srivastava[2,*] and *Arti Devi*[3]

1. Introduction

The first acetone butanol ethanol (ABE) production started in 1861. It was the French microbiologist Louis Pasteur who first discovered ABE fermentation. At the time of World War I, a huge volume of acetone were demanded because acetone was used in the production of cordite alternatives to gunpowder. Thus, there was an urgent need for a microorganism that might enhance acetone production. Strange and Chaim Weizmann (1919) succeeded in isolating bacteria *Clostridium acetobutylicum* from the soil matrix that has been used for acetone-butanol-ethanol fermentation, which in turn produced an enormous amount of acetone. After world War I, USA and Canada were the first ones to start large-scale industrial ABE production. Later, ABE production industries were set up in various countries like USSR, South Africa, China, Egypt, and Japan. In 1945, Japan started the production of butanol for aeroplane fuels derived from sugar plants. In 1950, n-butanol was produced through the petrochemical route. This was based on aldol condensation followed by dehydration

[1] School of Earth Sciences, Banasthali Vidyapith, Rajasthan.
[2] CSIR-National Botanical Research Institute, Lucknow.
[3] Department of Environmental Sciences, Central University of Jammu, J&K.
* Corresponding authors: patel7anju@gmail.com; drpankajk@gmail.com

and hydrogenation. The rapid growth of industrial methods for butanol production was known as oxo synthesis. Due to cheap oil prices, the ABE fermentation method of butanol extraction was abandoned (Uyttebroek et al. 2015) and the last factory based on the ABE process was closed in 1986 (South Africa). Although, once oil prices soaed high, many plants set up around the world (Slovakia, UK, France, Brazil, China, and the USA) have been using ABE fermentation. n-Butanol reduces more emission of CO, hydrocarbon, and nitrogen oxides than traditional petrol and Therefore, it has a positive impact on the environment. Thus, butanol is in the demand again, and there is an urgent need of renewable material as feedstock to get butanol to fulfil the increasing demand in energy sector.

In recent times, there has been great advancement in the utilization of sustainable and renewable energy resources at the global level and the prime reason behind this is the growth in energy consumption that has caused a depletion of fossil fuels and an increase in environmental issues at an alarming rate. Keeping today's scenario in mind, it is now of utmost importance to slowly move towards ecofriendly renewable resources that can help us in solving the dearth of supply of fuel, environmental problems, and issues related to climate change (Gao et al. 2012).

Eighty-six million tonnes bioethanol were produced in the year 2018 and a major portion (around 56%) were contributed by USA, followed by Brazil (28%), European Union (5%), China (4%), Canada, and India (2%) (Busic 2018). In the same year, the worldwide renewable energy was 53.0 EJ of the total energy generated. Global production of biofuel was approximately 160 billion liters, with 62% of bioethanol, followed by 26% of biodiesel, and the rest made up 12% (Global Bioenergy Statistics 2020).

Depending upon the primary raw material used, biofuels can be further classified as first (I)generation, second (II)generation, and third (III)generation fuel (Bharathiraja et al. 2017). The category of biofuels that are traditionally produced from food crops like grains, starch, vegetable oils, and sugars are the conventional biofuels and are known as the first-generation biofuels. When biofuels are produced from forestry and agricultural wastes like rice straw, wood chips, sawdust, and rice husk, they can be considered as sustainable fuels, that is, the second-generation biofuels that can easily be mingled with petroleum-based fuels for use in current internal combustion (IC) engines. Biofuel mainly comes from first-generation biomass, that is, the edible part of plants such as maize and molasses (Busic et al. 2018). However, second-generation biomass contribute a relatively small share to the total biofuel production.

Bioethanol, biomethane, jet fuel, and biodiesel are all known as third-generation biofuels or algae-based fuels. The third-generation biofuels have 10 times higher productivity (approximately 9000 L bio-fuels/hectare) as compared to the first- and second-generation biofuels

(Joshi et al. 2017). Utilization of renewable resources gained momentum in recent times to replace nonrenewable energy resources (Yao et al. 2021).

There is still a way to go before petrochemical products are entirely phased out, because for a complete phase out, entire petro-refineries need to be replaced by biorefineries (Chandel et al. 2018). To implement the biorefinery process, lignocellulosic biomass is what is mainly required from renewable resources. This lignocellulosic material is crucial for meeting the energy demand of the global economy (Ubando et al. 2020). Biomass of lignocellulosic origin is composed of cellulose, hemicellulose, oil starch, lignin, and protein. Lignocellulosic biomass is heterogenous and complex in nature (Bano and Irfan 2019). Due to variations in the nature of biomass composition, no standard method is currently available for the bio-conversion (Singhvi and Gokhale 2019).

2. Biobutanol as a fuel

According to an estimate, oil reserves present all around the globe will be diminished before 2042. To combat such a situation, finding alternatives of gasoline, such as renewable fuels, will eventually become a necessity rather than a choice in the proceeding decades (Shafiee et al. 2009).

Ethanol has been produced commercially from starchy and sugary materials and has been blended with gasoline since the 1970s. Though developing fuel-grade ethanol has many serious challenges, the biggest obstacle is infrastructure and equipment. Due to the current equipment used as well as the infrastructure of fuel distribution, it becomes difficult to expand the usage of ethanol beyond the 10% blend (Trindade and Santos 2017). Thus, if the percentage of ethanol in gasoline increased, entirely new infrastructure would be required to handle it.

Currently, butanol can easily be used in gasoline engines and no new infrastructure is needed for its distribution and storage. Hence, no modification is required in butanol production and it also has many advantages when compared to ethanol.

- The butanol-gasoline mixture is less likely to go through phase separation because butanol holds less water in solution as it is less soluble in water, which is not the case with ethanol-gasoline blends.
- Butanol-gasoline blends also facilitate the mixing of a large quantity of lower value hydrocarbon components with high vapor pressure with gasoline.
- Butanol has a relatively low vapour pressure that helps in limiting the emissions of unburned fuel by evaporation and also helps it to blend with gasoline at a low cost of vapour pressure.

Though butanol is still less popular than biodiesel and ethanol, it can prove to be a promising, renewable fuel with tremendous benefits when used in internal combustion engines (Nakata et al. 2006).

Biobutanol or butyl alcohol is a four-carbon alcohol that can also be manufactured by the fermentation of different types of feedstocks or biomass (Al-Shorgani et al. 2012b, Al-Shorgani et al. 2012a, Li et al. 2013, Becerra et al. 2015, Li et al. 2015) with the utilization of anaerobic as well as aerobic bacteria (Al Shorgani et al. 2015). Since biobutanol has the potential to be blended with diesel and contains more oxygen than biodiesel, it helps in reducing emissions from the engines. Biobutanol has a higher evaporation heat that helps in reducing the temperature of the internal engine, thus reducing the formation of NOx during combustion.

The production of bio-butanol through fermentation faces a lot of economic and technical challenges that include finding effective microorganisms in order to convert fermentable sugars to biobutanol (Gao et al. 2012). Thus, it becomes necessary to adopt alternate routes to produce biobutanol (Ogo et al. 2011, Santacesaria et al. 2012). The most promising way is the chemical route which usually takes place in simple steps to allow suitable ethanol conversion into n-butanol in the presence of a catalyst that helps in providing a higher yield and conversions (Gao et al. 2015).

3. Production route of biobutanol

The predominant method of biobutanol production is the fermentation of sugar, glycerol, or lignocellulosic material along with microorganisms belonging to the *Clostridiaceae* family (Yadav et al. 2014). ABE fermentation is also an alternative to crude oil. Figure 1 depicts how starch has been converted to butanol. This is basically a four-step process (Abo et al. 2019).

Step 1: Glucose is derived from the hydrolysis of carbohydrates to produce fatty acids and solvents using amylase enzyme (anaerobic fermentation) provided by *C. acetobutylicum*.

Step 2: Carbon from carbohydrates in form of pentose and hexose sugars are metabolized via the Embden-Meyerhof pathway to pyruvate.

$$1 \text{ mole of sugar} \longrightarrow 2 \text{ moles of pyruvate} \longrightarrow \begin{cases} 2 \text{ moles of ATP} \\ \textit{Net formation} \\ 2 \text{ moles of NADH} \end{cases}$$

Step 3: Pyruvate is converted to acetyl-CoA and carbon dioxide.

Step 4: In this stage, Acetyl CoA is been converted to intermediates products like acetaldehyde and butyraldehyde that ultimately form oxidized products like butanol and ethanol.

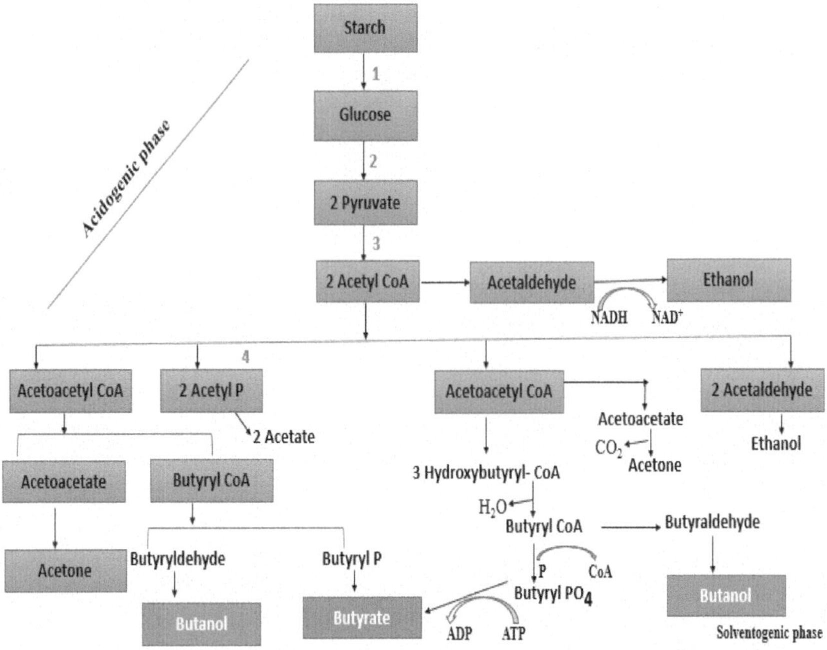

Figure 1. Starch converted to butanol.

Accumulation of organic acid is known as acidogenesis; during the abovementioned process, the pH of the culture reduces because of the metabolic switch from acidogenesis to solventogenesis (the formation of solvents) of *C. acetobutylicum* (Durre 2007). Organic acids are utilized for solventogenesis.

4. Feedstock for butanol production

4.1 First-generation biobutanol

One of the simplest ways to obtain biobutanol is by using sucrose-containing feedstocks (hexose sugars), that is, the hydrolysis of starch-rich crops like cassava, rice, maize, and wheat. First, the raw materials are hydrolyzed into dextrose, which is further converted into glucose utilizing the glucoamylase enzyme. Various researchers illustrated that first-generation biobutanol manufacturing can be enhanced using microorganisms. Different microbes associated with biobutanol have been illustrated in Table 1.

Table 1. First- and second-generation biomass and microorganisms associated with enhancing biobutanol production.

Feedstock	Fermentation Conditions (T, °C, & pH)	Microorganisms	Yield g/Lor g/g	Product distribution	References
First Generation					
Maize meal	37 and 6	*Clostridium beijerinckii* BA 101	26	Acetone & Butanol	Ezeji et al. 2004
Oil Palm Sap	37 and 6	*C. acetobutylicum* DSM 1731	14.4	Butanol	Komonkiat and Cheirsilp 2013
Glucose	37 and 4	*C. acetobutylicum* CICC 8022		Butanol	Chen et al. 2014
Cassava flour	37 and pH controlled	*C. acetobutylicum* DP 217	574.3	Acetone-Butanol-Ethanol	Li et al. 2014
Cassava Starch	37 and 5	*C. beijerinckityrobutyricum*	6.66/0.18	Butanol	Li et al. 2013
Second Generation					
Sugarbeet Juice		*C. beijerinkii* CCM 6182	0.37/0.40	Butanol	Shapovalov and Ashkinazi 2008
Palm Kernel Cake	30	*C. saccharoperbutylacetonicum*	7.12	Butanol	Shukor et al. 2016
Bamboo	37	*C. beijerinckii*		Butanol	Kumar et al. 2017
Corncobs	37	*C. saccharobutylicum* DSM 13864		Butanol	Gao and Rehmann 2014
Oil palm trunk fiber	37 and 6	*C. beijerinckii* TISTR 1461	10/0.41	Butanol	Komonkiat and Cheirsilp 2013
Rice straw	37 and 6.7	*C. sporogenes* BE01	5.52	Butanol	Gottumukkala et al. 2013
Barley liquor silage	37 and 6.5	*C. acetobutylicum* DSM 1731	9	Acetone-Butanol-Ethanol	Yang et al. 2014

4.2 Second-generation bio-butanol

In this method, a biofuel is derived from agricultural residues majorly sourced from non-edible biomass, for example, energy crops and grasses. Second-generation biofuels have an advantage over the first-generation biofuels because of their lack of competition with the food feed chain and as they are readily available. Second-generation biofuel is generally produced by biological or thermochemical processing to utilize feedstock (Cellulosic). Table 1 shows the various types of feedstock used in the production of second-generation biobutanol production. First- and second-generation biofuels require a large area as feedstock processing requires a long period of time and a lot of storage. It is important to note that the cons of the second-generation biofuel have been overcome by third-generation bio-butanol production.

4.3 Third-generation biobutanol

Algae are one of the promising raw materials for biobutanol production. They can be easily grown in seawater, wastewater, or freshwater, without taking much space and do not compete with conventional agriculture. Most species of algae have an enormous content of oil (around 50%), making them a suitable candidate for biodiesel production. The refuse left after extraction can be utilized for biobutanol production (Potts et al. 2012). Complex lignocellulosic materials are absent in algae; hence, simple hydrolysis (either acid or alkaline) is adequate to convert carbohydrates into sugars from the algal biomass (Ndaba et al. 2015).

Third-generation biobutanol also has certain challenges like inhibition of ABE fermentation during butanol production. Availability of feedstock biomass is also a challenge.

5. Lignocellulosic biomass for Biobutanol production

Lignocelluloses are an intricate combination of lignin, cellulose, and hemicellulose combined with extractives and inorganic materials. The chemical structure of cellulose, hemicellulose, and lignin is shown in Figure 2. Cellulose is the indispensable component of the lignocellulosic biomass, having a straight chain of repeated units of β-D-glucopyranose connected together by glycosidic bonds, which provide a stable structure to the cell wall. Cellulose chains bounded together by H-bonds and Van der Waal forces strengthen the parallel alignment, making a crystalline structure that limits the enzyme accessibility. The amorphous part of the cellulose structure binds with loose H-bonds, which make its lyses easier than in the crystalline form (Ashok Kumar et al. 2022). Hemicellulose is

Table 2. Algal biomass used for butanol production.

Feed Stock	Pre-treatment	Fermentation Conditions (T, °C, & pH)	Microorganisms	Yield g/L or g/g	Product distribution	References
Microalgae	Acid hydrolysis	37	*Clostridium saccharoperbutylacetonicum* N1-4	3.74	Butanol Sugar yield of 166.1 g/kg of dry algae	Castro et al. 2015
Microalgae	Methanol and microwave for 10 minutes	37	*C. acetobutylicum*	3.86	Butanol & 0.13 g/g carbohydrate	Cheng et al. 2015
Ulva lactuca (green sea weed)	Mildly alkaline	37 and 6.4	*C. acetobutylicum* and *C. beijerinckii*		Acetone and Ethanol	Van der wall et al. 2013
Macro algae		37	*C. beijerinckii* and *C. saccharoperbutylacetonicum*	4	Butanol	Potts et al. 2012
Wastewater algae		35 and over 6.5	*C. Saccharoperbutylacetonicum* N1-4	9.74	ABE	Ellis et al. 2012

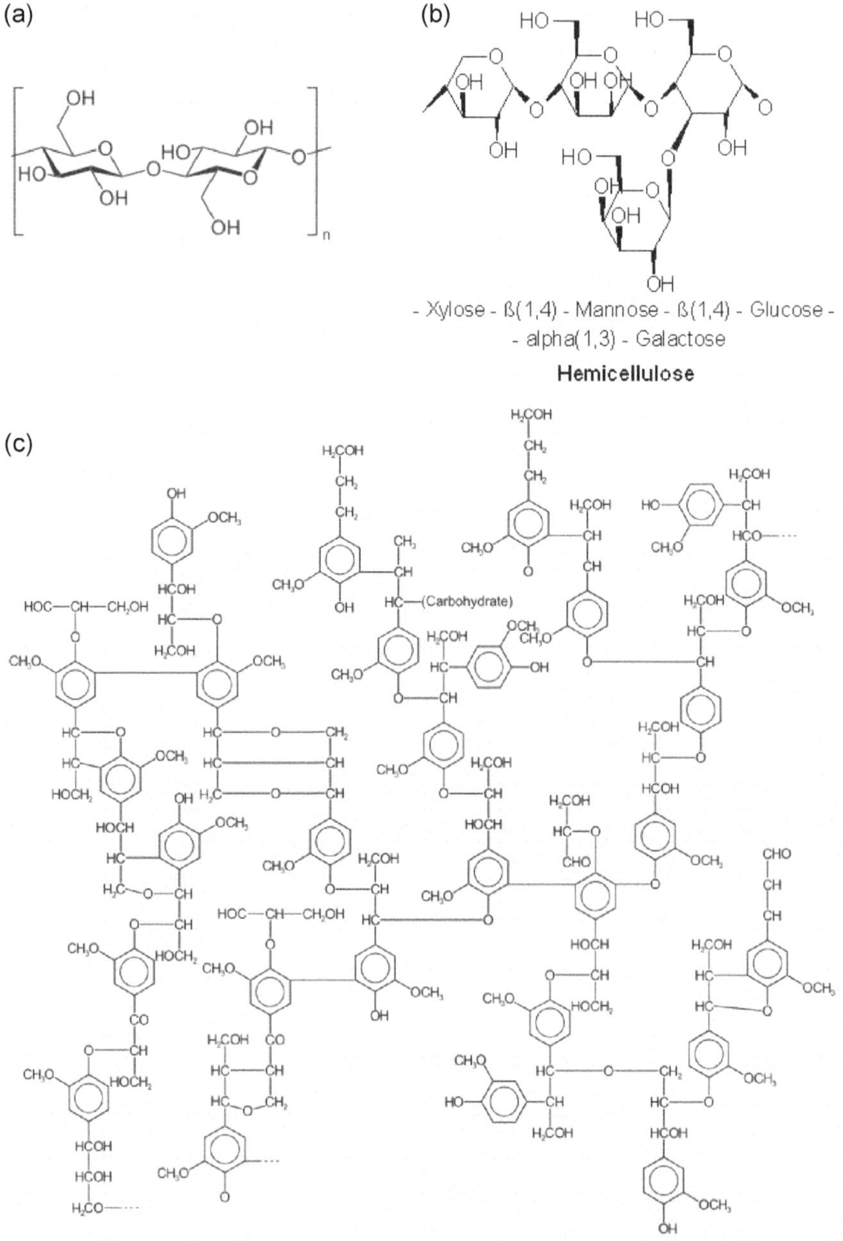

Figure 2. Structure of cellulose (a), hemicellulose (b) and lignin (c) (Wikipedia, creative common license).

the heteropolymer, consisting of C6 and C5 sugar units linked together with non-covalent bonding found under the cellulose fibers and between lignin, forming a complex structure. Two type of covalent interactions were found in hemicellulose, that is, diferulic acid bridges and lignin-glucuronic acid ester links. This interaction of covalent and non-covalent bonds inside the structure resists enzymatic action (Zeng et al. 2017). Lignin is a branched polymeric structure mainly composed of three phenylpropanoid units, that is, guaiacyl (G), *p*-hydroxyphenyl (H), and syringyl (S) obtained from precursors, that is, coniferyl, *p*-coumaryl, and sinapyl, respectively. The lignin polymer contains various ether and C-C bonds. Due to the presence of the complex lignin carbohydrate composite, it is difficult to extract lignin from its cell wall. Lignin composition varies in different bioresources based on the presence of G, H, and S subunits (Sheng et al. 2021).

Lignocellulosic materials commonly include municipal solid waste (MSW), forest and agricultural residues, crop residues, animal manure, and woodlot cuttings amongst others. Recently, lignocellulosic materials have risen as a propitious carbon source for acquiring energy and other valuable products like butanol. Amongst all the sources of lignocellulosic biomass, agricultural biomass can be contemplated as the sustainable source having high suitability for ABE-type fermentation as according to an estimate, agricultural waste has a sugar content of around 50% (Saini et al. 2015). Also, agricultural waste has a low cost and high availability (Karimi et al. 2006, Lee 1997).

Lignocellulosic materials need to be converted to fermentable hydrolysates for the production of butanol. The process suggested for ABE production from lignocellulosic materials comprises pretreatment, followed by hydrolysis, sterilization and fermentation of hydrolysates, and product recovery. Both cellulose and hemicellulose, which comprise lignocelluloses, can be utilized for ABE production as the solvent-producing *Clostridia* can effectively ferment hemicellulosic sugars as efficiently as it ferments glucose.

Pretreatment is a very important step as it helps in improving the enzymatic hydrolysis efficacy of celluloses and hemicelluloses rich in fermentable sugar (Cheng et al. 2012). Despite being an important step, pretreatment of lignocellulosic biomass has a limitation, that is, it generates undesirable products (inhibitors) in the lignocellulosic hydrolysates, which cause the inhibition of the growth of microorganisms (Ezeji et al. 2007, Baral and Shah 2014). It is important to neutralize or reduce the concentration of these inhibitors in order to facilitate the efficient fermentation of the microorganism.

Pretreatment and saccharification along with the techniques for removing the inhibitors are helpful in reducing the toxicity and thus are used to disintegrate lignocelluloses into lignin, cellulose, and hemicellulose.

6. Technology enhancement in commercial biofuel production

6.1 Feedstock availability

There should be a continuous supply of feedstock to meet the energy demand,. The composition of lignocellulosic material varies due to geographical region and environmental conditions. Su. et al (2020) reported around 170 billion metric tonnes/year production of biomass, and 10% agricultural and forestry residues contribute 233 billion litres of bioethanol. In India, 0.2 billion tonnes of agricultural residues and around 1.3×10^{10} metric tonnes of wood are produced annually. These numbers are even higher in the United States. The lignocellulosic biomass contributed by USA in 2018 was 1.3 billion tonnes per year. Biomass that contains high carbohydrate content yields a high percentage of ethanol; for example, switch grass contributes a higher percentage of ethanol than forest residues (spruce).

6.2 Biomass pretreatment methods

As mentioned earlier, lignocellulosic materials mainly composed of hemicellulose, cellulose, and lignin. They are recalcitrant and take time in their transformation to fuel. For the commercial production of bioethanol/butanol, pretreatment is a necessary step. Pretreatment should be cost-effective, having a low retention time, capable of utilizing a variety of feedstock, be ecofriendly, and generate low inhibitor along with producing a highly digestible carbohydrate-rich product (Alvira et al. 2010). Various methods are available for disintegration of lignin and cellulose so that its porosity will increase, which in turn, provides better enzymatic accessibility for biological conversion (Yamakawa et al. 2020). Some popular pretreatment methods are mentioned below:

i. Hydrothermal
ii. Steam explosion and high temperature
iii. Dilute acid
iv. Beta renewable – PROESA™ technology
v. Borregaard's BALI™
vi. Acid/Alkali treatment
vii. Microwave irradiation
viii. Ionic Liquids
ix. Supercritical fluids-based pretreatments
x. Low-temperature steep delignification
xi. Cosolvent-enhanced lignocellulosic fractionation
xii. Inorganic solvents and many more

Dilute acid pretreatment has been widely applied at commercial scale as it solubilizes the hemicellulose and enhances the surface area for enzymatic action. A drawback of this method is that it generates a number of inhibitors like levulinic acid, malic acid, formic acid, and HMF. These are detrimental to the enzymes. Thus, enzyme activity is hindered, which in turn affects the production of biofuel.

Some new and pioneering technologies like PROESA™ at Crescentino biorefinery, Italy, used this technology. In this process, integrated advanced steam explosion deals with 200,000 tonnes of lignocellulosic biomass (agriculture residues) and generates 40,000 tonnes of ethanol per year (Duque et al. 2021), at a cost of approximately $ 0.40/L of ethanol. Another technology named BALI™ utilized pretreated slurry (Sulfite treatment of 25 MT of spruce woods) undergoes enzymatic hydrolysis and utilizes CTec2/CTeec3 cellulase, generating 30,000 litres of bioethanol (Sjode et al. 2013).

Furthermore, Chundawat et al. (2020) used an ammonia salt solvent treatment and efficiently dissolved 80–85% of lignin. Combining mechanical, biochemical, physico-chemical pretreatment, saccharification, and fermentation releases more sugars and ethanol (Mikulski and Klosonski 2020).

6.3 Ionic Liquids

Ionic liquids are a group of mixtures of ionic species and liquids at room temperature. It is a green solvent having unique solvation properties with low vapor pressure (Morais et al. 2015). Certain ionic liquids have showed the ability to remove lignin, hemicellulose, and exposed pure cellulose for the next phase, that is, hydrolysis. Pretreatment with ionic liquids is very successful in the continuous mode along with a high input of biomass (Brandt et al. 2011). Some examples of ionic liquid pretreatment on lignocellulosic biomass is discussed in Table 3.

Some common ionic liquids favor dissolving cellulose over the lignin. To overcome this, filter out the undissolved solids and then regenerate the dissolved biomass rich in cellulose (Liu et al. 2020). The dissolution of cellulose is attributed to the strong hydrogen bond between equatorial hydroxyl groups of anions of ionic liquids and of cellulose.

The cost of ionic liquids is crucial while considering its application at the industrial level. The cost of imidazolium cations poses a big challenge for their commercial and large-scale use. However, certain ionic liquids can be fully recovered by adding glycerol or carboxylate ionic liquids (Clough et al. 2016)

Table 3. Examples of ionic liquids utilized on various lignocellulosic biomass.

Biomass	Ionic Liquids	Conditions	Lignin removed	References
Miscanthus grass	1,3-Dimethylimidazolium methyl sulfate (C_4C_1im) ($MeSO_4$)	(C_4C_1im) ($MeSO_4$) 80%, 120°C, 2 h	27.2%	Gschwend et al. 2018
Willow	1-Butyl-3-methylimidazolium acetate (C_4C_1im) ($MeCO_2$)	(C_4C_1im) (Me CO_2), 80%, 120°C, 22 h	17.43%	Weigand et al. 2017
Switch grass	1-Butyl-3-methylimidazolium acetate (C_4C_1im) ($MeCO_2$)	(C_4C_1im) (Me CO_2), 100%, 160°C, 3 h	65%	Williams et al. 2018
Sugarcane bagasse	1-Butyl-3-methylimidazolium	140°C, 25 min with ultrasonication	65.72%	Lie et al. 2020
Horn beam	1-butylimidazolium hydrogen sulfate	150°C, 3 h	91%	Semerci and Ersan 2021
Corn stover	Tetra-butyl-phosphonium 2 ethyl-hexanoate	80°C, 3 h	84%	Glinska et al. 2021
Pine	1-Butyl-3-methylimidazolium hydrogen sulfate (C_4C_1im) (HSO4)	(C_4C_1im) (HSO4) 80%, 120°C, 22 h	65.5%	Gschwend et al. 2019

6.4 India's biofuel policy

India has enriched its efforts to reach 20 percent blending of ethanol by 2025 and also maintaining the goal of blending ethanol 10–20 percent by 2022 and 2030, respectively (Sakthivel et al. 2018). The average blending rate in the year 2021 was 7.5% because of limited fuel pools. Due to government incentives and efforts, companies were able to generate 2.7 billion gallons of gasoline-ethanol in the year 2021.

Indian Oil Corporation announced plans to construct a 63-million-liter cellulosic ethanol unit utilising the dilute acid pre-treatment method. TATA has also planned to install a 100,000 -liters ethanol plant at Bargarh, Odisha, India (Bioenergy 2020).

7. Conclusion & limitations

Biobutanol has now become an intriguing advanced biofuel with exceptional character, making it suitable to be used as a 'drop-in' fuel. The production of biobutanol addresses sustainable development and can also help in reviving the ABE-fermentation industries. There has been considerable advancement in biobutanol production over the years.

Though there are a number of advances in ongoing research vis-à-vis evolving technologies for biobutanol production, there are still many challenges ahead related to the process, substrate, and strain.

Lignocelluloses are one of the propitious materials that are appropriate for the sustainable generation of biobutanol production at a large scale. After the proper pretreatment and hydrolysis stage of lignocellulosic material, the hemicellulose and the cellulose content present in the feedstock can be productively converted to biobutanol. The most pressing challenge for the economically efficient production of biobutanol from lignocellulosic biomass is its efficient conversion to hydrolysates with a pertinent concentration of stimulators, inhibitors, and sugars.

Butanol is one of the promising fuels as it is a pure alcohol with a long hydrocarbon chain, which makes it non-polar and having similar energy content as petrol. Thus, it can directly be utilized in various internal combustion (IC) engines with or without mixing it with petrol. Although limitations in ABE fermentation and pretreatment of lignocellulosic biomass limit its commercial production.

However, the major factor which inhibits the production of biobutanol at a large scale is the lack of consciousness among decision-makers and the opinion of the general public on the characteristics of biobutanol regarding its low yield of butanol when it obtained from the fermentation process. More research should be focused on biobutanol synthesis and a specially designed biomass approach should be adopted for cheaper biobutanol production.

References

Abo, B.O., Gao, M., Wang, Y., Wu, C., Wang, Q., and Ma, H. (2019). Production of butanol from biomass: recent advances and future prospects. Environ. Sci. Pollut. Res., 26(20): 20164–20182. https://doi.org/10.1007/s11356-019-05437-y.

Al-Shorgani, N.K.N., Kalil, M.S., and Yusoff, W.M.W. (2012a). Biobutanol production from rice bran and de-oiled rice bran by *Clostridium saccharoperbutylacetonicum* N1-4. Bioprocess BiosystEng., 35(5): 817–826.

Al-Shorgani, N.K.N., Kalil, M.S., Ali, E., Hamid, A.A., and Yusoff, W.M.W. (2012b). The use of pretreated palm oil mill effluent for acetone-butanol-ethanol fermentation by *Clostridium saccharoperbutylacetonicum* N1-4. Clean Techn. Environ. Policy, 14(5): 879–887.

Al-Shorgani, N.K.N., Kalil, M.S., Yusoff, W.M.W., and Hamid, A.A. (2015). Biobutanol production by a new aerotolerant strain of *Clostridium acetobutylicum* YM1 under aerobic conditions. Fuel, 158: 855–863.

Alvira, P., Tom´ as-Pejo, ´ E., Ballesteros, M., and Negro, M.J. (2010). Pretreatment technologies for an efficient bioethanol production process based on enzymatic hydrolysis: A review. Bioresour. Technol., 101(13): 4851–4861. https://doi.org/10.1016/j. biortech.2009.11.093.

Arancon, R.A.D., Lin, C.S.K., Chan, K.M., Kwan, T.H., and Luque, R. (2013). Advances on waste valorization: new horizons for a more sustainable society. Energy Sci. Eng., 1(2): 53–71.

Arianna Callegari, Silvia Bolognesi, Daniele Cecconet, and Andrea G. Capodaglio. (2020). Production technologies, current role, and future prospects of biofuels feedstocks: A state of-the-art review, Critical Reviews in Environmental Science and Technology, 50: 4, 384–436, DOI: 10.1080/10643389.2019.1629801.

Ashokkumar, V., Venkatkarthick, R., Jayashree, S., Chuetor, S., Dharmaraj, S., Kumar, G. Chen, W.H., and Ngamcharussrivichai, C. (2022). Recent advances in lignocellulosic biomass for biofuels and value-added bioproducts-A critical review. Bioresource Technology, 344: 126195.

Association, W.B. (2020). Global Bioenergy Statistics. World Bioenergy Association, http://www.worldbioenergy.org/uploads/201210 WBA GBS 2020.pdf.

Bano, A., and M. Irfan. (2019). Alkali pretreatment of cotton stalk for bioethanol. Bangladesh Journal of Scientific and Industrial Research, 54: 73–82.

Baral, N.R., and Shah, A. (2014). Microbial inhibitors: formation and effects on acetone-butanol-ethanol fermentation of lignocellulosic biomass. Appl. Microbiol. Biotechnol., 98: 9151–9172.

Becerra, M., Cerdán, M.E., and González-Siso, M.I. (2015). Biobutanol from cheese whey. Microb Cell Factories, 14(1): 27.

Bharathiraja, B., Jayamuthunagai, J., Sudharsanaa, T., Bharghavi, A., Praveenkumar, R., Chakravarthy, M., and Yuvaraj, D. (2017). Biobutanol e an impending biofuel for future: a review on upstream and downstream processing techniques, Renewable and Sustainable Energy Reviews, 68: 788–807, https://doi.org/10.1016/j.rser.2016.10.017.

Bioenergy, E. (2020). Current status of Advanced Biofuels demonstrations in Europe. ETIP Bioenergy Working Group 2 and ETIP-B-SABS2 project team. Final version 09/03/ 2020.

Brandt, A., Ray, M.J., To, T.Q., Leak, D.J., Murphy, R.J., and Welton, T. (2011). Ionic liquid pretreatment of lignocellulosic biomass with ionic liquid–water mixtures. Green. Chem. 13: 2489–2499.

Bušić, A., Marđetko, N., Kundas, S., Morzak, G., Belskaya, H., Ivančić Šantek, M., Komes, D., Novak, S., and Šantek, B. (2018). Bioethanol production from renewable raw materials and its separation and purification: A review. Food technology and biotechnology, 56(3): 289–311. https://doi.org/10.17113/ftb.56.03.18.5546.

Chandel, A.K., Garlapati, V.K., Singh, A.K., Antunes, F.A.F., and da Silva, S.S. (2018). The path forward for lignocellulose biorefineries: bottlenecks, solutions, and perspective on commercialization. Bioresour. Technol., 264: 370–381.

Cheng, C.L., Che, P.Y., Chen, B.Y., Lee, W.J., Lin, C.Y., and Chang, J.S. (2012). Biobutanol production from agricultural waste by an acclimated mixed bacterial microflora. Appl Energy, 100: 3–9.

Chundawat, S.P.S., Sousa, L.d.C., Roy, S., Yang, Z., Gupta, S., Pal, R., Zhao, C., Liu, S.-H., Petridis, L., O'Neill, H., and Pingali, S.V. (2020). Ammonia-salt solvent promotes cellulosic biomass deconstruction under ambient pretreatment conditions to enable rapid soluble sugar production at ultra-low enzyme loadings. Green Chem., 22(1): 204–218. https://doi.org/10.1039/C9GC03524A.

Duque, A., Alvarez,´ C., Dom´enech, P., Manzanares, P., Moreno, and Antonio D. Moreno. (2021). Advanced bioethanol production: from novel raw materials to integrated biorefineries. Processes, 9(2): 206. https://www.mdpi.com/2227-9717/9/2/206.

Durre, P. (2007). Biobutanol: an attractive biofuel. Biotechnol. J. 2: 1525–1534.

Ellis, J.T., Hengge, N.N., Sims, R.C., and Miller, C.D. (2012). Acetone, butanol, and ethanol production from wastewater algae. BioresourTechnol, 111: 491–495.

Ezeji, T.C., Qureshi, N., and Blaschek, H.P. (2007). Bioproduction of butanol from biomass: from genes to bioreactors. Curr. Opin. Biotechnol., 18(3): 220–227.

Gao, K., Li, Y., Tian, S., and Yang, X. (2012). Screening and characteristics of a butanol-tolerant strain and butanol production from enzymatic hydrolysate of NaOH-pretreated corn stover. World J. Microbiol. Biotechnol., 28(10): 2963–2971.

Gao, M., Tashiro, Y., Yoshida, T., Zheng, J., Wang, Q., Sakai, K., and Sonomoto, K. (2015). Metabolic analysis of butanol production from acetate in *Clostridium saccharoperbutylacetonicum* N1-4 using 13C tracer experiments. RSC Adv., 5(11): 8486–8495.

Glinska, ´K., Gitalt, J., Torrens, E., Plechkova, N., and Bengoa, C. (2021). Extraction of cellulose from corn stover using designed ionic liquids with improved reusing capabilities. Process Saf. Environ. Protect., 147: 181–191. https://doi.org/10.1016/j.psep.2020.09.035.

Gschwend, F.J.V., Malaret, F., Shinde, S., Brandt-Talbot, A., and Hallett, J.P. (2018). Rapid pretreatment of Miscanthus using the low-cost ionic liquid triethylammonium hydrogen sulfate at elevated temperatures. Green. Chem., 20: 3486–3498.

Gschwend, F.J.V., Chambon, C.L., Biedka, M., Brandt-Talbot, A., Fennell, P.S., and Hallett, J.P. (2019). Quantitative glucose release from softwood after pretreatment with lowcost ionic liquids. Green. Chem., 21: 692–703.

Hamid Amiri, Keikhosro Karimi, Chapter 6—Biobutanol Production, Editor(s): Majid Hosseini. pp. 109–133. *In:* Woodhead Publishing Series in Energy, Advanced Bioprocessing for Alternative Fuels, Biobased Chemicals, and Bioproducts, Woodhead Publishing, 2019, ISBN 9780128179413,https://doi.org/10.1016/B978-0-12-817941-3.00006-1.

Jin, C., Yao, M., Liu, H., Lee, C.F., and Ji, J. (2011). Progress in the production and application of n-butanol as a biofuel. Renewable and Sustainable Energy Reviews, 15(8): 4080–4106. doi: https://doi.org/10.1016/j.rser.2011.06.001.

Joshi, G., Pandey, J.K., Rana, S., and Rawat, D.S. (2017). Challenges and opportunities for the application of biofuel, Renewable and Sustainable Energy Reviews, 79: 850–866, https://doi.org/10.1016/j.rser.2017.05.185.

Karimi, K., Kheradmandinia, S., and Taherzadeh, M.J. (2006. Conversion of rice straw to sugars by dilute acid hydrolysis. Biomass Bioenergy, 30: 247–253.

Kim, D.Y., Vijayan, D., Praveenkumar, R., Han, J.I., Lee, K., Park, J.Y., Chang, W.S., Lee, J.S. and, You-Kwan. (2016). Cellwall disruption and lipid/astaxanthin extraction from microalgae: chlorella and haematococcus. Bioresource Technology, 199: 300–310, https://doi.org/10.1016/j.biortech.2015.08.107.

Lee, J. (1997). Biological conversion of lignocellulosic biomass to ethanol. J. Biotechnol., 56: 1–24.

Li, H.G., Luo, W., Gu, Q.Y., Wang, Q., Hu, W.J., and Yu, X.B. (2013). Acetone, butanol, and ethanol production from cane molasses using *Clostridium beijerinckii* mutant obtained by combined low-energy ion beam implantation and N-methyl-N-nitro-N-nitrosoguanidine induction. Bioresour. Technol., 137: 254–260.

Li, S., Guo, Y., Lu, F., Huang, J., and Pang, Z. (2015). High-level butanol production from cassava starch by a newly isolated *Clostridium acetobutylicum*. Appl. Biochem. Biotechnol., 177(4): 831–841.

Li, M., Jiang, H., Zhang, L., Yu, X., Liu, H., Yagoub, A.E.A., and Zhou, C. (2020). Synthesis of 5- HMF from an ultrasound-ionic liquid pretreated sugarcane bagasse by using a microwave-solid acid/ionic liquid system. Ind. Crop. Prod. 149: 112361. https://doi.org/10.1016/j.indcrop.2020.112361.

Lu, C., Yu, L., Varghese, S., Yu, M., and Yang, S.T. (2017). Enhanced robustness in acetone-butanol-ethanol fermentation with engineered *Clostridium beijerinckii* overexpressing adhE2 and ctfAB. Bioresour. Technol., 243: 1000–1008

Mikulski, D., and Kłosowski, G. (2020). Hydrotropic pretreatment on distillery stillage for efficient cellulosic ethanol production. Bioresour. Technol., 300: 122661. https://doi.org/10.1016/j.biortech.2019.122661.

Morais, A.R.C., da Costa Lopes, A.M., and Bogel-Łukasik, R. (2015). Carbon dioxide in biomass processing: contributions to the green biorefinery concept. Chem. Rev., 115: 3–27.

Nakata, K., Utsumi, S., Ota, A., Kawatake, K., Kawai, T., and Tsunooka, T. (2006). The effect of ethanol fuel on a spark ignition engine. In SAE Technical Paper. Warrendale, PA: SAE Internationaldoi: https://doi.org/10.4271/2006-01-3380.

Ndaba, B., Chiyanzu, I., and Marx, S. (2015). n-Butanol derived from biochemical and chemical routes: A review. Biotechnology Reports, 8: 1–9.

Ogo, S., Onda, A., and Yanagisawa, k. (2011). Selective synthesis of 1-butanol from ethanol over strontium phosphate hydroxyapatite catalysts. Appl. Catal., 402: 188–195.doi: https://doi.org/10.1016/j.apcata.2011.06.006.

Pfromm, P.H., Amanor-Boadu, V., Nelson, R., Vadlani, P., and Madl, R. (2010). Biobutanol vs. bio-ethanol: a technical and economic assessment for corn and switchgrass fermented by yeast or *Clostridium acetobutylicum*. Biomass Bioenergy, 34(4): 515–524.

Raj, T., Gaur, R., Dixit, P., Gupta, R.P., Kagdiyal, V., Kumar, R., and Tuli, D.K. (2016). Ionic liquid pretreatment of biomass for sugars production: Driving factors with a plausible mechanism for higher enzymatic digestibility. Carbohydr. Polym., 149: 369–381. https://doi.org/10.1016/j.carbpol.2016.04.129.

Saini, J. K., Saini, R., and Tewari, L. (2015). Lignocellulosic agriculture wastes as biomass feedstocks for second-generation bioethanol production: concepts and recent developments. Biotech, 5: 337–353.

Sakthivel, P., Subramanian, K.A., and Mathai, R. (2018). Indian scenario of ethanol fuel and its utilization in automotive transportation sector. Resour. Conserv. Recyc. 132: 102–120. https://doi.org/10.1016/j.resconrec.2018.01.012.

Santacesaria, E., Vicente, G.M., Di Serio, M., and Tesser, R. (2012). Main technologies in biodiesel production: State of the art and future challenges. Catalysis Today, 195: 2–13.

Sarchami, T., and Rehmann, L. (2014). Optimizing enzymatic hydrolysis of inulin from Jerusalem artichoke tubers for fermentative butanol production. Biomass Bioenergy, 69: 175–182.

Semerci, I., and Ersan, G. (2021). Hornbeam pretreatment with protic ionic liquids: cation, particle size, biomass loading and recycling effects. Ind. Crop. Prod., 159:, 113021. https://doi.org/10.1016/j.indcrop.2020.113021.

Shafiee, S., and Topal, E. (2009). When will fossil fuel reserves be diminished? Energy Policy, 37: 181189. https://doi.org/10.1016/j.enpol.2008.08.016.

Sheng, Y., Lam, S.S., Wu, Y., Ge, S., Wu, J., Cai, L., Huang, Z., Van Le, Q., Sonne, C., and Xia, C. (2021). Enzymatic conversion of pretreated lignocellulosic biomass: a review on influence of structural changes of lignin. Bioresource Technology, 324: 124631.

Singhvi, M.S., and Gokhale, D.V. (2019). Lignocellulosic biomass: hurdles and challenges in its valorization. App. Microbiol. Biotechnol., 103: 9305–9320.

Siwale, L., Kristof, L., Adam, T., Bereczky, A., Mbarawa, M., Penninger, A., and Kolesnikov, A. (2013). Combustion and emission characteristics of n-butanol/diesel fuel blend in a turbo-charged compression ignition engine. Fuel, 107: 409–418.doi: https://doi.org/10.1016/j.fuel.2012.11.083.

Sjode, ¨ A., Frolander, ¨ A., Lersch, M., and Rødsrud, G. (2013). Lignocellulosic biomass conversion by sulfite pretreatment. Patent EP2376642 B 1.

Su, T., Zhao, D., Khodadadi, M., and Len, C. (2020). Lignocellulosic biomass for bioethanol: Recent advances, technology trends, and barriers to industrial development. Curr. Opin. Green Sustain. Chem., 24: 56–60. https://doi.org/10.1016/j. cogsc.2020.04.005.

Trindade, W.R.d.S., and Santos, R.G.D. (2017). Review on the characteristics of butanol, its production and use as fuel in internal combustion engines. Renewable and Sustainable Energy Reviews, 69: 642–651. doi:https://doi.org/10.1016/j.rser.2016.11.213.

Ubando, A.T., Felix, C.B., and Chen, W.H. (2020). Biorefineries in circular bioeconomy: a comprehensive review. Bioresour. Technol., 299: 122585.

Uyttebroek, M., Vam Hecke, W., and Vanbroekhoven, K. (2015). Sustainability metrics of 1-butanol, Cat. Today, 239: 7–10. doi: https://doi.org/10.1016/j.cattod.2013.10.094.

Weigand, L., Mostame, S., Brandt-Talbot, A., Welton, T., and Hallett, J.P. (2017). Effect of pretreatment severity on the cellulose and lignin isolated from Salix using iono Solvpretreatment. Faraday. Discuss. 202: 331–349.

Williams, C.L., Li, C., Hu, H., Allen, J.C., and Thomas, B.J. (2018). Three way comparison of hydrophilic ionic liquid, hydrophobic ionic liquid, and dilute acid for the pretreatment of herbaceous and woody biomass. Front. Energy. Res., 6.

Yadav, S., Rawat, G., Tripathi, P., and Saxena, R.K. (2014). Dual substrate strategy to enhance butanol production using high cell inoculum and its efficient recovery by pervaporation. Bioresource Technology, 152: 377–383.

Yamakawa, C.K., Kastell, L., Mahler, M.R., Martinez, J.L., and Mussatto, S.I. (2020). Exploiting new biorefinery models using non-conventional yeasts and their implications for sustainability. Bioresour. Technol., 309 https://doi.org/10.1016/j. biortech.2020.123374.

Yao, Y., Xu, J.-H., and Sun, D.-Q. (2021). Untangling global levelised cost of electricity based on multi-factor learning curve for renewable energy: Wind, solar, geothermal, hydropower and bioenergy. J. Clean. Prod., 285: 124827. https://doi.org/10.1016/j. jclepro.2020.124827.

Zeng, Y., Himmel, M.E., and Ding, S.Y. (2017). Visualizing chemical functionality in plant cell walls. Biotechnology for Biofuels, 10(1): 1–16.

9

Advancement in Algal Biomass Based Biobutanol Production Technologies and Research Trends

Kulvinder Bajwa,[1,2,*] *Narsi R. Bishnoi,*[2]
Muhammad Yousuf Jat Baloch[3] and *S.P. Jeevan Kumar*[4]

1. Introduction

The rapid growth of the world population, as well as the rise of developing countries such as China and India, have resulted in a significant increase in energy demand (Harun et al. 2010, Kongjan et al. 2021). Currently, coal, natural gas, and petroleum supply approximately 90% of energy needs, and sustainable energy supplies must be found due to the depleting reserves of these fossil fuel resources (Chen et al. 2011, Demirbaş et al. 2010, Chen et al. 2013). As oil and fossil fuels run out, and climate change worsens, efforts to generate renewable energy in the form of biofuels must continue. Biofuel is described as the fuel derived from or generated by living organisms. The energy in biofuel comes from the carbon fixation process, in which CO_2 is converted into sugar, which is only present in

[1] University School for Graduates Studies, Chaudhary Devi Lal University, 125055, Sirsa, Haryana, India.
[2] Department of Environmental Science and Engineering, Guru Jambheshwar University of Science and Technology, Hisar, Haryana, India.
[3] College of New Energy and Environment, Jilin University, Changchun 130021, China.
[4] ICAR-Directorate of Floricultural Research, Pune-411036, Maharashtra, India.
* Corresponding author: kulvinderbajwa3@gmail.com

living creatures and plants. Biofuels are divided into four generations as shown in Figure 1 (Demirbaş 2011). First-generation biofuels are made directly from food crops like maize, wheat, and soybeans. These biofuels reduce CO_2 emissions and can aid in the resolution of domestic energy security issues. The most contentious problem, however, is that first-generation biofuels compete with human and animal sustenance (Olabi 2013). As a result, "second-generation biofuels" were created to address this issue. These are created from non-edible crops such as grass, wood, and other organic wastes (lignocellulosic biomass) (Gomez et al. 2008). However, a lack of arable land is a disadvantage for some second-generation biofuels. This problem is permanently overcome by generating "third-generation biofuels" sourced mostly from algae (Singh et al. 2011), whereas the fourth-generation uses metabolically altered algae to produce biofuels through increased photosynthesis (Lü et al. 2011, Shanmugam et al. 2021). Microalgae are a kind of photosynthetic organism that produces carbohydrates, lipids, and proteins in relatively high quantities in a matter of days with basic growth conditions (light, CO_2, N, P, K, and other inorganic nutrients) (Huang et al. 2016, Vidhya 2022). These components can be utilized to make biofuels and other commercially valuable by-products (Brennan and Owende 2010, Olabi 2013). Algal biofuels can be direct combustion (biomass), liquid fuels (bioethanol, biobutanol, bio-oil, and biodiesel), or gaseous fuels (biogas, biomethane, syngas, and biohydrogen). Microalgae may also create alternative biofuels

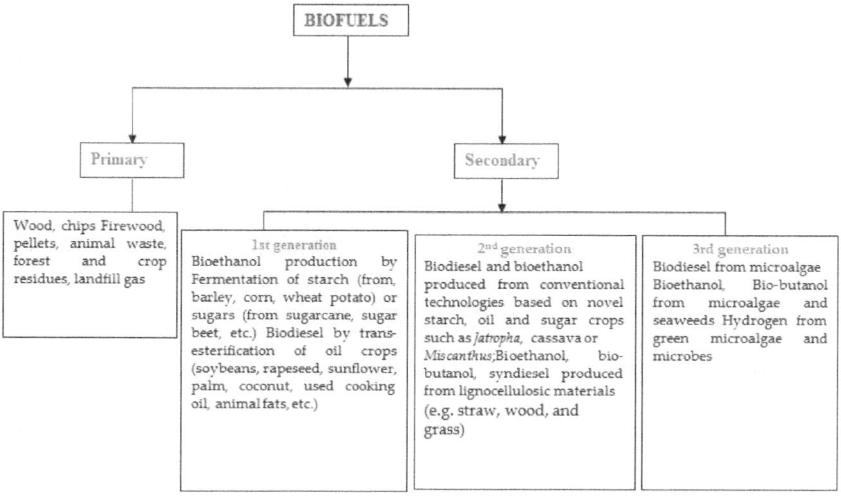

Figure 1. The flowchart of biofuel classification (Modified from Demibras 2011).

such as biobutanol, bioethanol, and biomethane by anaerobic fermentation of algal carbohydrates with various bacteria (Linhares et al. 2019, de Carvalho Silvello et al. 2022).

Biobutanol is one such potent biofuel, with characteristics similar to gasoline, allowing for easier public distribution using existing oil and gas infrastructure. Also, the development of third-generation biofuels derived from microalgae culture appears to be a viable renewable energy source (Elkatory et al. 2022). This is due to its inherent benefits over earlier ways of producing ethanol from crops and plant waste. Despite ongoing efforts, research on micro-algal biobutanol production is lacking (Veza et al. 2021). Working on both strengths may help the booming biofuel business, ultimately helping to meet global energy demand and cut CO_2 emissions (Yeong et al. 2018).

In this chapter, the authors discuss the concept of butanol, the comparability of butanol to other fuels, the principle of butanol production, the prospects, and the obstacles related with algal butanol. Thus, the purpose of this chapter is to provide technical guidance for the conversion of algae to biobutanol liquid biofuel, including the potential of algal biomass feedstock, ABE fermentation principles, and factors impacting butanol production, fermentation type, and reactor architecture.

2. Biobutanol: a potential alternative biofuel

Bio-butanol is a biomass-derived butanol that may be used as a fuel in a gasoline-powered internal combustion engine. Biobutanol is characterized by high miscibility, low volatility, high energy contents from 33.07 MJ Kg^{-1} (Klass 1998) to 36.1 MJ Kg^{-1} (Laza and Bereczky 2011), and a density of 810 Kgm^{-3} (Pfromm et al. 2010).

Biobutanol is envisioned as a viable alternative to conventional fuels. It is a better biofuel than biomethanol or bioethanol because it is a completely nonpolar, long hydrocarbon, and it is identical to gasoline and may be used in gasoline cars without any upgrades or mechanism changes (Hönig, et al. 2014). A comparison of various physicochemical properties of fuels are mentioned in Table 1. Hence, the heat of vaporization rate of butanol is similar to ethanol, so the butanol blended gasoline doesn't show a cold start problem and can be used as a 100% biobutanol fuel instead of gasoline (Pospíšil et al. 2014, Elkatory et al. 2022) as a result of its high energy density and molecular resemblance to gasoline. This makes it more suitable for in-place fuel engines: either in a diesel blend (excellent inter-solubility) or unmodified (Visioli et al. 2014). The fact that it possesses almost half the heat of vaporization of that of ethanol (Lee et al. 2008, Pospíšil et al. 2014), indicates its superiority over either ethanol or methanol when it comes to engine initiation at sub-zero temperatures

(Berni et al. 2013). Furthermore, it has lower vapor pressure and lower volatility, which facilitates easier storage and transport, and subsequently makes it less prone to problems like pipeline rupture, cavitation, and vapor lock (Jin et al. 2011, Ranjan et al. 2012). Besides being used as a fuel, it also has applications as a solvent in food and pharmaceutical industries. Undoubtedly, it possesses a superior application range over other biofuels.

3. Principle of acetone-butanol-ethanol (ABE) fermentation

Biobutanol (butanol) is generated commercially on a large scale via a fermentative method, and bacteria have shown their ability to use sugars in maize and other biomass to create butanol. However, because of the current move from first- and second-generation biofuels to third-generation biofuels for the reasons described earlier, butanol synthesis utilizing microalgae as a substrate has been researched. Because biomass largely comprises non-fermentable carbohydrates that must be converted to simple sugars, microbial fermentation of microalgae to butanol requires some chemical or mechanical treatment, similar to the ethanol production process (Wang et al. 2016). A simplified process flow diagram for manufacturing biobutanol from algal biomass is shown in Figure 2. The acetone-butanol-ethanol (ABE) fermentation method produces biobutanol by using microalgal biomass as a substrate and feeding it to the *C. acetobutylicum* bacteria for anaerobic digestion (Inui et al. 2008).

Major routes and products generated during *Clostridium* sp.-mediated butanol synthesis from carbohydrates is shown in Figure 3. Acidogenesis is the process by which the bacteria creates butyric and acetic acids, followed

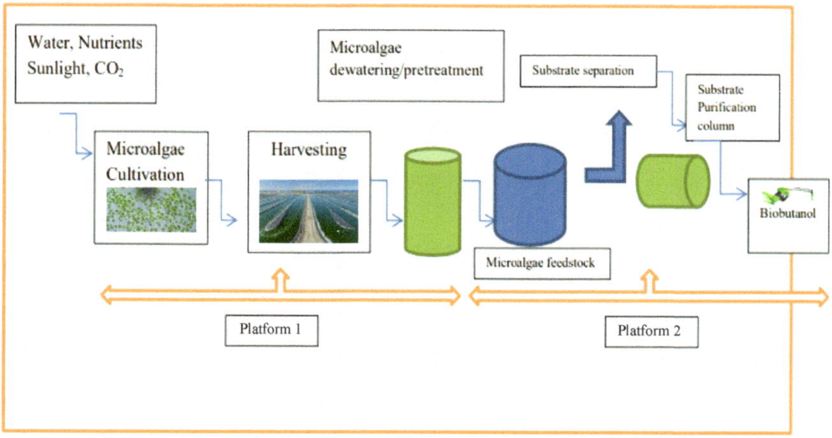

Figure 2. A simplified process flow diagram for manufacturing bio-butanol from algal biomass.

by solventogenesis, which produces butanol, acetone, and ethanol (Lee et al. 2008). Carbohydrate/starch, which is stored as cellulose or in plastids, is the principal microalgal ingredient involved in this process. The generation of acetate and butyrate lowers the pH of the culture, triggering a metabolic process that results in solvent formation. Because *C. acetobutylicum* is saccharolytic, it is a simpler method than bioethanol synthesis. This suggests that neither amylase nor the saccharification process has undergone any pretreatment (such as starch liquefaction). Because of the essentially simpler physiological structure of microalgae, minimum pretreatment is required, resulting in cost savings over other feedstock types. This results in a 3:6:1 ratio of acetone, butanol, and ethanol (Ranjan and Moholkar 2012, Dürre 2007). By upregulating ribulose 1,5-biphosphate carboxylase/oxygenase (Rubisco) in the 2-ketoacid (Ehrlich) pathway Atsumi et al. (2009) discussed about the first successful isobutanol (synchronised with isobutyraldehyde) generating strain *Synechococcus elongatus* PCC 7942 directly from CO_2. The altered strain containing *E. coli* alcohol dehydrogenase (adh) encoded by yqhD and ketoacid decarboxylase gene Figure 1 was cultivated. *Clostridium* sp.

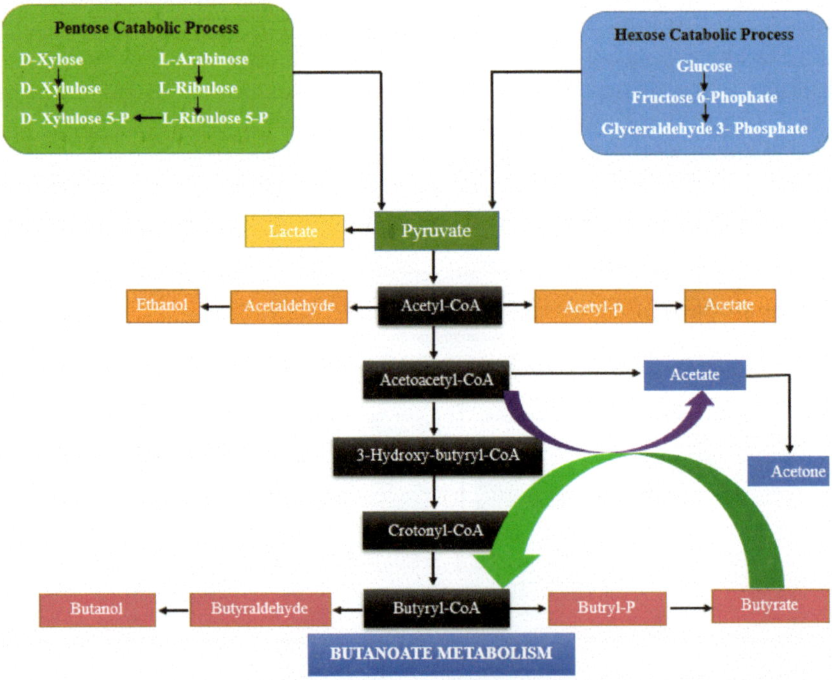

Figure 3. Major pathways and products formed during butanol production from sugars using *Clostridium* sp. (Modified by Shanmugam et al. 2021).

(Kivd) from *Lactobacillus lactis* generated 450 mg/L of isobutanol during the synthesis of butanol from carbohydrates.

However, microalgae have not been thoroughly explored as a substrate for large-scale ABE fermentation (de Carvalho Silvello et al. 2022). Batch fermentation, fed-batch fermentation, and continuous fermentation are the three most common techniques of fermentation. Ranjan and Moholkar (2012) examined the economics of these three fermentations and concluded that continuous fermentation is the best option for large-scale production because it eliminates the downtime costs associated with batch and fed-batch fermentation processes, which are primarily related to broth incubation and harvesting. At a pH of 4.5, Li et al. (2011) examined the performance of the three methods and found that continuous fermentation produced the most butanol and had the greatest butanol productivity. Butanol solvent synthesis was shown to be better when a pH of 4.5 was maintained, compared to the uncontrolled pH condition. The adaptation of bacterial transition between acidogenesis and solventogenesis phases is constantly influenced by time. As a result, it was suggested that for continuous fermentation, a low dilution rate be utilized to shorten the time duration.

4. Comparison of butanol with other biofuels as a source of fuel

Biobutanol is getting more attention these days because of the drawbacks of ethanol as a universal fuel and as a component in fuel blends (Shi et al. 2012). As a matter of fact, butanol has a larger calorific value than ethanol. In addition, it has a higher-octane number, better miscibility with gasoline and diesel, and a lower miscibility with water. Butanol can be stored in humid environments because of its higher hydrophobicity. The non-corrosive nature of this fuel allows it to be used in existing combustion engines in up to 30% (vol/vol) combinations with gasoline (Patakova et al. 2011, Mascal 2012). As a starting material for a variety of chemical processes, butanol has the potential to provide isoprene, isobutene, butane, and other compounds (Kolesinska et al. 2019). Butanol is currently mostly manufactured using oxo synthesis, Reppe synthesis, or crotonaldehyde hydrogenation at a cost of $7.4–8.4 billion per year. However, these items cannot be considered for use as alternative fuel components at this time due to budgetary constraints (Kumar et al. 2012).

5. Microorganisms involved in butanol production using microalgae and their efficiency in biobutanol production

Biobutanol is considered as renewable energy source to replace existing fossil fuels. Among various used feedstocks, various strains microalgae have been extensively used by researchers for biobutanol production. Since lipid-based microalgae biodiesel manufacturing has received much attention, carbohydrate-based microalgae and microalgae wastes resulting from biodiesel production should be considered (de Carvalho Silvello et al. 2022). Some research has been conducted on microalgae-based carbohydrate fermentation and the conversion of microalgal biodiesel leftovers to alternative biofuels such as ethanol (Brennan and Owende 2010, Daroch et al. 2013), butanol (Cheng et al. 2015b, Wang et al. 2016), and biogas (Lakaniemi et al. 2013).

Algal biomass, which includes both macroalgae and microalgae, has the ability to be transformed into biobutanol, an advanced liquid biofuel with the potential to replace petroleum-based fuels. *Clostridium* sp. ABE fermentation process, which is a complicated two-phase heterofermentation involving acidogenesis and solventogenesis, may be used to make biobutanol from biomass (Ge et al. 2022). Algae, being the third-generation biomass for biofuel production, is regarded the ideal feedstock for ABE production because to its quicker growth and great tolerance to unfavourable environmental conditions. Traditional ABE fermentation, on the other hand, is hampered by low productivity due to severe inhibition produced by butanol concentrations in fermentation broth more than 2% (w/v) (Kongjan et al. 2021). Biobutanol is synthesized through fermentation process of microalgae cell wall and other cellular parts as a source of carbon. Among the microalgal species, *Chlorella*, *Scenedesmus*, and *Chlamydomonas* are the best prospects for biobutanol synthesis due to their higher carbohydrate contents of 55%, 52%, and 53%, respectively, which can be easily digested for biobutanol production (Shanmugam et al. 2021). In a recent study, conducted by Kristiawan and Tambunan (2020), marine water algal strain was successfully used for butanol production (2.61%) through fermentation process, with the role of bacteria *Closteridium Acetobutylicum*. According to Onay (2020), effective biobutanol production with various acid hydrolysis treatment strategies H_2SO_4, HCl, and H_3PO_4 from *Nannochloropsis gaditana* grown under different light intensities. Narchonai et al. 2020 investigated the potential of the freshwater microalga *Chlorococcum humicola* isolated from the Temple Pond as a feedstock for *Clostridium acetobutylicum*-based biobutanol production. A recent study conducted by Figueroa-Torres et al. (2020) revealed that when *Chlamydomonas reinhardtii* hydrolysate was used, biobutanol yields were 10.31% and 10.75%, respectively. It was discovered that using microalgae biomass that had been grown in wastewater as a feedstock for ABE fermentation with *Clostridium saccharoper* butylacetonicum N1-4 was feasible by Ellis et al. (2012). According to the findings, adequate pretreatment and enzymatic hydrolysis considerably enhanced the ABE yield, while the addition of 1% glucose resulted in an increase of 1.6-fold in total ABE production. The maximum total ABE production yield and productivity were attained, with 0.311 g/g and 0.102 g/L/h, respectively, as the highest total ABE production yield and productivity. As shown in Table 1 which describes butanol production with different bacteria using microalgae as substrate. For the production of ABE, the selection of a strain is critical because it determines the fermentation performance and influences the methods used for feedstock pretreatment, hydrolysis, and solvent recovery. A large number of strains capable of ABE fermentation are currently recognized, and they can primarily be classified as genus *Clostridium strains* (e.g., *Clostridium acetobutylicum (C. acetobutylicum), Clostridium beijerinckii, Clostridium saccharoperbutylacetonicum, Clostridium saccharobutylicum*) and non-genus *Clostridium* strains (Li et al. 2019) *Clostridium* strains have been shown to be capable of utilizing both simple and complex carbohydrates, such as glucose, sucrose, and cellulose (Tracy et al. 2012). While there is a large range of *Clostridium* strains available, *C. acetobutylicum* (López-Contreras et al. 2000), *C. beijerinckii* (Qureshi et al. 2010), *C. saccharoperbutylacetonicum* (Tashiro et al. 2010), and *C. saccharobutylicum* (Liew et al. 2006) have all been shown to generate

Table 1. Butanol production with different bacteria using microalgae as substrate.

Fermenting bacteria	Microalgae as substrate	Pretreatment conditions	Butanol concentrations (g/L)	References
C. acetobutylicum	*Arthrospira platensi*	Thermal treatment 108°C for 30 min with 0.1 mM H_2SO_4	0.43	Efremenko et al. 2012
C. acetobutylicum B1787	*Nannochloropsis* sp.	Sulfuric acid pretreatment	10.9	Efremenko et al. 2012
C. acetobutylicum	Biodiesel residue of *C. sorokiniana*	2% Sulfuric acid at 121°C, followed by 2% NaOH at 121°C	3.86	Cheng et al. 2015a
C. acetobutylicum	*Chlorella vulgaris* JSC-6	1% NaOH followed by 3% H_2SO_4	13.1	Wang et al. 2016
C. acetobutylicum	*C. zofingiensis*	4% H_2O_2 pre-treatment	4.2	Onay 2020
C. acetobutylicum DSM 792	*C. reinhardtii*	Pulverized, Freeze-dried, Sterilized & acid hydrolyzed	0.89	Figueroa-Torres et al. 2020
C. acetobutylicum DSM 792	*C. reinhardtii*	Pulverized, Freeze-dried & Sterilized	1.031	Figueroa-Torres et al. 2020

solvent with relatively high yields during fermentation when the conditions are favourable.

6. Hurdles and challenges in butanol production with microalgal

ABE fermentation is a well-studied technique for producing biobutanol. Several studies have concentrated on ABE fermentation from cellulosic and non-cellulosic feedstock over the last decade. Nonetheless, there are barriers to commercialising ABE fermentation, including high feedstock costs and a reduced output profile by *Clostridium* species. Other than lignocelluloses as feedstocks must be used to lessen our reliance on plant biomass and move beyond the land use conflict between food and fuel (Shanmugam et al. 2021).

There major challenges in butanol production and possible solutions although butanol, as an advanced liquid fuel, has several advantages compared to ethanol, butanol process has several bottlenecks and challenges hindering its commercial production. Political and economic factors, among other things, have an impact on the development of biobutanol production. In several countries, such as Brazil and the United States, alcohol in the form of ethanol has been successfully commercialized and used as a biofuel. In contrast, biobutanol production is far behind the pace of ethanol production because it is heavily reliant on the advancement of ABE fermentation technology, which is currently facing a number of obstacles to commercialization (Veza et al. 2021). Here is a brief introduction to the main challenges and some of their possible solutions as mentioned above.

ABE is not a commercially profitable and competitive process without using an inexpensive and widely available substrate. The development of sophisticated ABE fermentation techniques was prompted by challenges with substrate and product inhibition in clostridial fermentation. ABE fermentation has been described as fed-batch fermentation, SSF, continuous fermentation, two-stage fermentation, multistage fermentation, high cell density fermentation, membrane biofilm fermentation, and fermentation utilizing immobilized cells. Lignocelluloses are suggested to be suitable substrate for ABE fermentation. Fed-batch fermentation with integrated product recovery was proven to generate improved butanol yields and productivity in ABE fermentation, but it is not a particularly attractive technology for lignocellulosic biomass hydrolysate due to the low sugar concentrations produced. The majority of fed-batch trials included glucose supplementation (Ezeji et al 2016, Gottumukkala et al. 2019). The resiliency of lignocellulosic biomass is as well recognized as its abundance. The abrasive character of lignocellulosic biomass necessitates energy and cost-intensive pretreatment and hydrolysis processes before fermentable sugars can be extracted for biofuel generation (Gottumukkala et al. 2019). Glycerol, a biodiesel industry waste, has been proposed as a viable substrate for ABE manufacturing (Khanna et al. 2013, Li et al. 2014).

Algae are also suggested as alternate substrates. Although algae biomass has significant advantages over other substrates, the generation of algae fuels is still in its early phases, and a number of serious issues must be addressed first (Shanmugam et al. 2021). The main impediment to meeting the potential of microalgae is the development of an economically viable technology for feedstock production (Pienkos et al. 2009). This can be greatly reduced by employing effective genome engineering approaches, as functional knowledge of the genes and proteins involved in central carbon metabolism could result in cellular network modification for increased biofuel generation (Radakovits et al. 2010).

7. Improvement strategies for biobutanol production

Butanol yields of 16–17 g/L can be produced employing the best native bacteria in traditional clostridial fermentation techniques. As a result, one of the primary focuses of research has been to generate strains that increase butanol output (Figure 4). The following topics have been studied: (i) bacterial strain selection, (ii) using less expensive carbon sources to

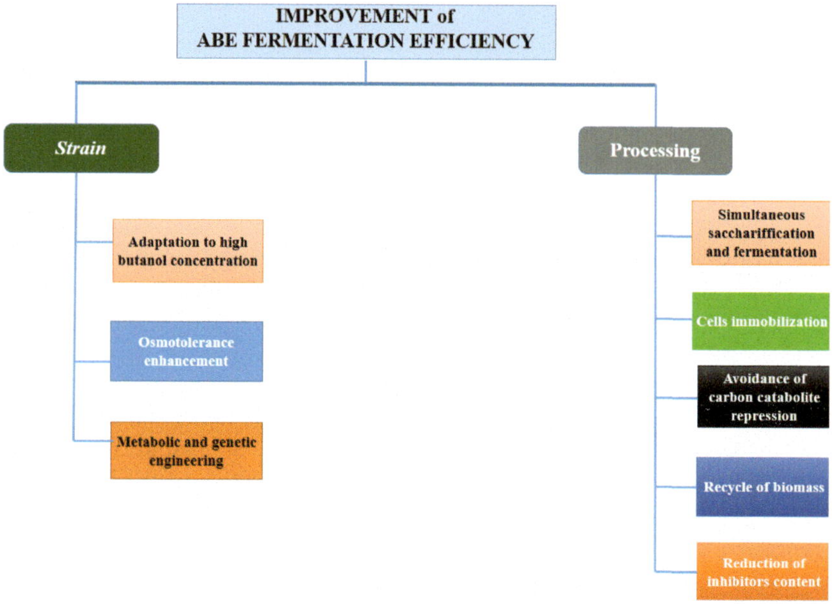

Figure 4. Improvement strategies for biobutanol production (Modified from Shanmugam et al. 2021).

create butanol, (iii) metabolic engineering methodologies, (iv) process development, and (v) biobutanol recovery techniques (Zhou et al. 2022).

7.1 Process improvement

To avoid the usage of high-cost glucose, wheat straw can be hydrolyzed into simple sugar and then fermented to butanol at the same time. The batch process for butanol production from various polysaccharide materials was tested by Qureshi et al. (2008a,b,c) using different combinations of pretreatment, including (i) fermentation with pretreated wheat straw; (ii) separate hydrolysis and fermentation of the wheat straw without removing the sediments; (iii) simultaneous hydrolysis and fermentation of the wheat straw without agitation; (iv) the final combination produced the maximum butanol production; and (v) the hydrolysate contained a wide range of monosaccharides, including glucose, xylose, arabinose, galactose, and mannose, among others (feed-batch fermentation's productivity was boosted, however, by sugar supplementation). Specific obstacles involved in the manufacture of biobutanol include high process costs associated with some feedstocks, product toxicity, and low product concentrations. Identifying the obstacles involved in converting lignocellulosic biomass to biobutanol and identifying significant process improvements will help to make biobutanol more commercially appealing (Kolesinska 2019).

7.2 Strain improvement

Clostridium strains that are wild-type have a poor solvent tolerance, slow growth, and low cell density during the solventogenic phase of growth, which severely limits the performance of ABE fermentation. These issues have been discussed in relation to mutagenesis, evolutionary engineering, and molecular engineering (Jang et al. 2012, Li et al. 2019). N-methyl-N′-nitro-N-nitrosoguanidine (MNNG), hydrogen peroxide, and nalidixic acid are only a few examples of the many mutagens utilized in mutagenesis engineering, although the direct action of MNNG appears to be the most effective. *C. beijerinckii* NCIMB 8052 was treated with MNNG to yield BA101, a mutant of *C. beijerinckii.* The technique of "mutagenesis followed-by selection" can also be used in evolutionary engineering to boost solventogenic performance even further. *C. acetobutylicum* T64 was created by simulating bio-evolution using *C. acetobutylicum* D64 as a starting point (Liu et al. 2013). Molecular engineering, which involves changing strains by the use of inactivated and/or overexpressed genes, was anticipated to be a more reasonable technique for addressing the difficulties of tediousness and instability associated with mutagenesis and evolution engineering. Numerous metabolic enzymes were either altered or overexpressed, including buk, ptb, and ack (which were all knocked out or down), as well as aad and adhE (which were both overexpressed and knocked out) (Liu et al. 2022) Many genes encoding metabolic enzymes have been knocked out or reduced (for, e.g., buk, ptb, and ack) and/or overexpressed in *Clostridium* strains (e.g., adhE and thiL) in order to modify the concentration, yield, and ratios of various solvents (Tashiro et al. 2010, Shen and Liao 2008, Li et al. 2019).

8. Future perspectives

Novel strains with promising butanol production capacities have been found in recent years, and a wide range of feedstock has been examined for their potential in ABE fermentation. Butanol manufacturing from macroalgae and food waste showed encouraging results. Although advanced pretreatment processes for boosting sugar yields and butanol fermentation have been documented, their commercial viability is contingent on operating costs and efficiency (Vidhya 2022). According to the results of the techno-economic analysis studies, pretreatment and hydrolysis are the major cost-contributing operations in terms of fixed capital investment and operating costs, and thus consolidated bioprocessing strategies may be advantageous for biobutanol production from lignocellulosic biomass. Butanol fermentation has a history dating back more than a century, and advances in metabolic engineering of production strains to boost butanol output, tolerance, selective blockage of co-solvent formation, and other factors have greatly advanced the sector. Furthermore, great progress has been made in the removal of solvents during fermentation by improved separation techniques such as reverse osmosis, pervaporation, perstraction, liquid-liquid extraction, gas stripping, and so on. Because butanol fermentative technology has been around for a long time, any future adjustments would involve making it capable of handling renewable second-generation feedstock. Although lignocellulose-derived sugar is a very effective substrate, the technology for producing sugar from biomass at a low cost must mature. *Clostridial* strains utilised for butanol fermentation have the ability to withstand C5 sugars, which gives them an edge over yeast-*Saccharomyces cerevisiae* used for bioethanol production. Aside from lignocellulose, additional key biobutanol feedstock choices include feedstock that does not place demand pressure on food or feed. While more difficult to manage, industrial wastes are also a viable solution. In any scenario, large-scale operations must be proven, as well as the techno economics and life cycle analysis. *C. carboxydivorans'* synthesis of butanol from carbon monoxide is one of the most intriguing R&D results. In addition to economic benefits, successful waste valorization to fuel and chemicals provides considerable environmental benefits. Biobutanol enterprises might be developed as integrated facilities with paper and pulp mills, sugarcane plants, and other facilities where biobutanol is not the only product and feedstock is easily accessible. According to environmental assessments, biobutanol utilized in transportation will have a considerable beneficial influence on greenhouse gas reduction (Zhou et al. 2022).

9. Conclusion

Butanol, a competitor to most of the other biofuels and petroleum-based products, can be produced biologically via ABE processes. It can be mixed with gasoline at any concentrations and easily transported without the risk of absorbing moisture. Moreover, butanol contains higher energy and has a lower corrosion rate than ethanol. The ABE production process is still accompanied with several challenges such as low concentration and difficulty in separation of products, higher feedstock consumption rate (that is, lower production yield), and sensitivity to substrate composition, inhibitors, and to the presence of oxygen. In summary, more research is necessary to revamp the old ABE production processes or the current butanol production processes to better reflect the requirements of the butanol production in the future.

References

Abo, B.O., Gao, M., Wang, Y., Wu, C., Wang, Q., and Ma, H. (2019). Production of butanol from biomass: recent advances and future prospects. Environmental Science and Pollution Research, 26(20): 20164–20182.

Atsumi, S., Higashide, W., and Liao, J.C. (2009). Direct photosynthetic recycling of carbon dioxide to isobutyraldehyde. Nature Biotechnology, 27(12): 1177–1180.

Berni, M., Dorileo, I., Prado, J., Forster-Carneiro, T., and Meireles, M. (2013). Advances in biofuel production. Biofuels Production, 11–58.

Brennan, L., and Owende, P. (2010). Biofuels from microalgae—a review of technologies for production, processing, and extractions of biofuels and co-products. Renewable and Sustainable Energy Reviews, 14(2): 557–577.

Chen, C.Y., Yeh, K.L., Aisyah, R., Lee, D.J., and Chang, J.S. (2011). Cultivation, photobioreactor design and harvesting of microalgae for biodiesel production: a critical review. Bioresource Technology, 102(1): 71–81.

Chen, C.Y., Zhao, X.Q., Yen, H.W., Ho, S. H., Cheng, C.L., Lee, D.J., and Chang, J. S. (2013). Microalgae-based carbohydrates for biofuel production. Biochemical Engineering Journal, 78: 1–10.

Cheng, H.H., Whang, L.M., Chan, K.C., Chung, M.C., Wu, S.H., Liu, C.P., and Lee, W.J. (2015). Biological butanol production from microalgae-based biodiesel residues by *Clostridium acetobutylicum*. Bioresource Technology, 184: 379–385.

Cheng, Hai-Hsuan, Liang-Ming Whang, Kun-Chi Chan, Man-Chien Chung, Shu-Hsien Wu, Cheng-Pin Liu, Shih-Yuan Tien, Shan-Yuan Chen, Jo-Shu Chang, and Wen-Jhy Lee. (2015). Biological butanol production from microalgae-based biodiesel residues by *Clostridium acetobutylicum*. Bioresource Technology, 184: 379–385.

Chong, C.T., and Hochgreb, S. (2011). Measurements of laminar flame speeds of acetone/ methane/ air mixtures. Combust Flame; 158: 490–500.

Daroch, M., Geng, S., and Wang, G. (2013). Recent advances in liquid biofuel production from algal feedstocks. Applied Energy, 102: 1371–1381.

de Carvalho Silvello, M.A., Gonçalves, I.S., Azambuja, S.P.H., Costa, S.S., Silva, P.G. P., Santos, L.O., and Goldbeck, R. (2022). Microalgae-based carbohydrates: A green innovative source of bioenergy. Bioresource Technology, 344: 126304.

Demirbaş, A. (2010). Social, economic, environmental and policy aspects of biofuels. Energy Education Science and Technology Part B: Social and Educational Studies.

Demirbas, A. (2011). Competitive liquid biofuels from biomass. Appl. Energy, 88: 17–28.

Doğan, O. (2011). The influence of n-butanol/diesel fuel blends utilization on a small diesel engine performance and emissions. Fuel, 90: 2467–72.

Dürre, P. (2007). Biobutanol: an attractive biofuel. Biotechnology Journal: Healthcare Nutrition Technology, 2(12): 1525–1534.

Efremenko, E.N., Nikolskaya, A.B., Lyagin, I.V., Senko, O.V., Makhlis, T.A., Stepanov, N.A., Maslova, O.V., Mamedova, F. and Varfolomeev, S.D. 2012. Production of biofuels from pretreated microalgae biomass by anaerobic fermentation with immobilized *Clostridium acetobutylicum* cells. Bioresource Technology, 114: 342–348.

Elkatory, M.R., Hassaan, M.A., and El Nemr, A. (2022). Algal biomass for bioethanol and biobutanol production. In Handbook of Algal Biofuels (pp. 251–279). Elsevier.

Ellis, J.T., Hengge, N.N., Sims, R.C., and Miller, C.D. (2012). Acetone, butanol, and ethanol production from wastewater algae. Bioresource Technology, 111: 491–495.

Ezeji, T.C., Qureshi, N., and Blaschek, H.P. (2004). Acetone butanol ethanol (ABE) production from concentrated substrate: reduction in substrate inhibition by fed-batch technique and product inhibition by gas stripping. Applied Microbiology and Biotechnology, 63(6): 653–658.

Figueroa-Torres, G.M., Mahmood, W.M.A.W., Pittman, J.K., and Theodoropoulos, C. (2020). Microalgal biomass as a biorefinery platform for biobutanol and biodiesel production. Biochemical Engineering Journal, 153: 107396.

Ge, S., Brindhadevi, K., Xia, C., Khalifa, A.S., Elfasakhany, A., Unpaprom, Y., and Van Doan, H. (2022). Enhancement of the combustion, performance and emission characteristics of spirulina microalgae biodiesel blends using nanoparticles. Fuel, 308: 121822.

Gomez, L.D., Steele-King, C.G., and McQueen-Mason, S.J. (2008). Sustainable liquid biofuels from biomass: the writing's on the walls. New Phytol., 178: 473–85.

Gottumukkala, L.D., Mathew, A.K., Abraham, A., and Sukumaran, R.K. (2019). Biobutanol production: microbes, feedstock, and strategies. In Biofuels: Alternative Feedstocks and Conversion Processes for the Production of Liquid and Gaseous Biofuels (pp. 355–377). Academic Press.

Graham, L.A., Belisle, S.L., and Baas, C.L. (2008). Emissions from light duty gasoline vehicles operating on low blend ethanol gasoline and E85. Atmos. Environ., 42: 4498–516.

Harun, R., Danquah, M. K., and Forde, G.M. (2010). Microalgal biomass as a fermentation feedstock for bioethanol production. Journal of Chemical Technology & Biotechnology, 85(2): 199–203.

Hönig, V., Kotek, M., and Mařík, J. (2014). Use of butanol as a fuel for internal combustion engines. Agronomy Research, 12(2): 333–340.

Huang, H.J., Ramaswamy, S., and Liu, Y. (2014). Separation and purification of biobutanol during bioconversion of biomass. Separation and Purification Technology, 132: 513–540.

Huang, W., Qu, Z., Chen, W., Xu, H., and Yan, N. (2016). An enhancement method for the elemental mercury removal from coal-fired flue gas based on novel discharge activation reactor. Fuel, 171: 59–64.

Inui, M., Suda, M., Kimura, S., Yasuda, K., Suzuki, H., Toda, H., Yamamoto, S., Okino, S., Suzuki, N., and Yukawa, H. (2008). Expression of *Clostridium acetobutylicum* butanol synthetic genes in *Escherichia coli*. Applied Microbiology and Biotechnology, 77(6): 1305–1316.

Jang, Y.S., Malaviya, A., Cho, C., Lee, J., and Lee, S.Y. (2012). Butanol production from renewable biomass by clostridia. Bioresource Technology, 123: 653–663.

Jin, C., Yao, M., Liu, H., Chia-fon, F. L., and Ji, J. (2011). Progress in the production and application of n-butanol as a biofuel. Renewable and Sustainable Energy Reviews, 15(8): 4080–4106.

Jones, D.T., and Woods, D.R. (1986). Acetone-butanol fermentation revisited. Microbiological Reviews, 50(4): 484–524.

Khanna, S., Goyal, A., and Moholkar, V.S. (2013). Production of n-butanol from biodiesel derived crude glycerol using *Clostridium pasteurianum* immobilized on Amberlite. Fuel, 112: 557–561.

Klass, D.L. (1998). Biomass for renewable energy, fuels, and chemicals. Elsevier.

Kolesinska, B., Fraczyk, J., Binczarski, M., Modelska, M., Berlowska, J., Dziugan, P., and Kregiel, D. (2019). Butanol synthesis routes for biofuel production: trends and perspectives. Materials, 12(3): 350.

Kongjan, P., Usmanbaha, N., Khaonuan, S., Jariyaboon, R., Sompong, O., and Reungsang, A. (2021). Butanol production from algal biomass by acetone-butanol-ethanol fermentation process. In Clean Energy and Resources Recovery (pp. 421–446). Elsevier.

Kristiawan, O., and Tambunan, U.S.F. (2020). Biobutanol production from microalgae *Nannochloropsis* sp. bomasses by *Clostridium acetobutylicum* fermentation. Scientific Contributions Oil and Gas, 43(2): 91–98.

Kumar, M., Goyal, Y., Sarkar, A., and Gayen, K. (2012). Comparative economic assessment of ABE fermentation based on cellulosic and non-cellulosic feedstocks. Applied Energy, 93: 193–204.

Lakaniemi, A.M., Tuovinen, O.H., and Puhakka, J.A. (2013). Anaerobic conversion of microalgal biomass to sustainable energy carriers–a review. Bioresource Technology, 135: 222–231.

Laza, T., and Bereczky, Á. (2011). Basic fuel properties of rapeseed oil-higher alcohols blends. Fuel, 90(2): 803–810.

Lee, S.Y., Park, J.H., Jang, S. H., Nielsen, L.K., Kim, J., and Jung, K.S. (2008). Fermentative butanol production by *Clostridia*. Biotechnology and Bioengineering, 101(2): 209–228.

Li, J., Baral, N.R., and Jha, A.K. (2014). Acetone–butanol–ethanol fermentation of corn stover by *Clostridium* species: present status and future perspectives. World Journal of Microbiology and Biotechnology, 30(4): 1145–1157.

Li, S.Y., Srivastava, R., Suib, S.L., Li, Y., and Parnas, R. S. (2011). Performance of batch, fed-batch, and continuous A–B–E fermentation with pH-control. Bioresource Technology, 102(5): 4241–4250.

Li, Y., Tang, W., Chen, Y., Liu, J., and Chia-fon, F.L. (2019). Potential of acetone-butanol-ethanol (ABE) as a biofuel. Fuel, 242: 673–686.

Liew, S.T., Arbakariya, A., Rosfarizan, M., and Raha, A.R. (2006). Production of solvent (acetone-butanol-ethanol) in continuous fermentation by Clostridium *saccharobutylicum* DSM 13864 using gelatinised sago starch as a carbon source. Malaysian Journal of Microbiology, 2(2): 42–50.

Linhares, F.G., Lima, M.A., Mothe, G.A., De Castro, M.P.P., Da Silva, M.G., and Sthel, M.S. (2019). Photoacoustic spectroscopy for detection of N 2 O emitted from combustion of diesel/beef tallow biodiesel/sugarcane diesel and diesel/beef tallow biodiesel blends. Biomass Conversion and Biorefinery, 9(3): 577–583.

Liu, X.B., Gu, Q.Y., and Yu, X.B. (2013). Repetitive domestication to enhance butanol tolerance and production in *Clostridium acetobutylicum* through artificial simulation of bio-evolution. Bioresource technology, 130, 638-643.

Liu, X., Xie, H., Roussou, S., and Lindblad, P. (2022). Current advances in engineering cyanobacteria and their applications for photosynthetic butanol production. Current Opinion in Biotechnology, 73, 143–150.

López-Contreras, A.M., Claassen, P.A., Mooibroek, H., and De Vos, W.M. (2000). Utilisation of saccharides in extruded domestic organic waste by *Clostridium acetobutylicum* ATCC 824 for production of acetone, butanol and ethanol. Applied Microbiology and Biotechnology, 54(2): 162–167.

Lü, J., Sheahan, C., and Fu, P. (2011). Metabolic engineering of algae for fourth generation biofuels production. Energy & Environmental Science, 4(7): 2451-2466.

Mascal, M. (2012). Chemicals from biobutanol: technologies and markets. Biofuels, Bioproducts and Biorefining, 6(4): 483–493.

Narchonai, G., Arutselvan, C., LewisOscar, F., and Thajuddin, N. (2020). Enhancing starch accumulation/production in *Chlorococcum humicola* through sulphur limitation and 2, 4-D treatment for butanol production. Biotechnology Reports, 28, e00528.

Olabi, A.G. (2013). State of the art on renewable and sustainable energy. Energy, 61: 2–5.

Onay, M. (2020). Enhancing carbohydrate productivity from *Nannochloropsis gaditana* for bio-butanol production. Energy Reports, 6: 63 67.

Patakova, P., Maxa, D., Rychtera, M., Linhova, M., Fribert, P., Muzikova, Z., and Melzoch, K. (2011). Perspectives of biobutanol production and use. Biofuel's Engineering Process Technology, 11: 243–261.

Pfromm, P.H., Amanor-Boadu, V., Nelson, R., Vadlani, P., and Madl, R. (2010). Bio-butanol vs. bio-ethanol: a technical and economic assessment for corn and switchgrass fermented by yeast or *Clostridium acetobutylicum*. Biomass and Bioenergy, 34(4): 515–524.

Pienkos, P.T., and Darzins, A.L. (2009). The promise and challenges of microalgal-derived biofuels. Biofuels, Bioproducts and Biorefining: Innovation for a Sustainable Economy, 3(4): 431–440.

Pospíšil, M., Šiška, J., and Šebor, G. (2014). BioButanol as fuel in transport, Biom [online].[cit. 2014-17-01]. Available at www: biom. cz.

Qureshi, N., Ezeji, T.C., Ebener, J., Dien, B.S., Cotta, M.A., and Blaschek, H.P. (2008a). Butanol production by *Clostridium beijerinckii*. Part I: use of acid and enzyme hydrolyzed corn fiber. Bioresource Technology, 99(13): 5915–5922.

Qureshi, N., Saha, B.C., Hector, R.E., Hughes, S.R., and Cotta, M.A. (2008b). Butanol production from wheat straw by simultaneous saccharification and fermentation using *Clostridium beijerinckii*: Part I—Batch fermentation. Biomass and Bioenergy, 32(2): 168–175.

Qureshi, N., Saha, B.C., Hector, R.E., Hughes, S.R., and Cotta, M.A. (2008c). Butanol production from wheat straw by simultaneous saccharification and fermentation using *Clostridium beijerinckii*: Part I—Batch fermentation. Biomass and Bioenergy, 32(2): 168–175.

Qureshi, N., Saha, B.C., Dien, B., Hector, R.E., and Cotta, M.A. (2010). Production of butanol (a biofuel) from agricultural residues: Part I–Use of barley straw hydrolysate. Biomass and Bioenergy, 34(4): 559–565.

Radakovits, R., Jinkerson, R.E., Darzins, A., and Posewitz, M.C. (2010). Genetic engineering of algae for enhanced biofuel production. Eukaryotic Cell, 9(4): 486–501.

Ramaswamy, S., Huang, H.J., and Ramarao, B.V. (2013). Separation and purification technologies in biorefineries (p. 1). John Wiley & Sons Incorporated.

Ranjan, A., and Moholkar, V.S. (2012). Biobutanol: science, engineering, and economics. International Journal of Energy Research, 36(3): 277–323.

Shanmugam, S., Hari, A., Kumar, D., Rajendran, K., Mathimani, T., Atabani, A.E., and Pugazhendhi, A. (2021). Recent developments and strategies in genome engineering and integrated fermentation approaches for biobutanol production from microalgae. Fuel, 285; 119052.

Shen, C.R., and Liao, J.C. (2008). Metabolic engineering of *Escherichia coli* for 1-butanol and 1-propanol production via the keto-acid pathways. Metabolic Engineering, 10(6): 313–320.

Shi, S., Yue, C., Wang, L., Sun, X., and Wang, Q. (2012). A bibliometric analysis of anaerobic digestion for butanol production research trends. Procedia Environmental Sciences, 16: 153–158.

Siddiki, S.Y.A., Mofijur, M., Kumar, P.S., Ahmed, S.F., Inayat, A., Kusumo, F., and Mahlia, T.M.I. (2022). Microalgae biomass as a sustainable source for biofuel, biochemical and biobased value-added products: An integrated biorefinery concept. Fuel, 307: 121782.

Sileghem, L., Alekseev, V.A., Vancoillie, J., et al. (2013) Laminar burning velocity of gasoline and the gasoline surrogate components iso-octane, n-heptane and toluene. Fuel, 112: 355–65.

Singh, A., and Olsen, S.I. (2011). A critical review of biochemical conversion, sustainability and life cycle assessment of algal biofuels. Appl. Energy, 88: 3548–55.

Tashiro, Y., and Sonomoto, K. (2010). Advances in butanol production by clostridia. Current Research, Technology and Education Topics in Applied Microbiology and Microbial Biotechnology, 2: 1383–94.

Tracy, B.P., Jones, S.W., Fast, A.G., Indurthi, D.C., and Papoutsakis, E.T. (2012). Clostridia: the importance of their exceptional substrate and metabolite diversity for biofuel and biorefinery applications. Current Opinion in Biotechnology, 23(3): 364–381.

Veloo, P.S., Wang, Y.L, Egolfopoulos, F.N., and Westbrook, C.K. (2010) A comparative experimental and computational study of methanol, ethanol, and n-butanol flames. Combust Flame, 157: 1989–2004.

Veza, I., Said, M.F.M., and Latiff, Z.A. (2021). Recent advances in butanol production by acetone-butanol-ethanol (ABE) fermentation. Biomass and Bioenergy, 144: 105919.

Vidhya, C.V. (2022). Microalgae—the ideal source of biofuel. In Biofuels and Bioenergy (pp. 389–405). Elsevier.

Visioli, L.J., Enzweiler, H., Kuhn, R.C., Schwaab, M., and Mazutti, M.A. (2014). Recent advances on biobutanol production. Sustainable Chemical Processes, 2(1): 1–9.

Wallner, T., Ickes, A., and Lawyer, K. (2013). Analytical assessment of C2-C8 alcohols as sparkignition engine fuels. Proceedings of the FISITA 2012 World Automotive Congress. Springer, pp. 15–26.

Wang, Y., Guo, W., Cheng, C.L., Ho, S.H., Chang, J.S., and Ren, N. (2016). Enhancing bio-butanol production from biomass of Chlorella vulgaris JSC-6 with sequential alkali pretreatment and acid hydrolysis. Bioresource Technology, 200: 557–564.

Yazdani, P., Zamani, A., Karimi, K., and Taherzadeh, M.J. (2015). Characterization of *Nizimuddinia zanardini* macroalgae biomass composition and its potential for biofuel production. Bioresource Technology, 176: 196–202.

Yeong, T.K., Jiao, K., Zeng, X., Lin, L., Pan, S., and Danquah, M.K. (2018). Microalgae for biobutanol production–Technology evaluation and value proposition. Algal Research, 31: 367–376.

Zhou, Y., Liu, L., Li, M., and Hu, C. (2022). Algal biomass valorisation to high-value chemicals and bioproducts: recent advances, opportunities and challenges. Bioresource Technology, 344: 126371.

10

Pretreatment and Hydrolysis of Biomaterials for Butanol Production

Arti Devi,[1] *Anita Singh,*[1,*] *Somvir Bajar*[2] and *Deepak Pant*[3]

1. Introduction

Energy plays a crucial role in the development and the growth of a nation. The non-renewability of fossil fuels and variation in the crude oil prices hastens the research in finding better renewable energy alternatives. Biofuels are a renewable, environment-friendly alternative that has the ability to fulfill the world's energy demand. Among biofuels, butanol is one of the liquid biofuels produced by the biological fermentation method that can be used in the internal combustion (IC) engines directly or in blend with petrochemicals. High cetane number, energy content, viscosity, and low flammability of butanol made it more suited to IC engines than ethanol (Li et al. 2019a). It has been mentioned by Dutta et al. (2021) that one liter of butanol can substitute 0.662 liter of gasoline. It can also reduce soot emissions when blended with petrochemicals. Butanol is produced by the process of acetone–butanol–ethanol (ABE) fermentation (Figure 1)

[1] Department of Environmental Sciences, Central University of Jammu, Jammu-180011, Jammu and Kashmir, India.
[2] Department of Environmental Science and Engineering, J.C. Bose University of Science & Technology, YMCA, Faridabad-121006, Haryana (India).
[3] Separation and Conversion Technology, Flemish Institute for Technological Research (VITO), Boeretang 200, Mol, 2400, Belgium.
* Corresponding author: anitasaharan@gmail.com

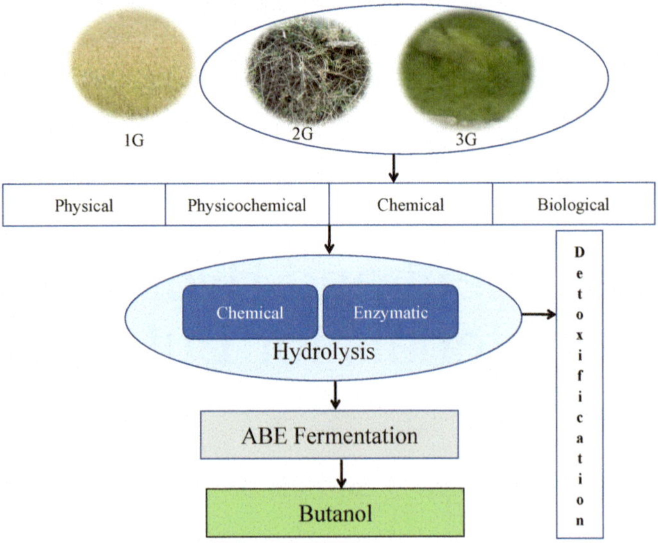

Figure 1. Schematic of the acetone-butanol-ethanol (ABE) fermentation process.

usually takes place with the help of bacteria namely *Clostridia*. Various substrates like edible crops, lignocellulosic biomass, and algal biomass that are categorized into three generation, respectively have been used for its production through biological route. Edible crops have directly affected food prices and its food quality as well as the soil for its growth. Lignocellulosic biomass is a low-cost resource but needs treatment before the fermentation process. Likewise algal biomass also possess carbohydrates, lipids, proteins, etc., in their cell wall and have advantage over other two generation feedstocks as there is no need of arable land and have less carbon inputs (Patil et al. 2019). In this chapter, pretreatment and hydrolysis of biomass has been discussed in detail which is necessary steps for the production of butanol from second- and third-generation feedstocks. Also, the challenges and prospects related to its production have also been mentioned.

2. Feedstock

Raw material is the most essential factor in production process due to its availability and economic utility. Non-lignocellulosic (food crops), lignocellulosic, and algal feedstocks are referred to as first-, second-, and third-generation substrates for butanol production. Non-lignocellulosic feedstock mainly comprise starch or sugar substrates that create conflict between food and fuel. Various non-lignocellulosic raw materials are cheese whey, corn, wheat, fruit juices, etc. Likewise, algal biomass (microalgae and macroalgae) and lignocellulosic biomass are also utilized for ABE fermentation. Microalgae are unicellular and have a high content of lipids but also contain carbohydrates in its cell wall while macroalgae are seaweeds and rich in carbohydrates with less fats and protein content in contrast. One major challenge related to production from algal biomass is the high-cost harvesting systems. Lignocellulosic biomass is abundantly available and comprise cellulose, hemicellulose, and lignin. Several crop residues like corn stover, wheat straw, rice straw, and sugarcane bagasse are used to obtain butanol. The advantage of Lignocellulosic biomass (LCB) is its low cost, abundant availability, and environment-friendly nature (Gottumukkala et al. 2019). However, to get butanol from lignocellulosic biomass and algal biomass, it should be pretreated and hydrolyzed to release sugars for the fermentation process, which is discussed in detail in the upcoming sections.

3. Pretreatment

It is the crucial step to open the recalcitrant structure of lignocellulosic biomass or hydrolyze the algal biomass to make it accessible for fermentation process. It decreases the crystallinity of the complex structure of the biomass. Various pretreatment methods like physical, chemical, physiochemical, and biological has been utilized by the researchers to produce butanol from it. While selecting the pretreatment method, one should consider the economics and energy consumption of that particular method. Effective pretreatment method is the one that possess all the characteristics, that is, (1) it should provide high yields with variable feedstocks, (2) less or no production of toxic compounds or inhibitors, (3) effective recovery of lignin, (4) reduced energy consumption and cost, (5) compatible with further processes, and (6) diminishes the crystallinity of cellulose and also improves the enzymatic hydrolysis by increasing its surface area (Jędrzejczyk et al. 2019).

3.1 Physical method

This method affects the physical properties of the biomass, that is, it reduces the particle size, increases the surface area, and reduces the crystallinity index. This method includes the temperature and pressure variations which cause the biomass structure to change. Some chemicals are also used to facilitate the process further.

3.1.1 Mechanical comminution

Mechanical pretreatment reduces the size of the particle, and enhances the surface area and bulk density of lignocellulosic biomass. Size diminution reduces the degree of polymerization and crystallinity index of the cellulose to increase its digestibility. Enhanced bulk density supports the handling, storage, and transport of the biomass with ease. Decreased particle size and increased surface area makes physical and chemical processes easier because of phase boundary formation among the biomass and the chemicals and removal of heat transfer limitation. Several methods like grinding, chipping, and milling have been used to reduce the size. This method is advantageous in size and crystallinity reduction but consumes high energy and power leads to the high cost of the process (Anu et al. 2020). So, complete knowledge about the feedstock mechanics is necessary to select the method and equipment required for its processing. It will help to create a balance between the cost and the efficacy of the process (Jędrzejczyk et al. 2019).

3.1.2 Extrusion

It is a thermo-mechanical process that modifies the biomass physically due to the mixing and shearing at high temperatures. Extruders were used for this process, that can be distinguished in to two types: single screw extruders and twin screw extruders. A single screw extruder is made of one single unit whereas twin screw extruders are made of small pieces called "screw elements" assembled in a shaft. The latter is of two kinds: counter-rotating and co-rotating. The shearing and plasticizing effect is radial in counter-rotating and axial in co-rotating (Raquez et al. 2008). The screw elements are mainly of three types, that is, kneading block, reverse, and conveying screw. Each one has different forms responsible for the production of different effects (transport, mixing, and shearing) on biomass. The combined effect of all these three screw elements disrupts the biomass, provides better flowability, and complete mixing. The positioning of these screw elements called screw configuration/profile is one of the important factors in the extrusion process. It is responsible for the distribution of residence time and mechanical forces generated in the extruder. This factor is also responsible for setting other operational variables like liquid-solid ratio and particle size of feedstock. Temperature also plays n important role and typically ranges between 40–200°C and has positive impact on the enzymatic hydrolysis process. The process of extrusion reduces the particle size, enhances the specific surface area, and changes the crystallinity index (Duque et al. 2017). Recently, the bioextrusion method has been developed as a novel technology to pretreat biomass. This process involves the use of enzyme during the extrusion process (Gatt et al. 2018). The advantage of this method is as it gives better monitoring and control of all process variables and its adaptability to process modification. The drawback of this process is partial degradation of hemicellulose and lignin-carbohydrate complex (Zheng and Rehmann 2014).

3.1.3 Irradiation

To disintegrate the biomass structure for hydrolysis process, several irradiations like microwave (MW), gamma (γ), ultrasound, and electron beam has been utilized. The fundamental principle of irradiation method includes the generation of short-lived ions and radicals by taking part in cross-linking processes and fragments the biomass. MW radiation causes a heating effect or thermal effect. This thermal effect induces temperature and pressure increase, which disrupts the biomass components and also resulted into the relocation of the crystalline cellulose (Haldar and Purkait 2021). MW-induced thermal effect occurs when both oscillating electric field and MW force dipole is in alignment with each other brought the hydrogen bond (H-bond) disruption. This H-bond breakdown leads to cell wall disruption, cellulose chain fractionation, and change in cellulose configuration that lead to increased hydrolysis (Bundhoo and Zumar 2018b). Similarly, ultrasound referred to as sound waves having frequencies > 20 KHz move in a series of compression and rarefaction motion. When these waves travel in a rarefaction (lower pressure) area, it leads to the formation of bubbles gas and vapors whose diameter increased continuously until when it collapses and creates a phenomenon called "cavitation". Due to the implosion of these gas bubbles, a huge amount of energy is released, which increases the temperature and pressure conditions and are called "hotspots". Hotspot formations depolymerize the cellulose and cause cell wall and membrane lysis. The cavitation phenomenon forms oxidative radicals that oxidize the organic molecule and support its solubilization (Bundhoo et al. 2018a). γ-rays and electron beams are also gaining immense attention among other irradiation techniques to pretreat the biomass for biofuel production. Irradiation techniques are efficient but the high energy input requirement and inhibitory compound generation are the factors that need to be reconsidered for its large-scale implementation (Harkat and Purkait 2021).

3.1.4 Pyrolysis

It is thermo-chemical conversion of biomass in which biomass is fragmented at a high temperature in absence of oxygen to yield liquid fuels. The biomass on pyrolysis results in organic vapors that contain hemicellulose, cellulose and lignin, bio-char, and gaseous product. Cellulose pyrolysis leads to the breakdown of glycosidic bonds by dehydration and fractionation of "anhydroglucose" units. Dehydration and disintegration of sugar molecules results in char formation (Dhyani and Bhaskar 2018). It has been reported by Yang et al. (2007) that the temperature range for decomposition of hemicellulose, cellulose, and lignin are 220–315°C, 314–400°C, and 160 to 900°C, respectively. At this temperature range, lignin gives 40% of the solid residues. Based on the temperature and product requirement, it is categorized in to three types: fast, intermediate, and slow. Fast pyrolysis occurs at high temperatures for a short residence time and yields more liquid product while slow pyrolysis occurs at a low temperature for a long period of time and results in more solid residues (char). Less bio-char forms at a high temperature. Intermediate pyrolysis temperature ranges about 300–500°C and provides liquid product but at a low rate (Zadeh et al. 2020). The linkages between the three components of lignocellulosic biomass also affect the behavior of pyrolysis process and its products (Kumar et al. 2020). So, pyrolysis conditions like temperature, pressure, residence period, catalyst, etc., can be adjusted as per the product requirement.

3.2 Physico-chemical pretreatment

Application of this method on biomass modified it both physically and chemically using different techniques like steam explosion, ammonia treatment, liquid hot water, etc., detailed in the section below.

3.2.1 Steam explosion

This process is also called as autohydrolysis in which biomass is fractionated at high temperature(160–260°C) and pressure (0.69–4.83 MPa) conditions. High pressure saturated steam was initially used but sudden pressure reduction causes decompression of biomass that results in the disruption of lignin and degradation of hemicellulose content. The process efficacy depends on temperature, moisture content, residence time, and biomass particle size. This process has been demonstrated and applied at the pilot scale mainly for agricultural residues and hardwood. The advantage of this process is its cost efficiency and less energy requirement. However, the inhibitory compounds generation lowers the saccharification yield (Dheeran and Reddy 2018). Bhatia et al. (2021) utilized steam explosion (SE) and ionic liquid (ILs) pretreatment for *Miscanthus* for production of oligosaccharides and biofuel production. SE and combined SE and IL showed highest glucans recovery (about 86–91%) compared to the untreated biomass while IL and combined SE and IL resulted in solubilization of xylan (~ 47–88%), lignin (~ 64–74%), and acetyl groups (~ 17–100%).

3.2.2 Carbon dioxide (CO_2) explosion

This method is similar to steam explosion in terms of using supercritical CO_2 fluid at high pressure steam and enhanced the rate of enzymatic hydrolysis while being different from it as it does not produce inhibitory compounds and is cost intensive. This pretreatment method particularly increases the rate of enzymatic hydrolysis of the biomass and contains a good amount of moisture content as a rise in yield of hydrolysis is proportional to the biomass moisture content. Unavailability of the adequate data of techno-economic study in literature indicates the need of further studies associated to its reactor design for pilot scale (Anu et al. 2020). Serna et al. (2016) used supercritical CO_2 to pretreat rice husk at operating conditions of 80ºC, 270 bar, 75% moisture content with 2:1 water: ethanol mixture ratio for about 10 minutes resulted 90% lignin removal and 7 g/g of sugar yield. Supercritical CO_2 pretreatment was used for green coconut fiber to enhance its cellulose enzymatic hydrolysis. Decreased content of phenolic and wax, and a diminished degree of H-bond between lignin and cellulose after pretreatment were observed by SEM images and FTIR spectra (Putrino et al. 2020).

3.2.3 Ammonia pretreatment

This treatment method is basically divided in to two broad categories: aqueous and anhydrous. Aqueous ammonia pretreatment includes soaking in aqueous ammonia (SAA) and ammonia recycle percolation (ARP) while the anhydrous method involves ammonia fiber explosion (AFEX) and extractive ammonia (EA). The SAA method was earlier used in preparation of cattle feed but is now modified to pretreat cellulosic biomass while the ARP method includes biomass treatment with an ammonia solution (aqueous) in a flow through reactor. ARP requires less ammonia and water loadings compared to SAA and obtains 70–90% lignin and more than 50% hemicellulose removal during the process. However, the ammonia and water recycling increases the processing cost and energy requirement (Zhao et al. 2020). Aqueous ammonia methods involve the formation of ammonium hydroxide which disintegrates lignin and hemicellulose by splitting the alkali labile bonds and also swells the cell wall partially. The AFEX method was first originated by Bruce Dale in 1980 and utilized ammonia solutions (anhydrous) for the pretreatment of grasses identical to other methods of ammonia methods for wood pulping with some operational differences. The process was modified subsequently over the last few decades by Dale and his coworkers (Chundawat et al. 2013, Oconnor 1972). In the AFEX process, the anhydrous ammonia solution was exposed to biomass with variable water loadings at moderate temperature (60–100ºC) and pressure

(1.7–2.1 MPa) for a period of time (Kim et al. 2016). The biomass undergoes depressurization due to the release of ammonia from the reactor. In this method, ammonia permeates the cell wall and caused ammonolytic and hydrolytic reactions on the ester linkages of lignin carbohydrate complex inside the cell wall. There is no generation of inhibitors in AFEX so it does not need additional detoxification and also the inputs of water and ammonia is lower than required in the aqueous ammonia processes. EA is a new technique somewhat similar to the organosolv pretreatment developed to operate fractionation-based biorefinery in which liquid anhydrous ammonia is used at elevated liquid to solid loading at high pressure, and low water loading. This method extracts lignin selectively and also converts native recalcitrant cellulose I to degradable allomorph cellulose III. The formation of cellulose III during the anhydrous ammonia method is due to the infiltration of liquid ammonia in the crystalline region of cellulose and formed an ammonia-cellulose complex by substituting hydrogen bonds in the cellulose chains (Da Costa Sousa et al. 2016a, Da Costa Sousa et al. 2016b). This NH_3-cellulose brought remarkable change in the biomass structure and includes a reduction in the crystallinity and microcrystalline region size. After ammonia removal, recrystallization of the cellulose chains form cellulose III with a variable degree of crystallinity based on the pretreatment conditions (Zhao et al. 2020). In totality, it has been noticed that the aqueous ammonia method enhances enzymatic hydrolysis of biomass by lignin and hemicellulose removal while the anhydrous ammonia method increased the enzymatic hydrolyzis by causing ultrastructural and physicochemical changes in the biomass and transform the recalcitrant cellulose to its digestible form.

3.2.4 Liquid hot water

Also called as "hydrothermolysis", this process involves the biomass degradation by liquid hot water at temperature of about 150–240°C for a short residence period (≤ 50 minutes). Water under pressure penetrates and ruptures the cell wall, resulting into two products, that is, liquid hydrolysate (contains hemicellulose, acetic acid, furfural, minerals) and solid residues (contains cellulose, lignin, and few hemicellulose residuals). Pretreated biomass has increased pore size and surface area and contains cellulose fibrils and lignin particles on the cellulose surface. The carbohydrate content increases with a rise in temperature upto a certain temperature limit after which the temperature rise results in the degradation of the carbohydrates. A temperature of more than 230°C reduces the pretreated biomass pore size and surface area and restricts the enzymatic hydrolysis. The hydrolysis rate also depends on the type of feedstock used in the process (Bensah and Mensah 2018). The advantage of this process is that there is no requirement of acids or solvent and neutralization but inhibitor removal is one of the issues created during the process. Yan et al. (2021) performed liquid hot water pretreatment on four biomasses, that is, rubber wood, sugarcane bagasse, sorghum stalk, and cassava stalk. Using rubber wood as a substrate results the highest reducing sugar and hydrolysis yield of 3.92 g/l and 13.1%, respectively. Orange waste has been subjected to hydrothermolysis for butanol production at 140°C for 30 minutes and resulted in a butanol yield of about 42.3 g/kg of orange waste (Saadatinavaz et al. 2021).

3.2.5 Wet oxidation

It is mostly used for dried biomass, which is pretreated at a high temperature and pressure in the presence of an oxidizing agent (air or O_2) and water. An oxidizing agent oxidizes the compounds dissolved in water. The biomass is solubilized due to the occurrence of the reactive oxidation reactions. This method is effective for biomass pretreatment but its commercialization is restricted due to the large expense of the catalyst, oxidizing agent, equipments required, and also because of inhibitor formation (Vivek et al. 2019, Devi et al. 2021). Wet oxidation was applied on cassava peels, ulva algae, and water hyacinth at two different operating conditions, that is, 125°C, 1.5 bar for 45 minutes and 130°C, 1.8 bar for 75 minutes. Seventy-five percent pretreatment efficiency was showed by the water hyacinth at operating conditions of 125°C, 1.5 bar for 45 minutes (Ahou et al. 2020).

3.3 Chemical pretreatment

Chemical treatment of biomass disrupts it according to the properties of the chemical used (acid, alkali, oxidant, etc.) in the pretreatment process (Devi et al. 2022). The optimum conditions and mechanism of the different chemical methods have been delineated below.

3.3.1 Acid pretreatment

Acid pretreatment of biomass is an effective method in solubilizing the hemicellulose and cellulose. It is performed with diluted and concentrated acids. Acid, at a high concentration, yields high sugars at a moderate temperature but inhibitor formation (hydroxymethyl furfural, acetic acid, and furfural) during the process and equipment corrosion increases the cost of the process. Dilute acids modify the biomass structure and also solubilize the hemicellulose completely in the form of monomeric or oligomeric sugars. Dilute acid pretreatment can be carried out at a high temperature (> 160°C) for a short period or low temperature (< 160°C) for a longer time period. High temperature hydrolyze the sugars but also forms inhibitory compounds that are not good for downstream processing. Also neutralization of the acid, its recovery, salts disposal, and detoxification of the hydrolysate enhance the operating cost of this method. Gypsum formation during neutralization also creates an environmental problem (Jung and Kim 2015). Jang and Choi (2018) utilized 75% concentrated sulfuric acid for biomass pretreatment for a techno-economic analysis for butanol production indicated that pretreatment and hydrolysis constitutes about 27.6% of the fixed capital cost. Li et al. (2019b) uses diluted acid pretreatment (0.04N HCl) for butanol production from lignocellulosic biomass (10% solid loading of corn fiber, cotton stalk, soybean hull. and sugarcane bagasse) resulted in a glucose content of 53.7%, 44.4%, 58% and 32.5%, respectively after acid pretreatment.

3.3.2 Alkaline pretreatment

This method initially swells the biomass and breaks the ester and glycosidic chains that open the cellulose crystalline structure, solubilizes the hemicellulose, and delignifies the biomass. Different alkalis have been used for this process, that is, sodium hydroxide, potassium hydroxide, calcium hydroxide, etc. However, the limitation of this process involves the neutralization of the slurry and is not very effective in the case of biomass with high lignin content (Anu et al. 2020). Valles et al. (2021) uses sodium hydroxide for the pretreatment of rice straw to enhance its conversion for butanol production. The outcomes of the study reported the maximum butanol titer of 10.1 g/l after 72 h with almost complete glucose uptake. Chi et al. (2019) applied 1% (w/v) NaOH for pretreatment of rice straw for ABE fermentation resulted in a 68.6% of lignin removal and showed less hemicellulose loss.

3.3.3 Ozonolysis

The method involves the use of ozone for lignin removal from lignocellulosic biomass. This method is specific and selective as it does not affect the cellulose and hemicellulose content thereby increases the enzymatic hydrolysis. The method depends on various parameters, that is, ozone gas concentration, water content in reactor, and material particle size. The biomass is fractionated by oxidation caused by ozone and the hydroxyl radical formed during the dissociation of ozone in the presence of water (Devi et al. 2021). Shamjuddin et al. (2021) analyzed the kinetics and dynamics of the ozonolysis pretreatment of empty fruit bunch for sugar production. Results reported the maximum lignin degradation of 78% and glucose yield of 12% by weight after hydrolyzing the ozonated substrate.

3.3.4 Organic solvent pretreatment

This process involves the use of organic solvents (methanol, ethanol, ethylene glycol, acetone, etc.) in presence of organic acids or alkali (salicyclic acid, oxalic acid, etc.) as a catalyst for biomass pretreatment. Cleavage of lignin and hemicellulose bond takes place during the process which improved the overall surface area of cellulose (Mankar et al. 2021). The organic solvent pretreatment (65% of ethanol with 1% sulfuric acid) used on pine wood chips at 17°C for 1 h in a reactor for butanol fermentation. The total reducing sugar concentration in a hydrolysate was about 74.3 g/l which is three-fold that of the initial concentration (Li et al. 2018). The organic fraction of municipal solid waste (OFMSW) has been pretreated with 75% or 85% (v/v) of ethanol-water mixture and 0.5% of sulfuric acid (w/w) at 120°C, 150°C and 180°C for 0, 30, and 60 minutes. The highest yield of ABE fermentation of 155.73 g/kg of OFMSW was achieved at 120°C for 30 minutes using 85% ethanol (Farmanbordar et al. 2018).

3.3.5 Ionic liquids (ILs)

ILs are the salts made from the combination of organic cations and anions. It is non-volatile, non-flammable, possess good thermal and chemical stability, and have wide applications for biomass processing. The choice of cations and anions are very significant as the properties of ILs depend on it. They can fractionate the biomass in to its components by disrupting its non-covalent interactions but there are also some limitations for their large-scale processing as they are costly and the removal of ILs from the sugar mixture is quite challenging (Galbe and Wallberg 2019). Protic ionic liquid (monoethanolammonium acetate) has been used for the pretreatment of sugarcane bagasse, resulting in hemicellulose and cellulose yields of 49% and 78.6%, respectively, along with 60% lignin removal (Nakasu et al. 2021).

3.3.6 Deep eutectic solvents (DESs)

DESs are the green solvents prepared by Abbott et al. (2003) for the first time by mixing the H-bond donor (HBD) and acceptor (HBA) in an appropriate ratio. The reaction mechanism between the biomass and the DESs depends on the properties of HBD and HBA used and their molar ratios. The preparation of DESs is quite simple and does not need very complex purification techniques. Compared to the conventional solvents and ILs, these are non-toxic, biodegradable, and economical. Similar to ILs, its physico-chemical properties also have wider tuneability. Due to its unique properties, it efficiently delignified the biomass structure (Xu et al. 2021). Teh et al. (2019) developed a choline-based DES for concentrating hemicellulose in an oil palm empty fruit bunch. Results showed lignin, hemicellulose, and cellulose content of 0.98%, 83.82%, and 6.60% in DESs (choline: formic acid; 1:1.5) pretreated empty fruit bunch which is notably different from an untreated one. DESs were prepared from choline chloride and lactic acid by Liu et al. (2019) in the molar ratio of 1:9 to pretreat *Phyllostachys pubescens* and resulted in 94.39% lignin removal and 91% of cellulose recovery.

3.4 Biological pretreatment

In this method, the natural ability of a microorganism has been exploited to degrade lignocellulosic biomass and cell wall components of microalgae. It is divided into two categories: microbial (fungi, bacteria and microbial consortia) pretreatment and enzymatic pretreatment. It can be done by submerged cultivation or solid state cultivation systems by allowing the microbial cells to grow on biomass. The use of fungi is only limited to lignocellulosic biomass as no reports were found to treat microalgae with fungal cells (Carrilo-Reyes et al. 2016). Fungi has ability to degrade lignin by producing hydrolytic enzymes but it is a time-consuming process while bacteria has an advantage in terms of time consumption as the growth and microbial activity of bacteria is faster than fungi. The potential of lignin- degrading bacteria (LDB) is less compared to fungi. LDB mainly belongs to *actinomyces*, γ-proteobacteria and α-proteobacteria (Bugg et al. 2011). Hydrolytic enzymes by hydrolytic bacteria are important for microalgae pretreatment. Hydrolytic bacteria can also pretreat biomass mainly related to *Clostridium* sp., *Cellulomonas* sp., *Streptomyces* sp. and *Bacillus* sp., etc. Microbial consortia acting on biomass synergistically degrades the complex biomaterial that might be difficult to decompose individually. This method has advantage over the sole microbial treatment such as increased hydrolysis efficacy and productivity, enhanced adaptability, and better substrate utilization. However, maintaining optimum conditions for the synergistic action of microbial species involved is one of the challenges. The enzymatic method includes the use of lignolytic or hydrolytic enzymes in its crude or pure form for biomass pretreatment (Zabed et al. 2019). This method is effective for lignocellulosic biomass and microalgae decomposition but has some challenges like long-time consumption, non-selective lignin degradation, high cost of enzymes, and generation of various other products during the growth and metabolism of microorganism. These byproducts affect the overall efficiency of the bioconversion process (Sharma et al. 2019). Mondal et al. (2022) produced enzymes from fungal strains, that is, *Aspergillus niger* SKN1 and *Trameteshirsuta* SKH1 and used it for saccharification of wheat bran, sugarcane bagasse, and orange peel. Some more examples of 1G and 2G substrates pretreatment have been mentioned in Table 1.

Table 1. Some examples presented in literature on pretreatment of lignocellulosic and algal biomass.

S. no.	Pretreatment	Substrate	Operating conditions	Results	References
1.	Hot water pretreatment	Rubber wood	200°C at a rate of 5°C/minute for 30 minutes	Total reducing content is 3.92 g/l and hydrolysis yield of 13.1%	Yan et al. 2021
2.	Hot water pretreatment	Sugarcane bagasse	200°C at a rate of 5°C/minute for 30 minutes	Total reducing content is 2.98 g/l and hydrolysis yield of 9.92%	Yan et al. 2021
3.	Hot water pretreatment	Sorghum stalk	200°C at a rate of 5°C/minute for 30 minutes	Total reducing content is 2.70 g/l and hydrolysis yield of 9.0%	Yan et al. 2021
4.	Hot water pretreatment	Cassava stalk	200°C at a rate of 5°C/minute for 30 minutes	Total reducing content is 2.96 g/l and hydrolysis yield of 9.86%	Yan et al. 2021
5.	Deep Eutectic solvents (Choline chloride: lactic acid: 1:15)	Empty fruit bunch	Solid to liquid ratio is 1:10, 120°C for 8 h	61% (wt.) lignin removal	Tan et al. 2019
6.	Ultrasonic assisted pretreatment	*Saccharina japonica* (seaweed)	Biomass powder mixed with 0.1 M H_2SO_4 and extracted with ultrasonic apparatus at 40 KHz and 200 W power for 1h	Highest butanol yield and productivity is 0.26 g/g and 0.19 g/l.h, respectively	Fu et al. 2021
7.	Acid pretreatment	Algae	1M H_2SO_4 at 90°C heat for 30 minutes	Total ABE yield with pretreated was 0.244 g/g	Ellis et al. 2012
8.	Sequential alkali and acid pretreatment	*Chlorella vulgaris* JSC-6	1% NaOH and 3% H_2SO_4	Butanol yield is 13.1 g/l	Wang et al. 2016
9.	Acid pretreatment	Pine wood	0–2.5 % H_2SO_4 at 121°C for 1 h	Xylose yield was increased by 37.6%; saccharification percentage is 29.6% and butanol concentration of 11.6 g/l	Nanda et al. 2014
10.	Hybrid method (Dilute acid and aqueous ammonia)	Corn stover	1% dilute H_2SO_4 acid at 121°C for 120 minutes and 10% aqueous ammonia at 80 °C for 24 h	Lignin reduction rate increased (86.77%) from single method (45.84%) and produce 10.89 g/l butanol	Xiao et al. 2019

11.	Ultrasound-assisted dilute acid pretreatment	*Puerariae slag*	1% H_2SO_4 in 1:8 ratio at 126°C for 70 minutes and ultrasound at 40 KHz and 50 W power	Reducing sugar of 0.69 g/g of raw material and yield of 0.19 g/g of butanol obtained	Zhou et al. 2020
12.	Steam explosion	Vinegar residues	Saturated steam of 2.5 MPa, 222.9°C for 2 minutes	29.47% glucans conversion, 71.62% xylan conversion obtained	Xia et al. 2020
13.	Mechanical comminution	Spruce wood	Milled, chipped biomass (20 mm) treated with H_2O_2-acetic acid (1:1) at 80°C for 4 h and 0.15% (w/v) H_2SO_4 at 190 °C for 10 minutes	Butanol yield of 126.5 g/kg obtained	Yang et al. 2018
14.	Steam explosion	*Pinus radiate* sawdust	215°C and 180°C for 2 minutes and 21.5 minutes, respectively	Yields of glucose and xylose is 1.31% and 1.99% (215°C) and for 180°C is 0.97% and 1.45%	MacAskill et al. 2018
15.	Liquid Hot water	Sugarcane straw	195°C at 200 rpm for 10 minutes	80% of cellulose were preserved and 82% hemicellulose were solubilized	Pratto et al. 2020
16.	Autohydrolysis	Sorghum bagasse	8% (w/v) solid: liquid ratio at 150, 180, or 210°C for 30, 60, or 90 min	79% of hemicellulose and 20% of cellulose obtained at 210°C	Mirfakhar et al. 2020

4. Hydrolysis

Hydrolysis of biomass for butanol production can be done by chemicals or enzymes. Chemical hydrolysis is utilized mainly for hydrolyzing hemicellulose into fermentable sugars as the glycosidic bonds in it are more reactive than cellulose while enzymatic hydrolysis utilized to hydrolyze cellulose. In cellulose hydrolysis, three main enzymes are involved: endo-1,4-β-glucanase, exo-1,4-β-glucanase and β-glucosidase. Endoglucanase breaks the cellulose chain length by acting on its glycosidic bonds. Exoglucanase act on the reducing and non-reducing ends released cellobiose as major product. β-glucosidase acts on cellobiose and releases glucose to complete the hydrolysis process (Singh et al. 2021). The breakdown of cellobiose to glucose is the rate determining step. Cellulosic butanol is produced by *Clostridia* that solubilizes it to its fermentable sugars. *Clostridia* possess the characteristics due to which it utilizes the cellobiose efficiently which is not much utilized by *Saccharomyces* yeast in ethanol production and that can increase the cost of the overall process due to additional enzyme cost. This cost is saved in the case of cellulosic butanol hydrolysis. Enzymatic hydrolysis should be tunable with the growth of *Clostridia* for better efficiency of butanol production process. The concentration of oligomers and its type are the important factors that impact the butanol production. It has been observed that a sugar concentration less than threshold cannot shift the process from acidogenesis phase to the solventogenesis phase (Ezeji et al. 2005). *Clostridia* also possesses the ability to produce hemicellulolytic enzymes that degrades hemicellulose partially. It was reported by Shallom and Shoham (2003) that the presence of 14 hemicellulose genes in *C. acetobutylicum*, namely xylanases (3 in number), xylanosidase (4), arabinases (2), mannoses (4), and an acetyl xylanesterases (1). The chemical hydrolysis of hemicellulose can be done in both acidic and alkaline conditions. At optimum conditions of pH, temperature, and residence time, a complete recovery of hemicellulose is possible in the chemical hydrolysis process. Hemicellulose depolymerizes the hydrolysis of acetyl functional groups. The hydrolytic reactions occur in the hemicellulose hydrolysate due to the presence of ester groups, which leads to the formation of acetic acid. Other than oligomers, chemical hydrolysis are also accompanied with degradation products like hydroxymethyl furfural and furfural. Also, lignin degradation results in phenolic products, which affects the butanol efficiency (Amiri and Karimi 2018).

Another concept for cost effective butanol production involves the co-utilization of six-carbon and five-carbon sugars by *Clostridia*. When both sugars are present in a medium, *Clostridia* utilizes it by using catabolite repression mechanism in which the use of least preferred sugar is delimited by the most preferred sugar (Amiri and Karimi 2018). Enzymatic hydrolysis of pretreated wheat straw by a cocktail of cellulase and xylanases at 50°C and 150 rpm for 48 h at different biomass concentration (6%, 7.5%, 9%, and 10.5% w/v) and enzyme loading (5 FPU+10 IU, 10 FPU+20 IU, 15 FPU+30 IU and 20 FPU+40 IU cellulase + xylanase). Glucose and xylose concentration were increased from 23.86 g/l to 26.16 g/l and 13.49 g/l to 15.98 g/l, respectively, when enzyme loading increased from 5 FPU+10 IU to 10 FPU+20 IU. No further significant changes were observed after a further increase in enzyme loadings. A maximum butanol titer (13.14 g/l) was obtained at 15 FPU+30 IU cellulase + xylanases per gram of dried biomass (Qi et al. 2019). Saccharification of untreated apple pomace, and acid and alkali treated apple pomace were carried out at 13 FPU cellulase per gram of biomass at 50°C, 120 rpm for 72 h, resulted in the high glucose yields (16.2 and 14.8 g/L) with acid and alkali pretreated biomass compared to the untreated sample (Jin et al. 2019). The pretreated rice straw was hydrolyzed at 50°C, 150 rpm for 72 h by Valles et al. (2020) with solid loading and enzyme loading of 10% (w/v) and FPU g-dw-1, resulting in a glucose, xylose, and arabinose concentration of 17.68 g/l, 6.10 g/l and 0.39 g/l, respectively and butanol productivity of 0.040 g/l/h in separate hydrolysis and fermentation process. Seifollahi and Amiri (2019) conducted the simultaneous co-saccharificationand fermentation of cellulose oligomers obtained from chemical hydrolysis (1% dilute phosphoric acid at 120°C for 1 h) of hemicellulose, which resulted in 24.1 g/l ABE concentration. Pretreated eucalyptus sawdust was exposed to cellulase enzyme at operating conditions of 50°C, 150 rpm for 72 h with solid and enzyme loadings of 16% (w/w) and 25 FPU, respectively. Eighty-eight percent of the glucose concentration was achieved after enzymatic hydrolysis (Cebreiros et al. 2019).

5. Challenges and prospects

The challenges involved in the butanol production process include the economics of the process, low titer, and yield of the butanol. Near about 66% of the cost comes from the substrate utilized for fermentation (especially in case of 1G butanol) (Jiang et al. 2015). In 2G butanol production, pretreatment of the substrate to fermentable sugar is also expensive, releasing mixed sugars in the hydrolysate and generates inhibitory compounds during the process. Due to the carbon catabolite repression mechanism of the microorganism (*Clostridia*), the prime sugar present in the hydrolysate suppresses the least sugar utilization that affects the efficiency of the process. Future prospect includes the development of most appropriate and cost-effective pretreatment method for butanol production. Developing strains by adaptive evolution or genetic engineering with such characters that utilize both pentose and hexose sugars efficiently. Compared to ethanol, butanol fermentation consumes the substrate almost twice than in ethanol fermentation. Other problems like toxicity of the solvent and other product generation (ethanol, acetone) along with butanol need to be addressed to improve the overall yield and titer of butanol production. Metabolic-engineered strains have developed high solvent tolerance and are non-acetone forming, but still need to be tested over a period of time under different operating conditions to improve them further to get more industrially viable strains (Birgen et al. 2019, Patil et al. 2019).

6. Conclusion

Butanol is an advanced biofuel that possess properties which made it better than ethanol and methanol. For its production from biomass resources, pretreatment and hydrolysis are the necessary steps but some challenges like inhibitor generation, cost of the substrate, and its conversion process, etc., made its commercialization difficult. For its large-scale production, it has been concluded that low-cost substrate, more tolerant strains, more specific pretreatment methods, advanced fermentative tools, and more integrated approaches are required.

References

Abbott, Andrew, P., Glen Capper, David L. Davies, Raymond K. Rasheed, and Vasuki Tambyrajah. (2003). Novel solvent properties of choline chloride/urea mixtures. Chemical Communications, 1: 70–71.

Ahou, YaoviSylvestre, Margareta Novia AsihChristami, SaryAwad, Cindy RiantiPriadi, Lamine Baba-Moussa, SetyoSarwantoMoersidik, and Yves Andres. (2020). Wet oxidation pretreatment effect for enhancing bioethanol production from cassava peels, water hyacinth, and green algae (Ulva). In AIP Conference Proceedings, 2255(1): 030039. AIP Publishing LLC.

Amiri, Hamid, and Keikhosro Karimi. (2018). Pretreatment and hydrolysis of lignocellulosic wastes for butanol production: challenges and perspectives. Bioresource Technology, 270: 702–721.

Anu, Kumar, Anil, Alexander Rapoport, Gotthard Kunze, Sanjeev Kumar, Davender Singh, and Bijender Singh. (2020). Multifarious pretreatment strategies for the lignocellulosic substrates for the generation of renewable and sustainable biofuels: A review. Renewable Energy.

Bensah, Edem C., and Moses Y. Mensah. (2018). Emerging physico-chemical methods for biomass pretreatment. Fuel Ethanol Production from Sugarcane.

Bhatia, Rakesh, Jai B. Lad, Maurice Bosch, David N. Bryant, David Leak, Jason P. Hallett, Telma T. Franco, and Joe A. Gallagher. (2021). Production of oligosaccharides and biofuels from Miscanthus using combinatorial steam explosion and ionic liquid pretreatment. Bioresource Technology, 323: 124625.

Birgen, Cansu, Peter Dürre, Heinz A. Preisig, and Alexander Wentzel. (2019). Butanol production from lignocellulosic biomass: revisiting fermentation performance indicators with exploratory data analysis. Biotechnology for Biofuels, 12(1): 1–15.

Bugg, Timothy D.H., Mark Ahmad, Elizabeth M. Hardiman, and Rahul Singh. (2011). The emerging role for bacteria in lignin degradation and bio-product formation. Current opinion in biotechnology 22(3): 394–400.

Bundhoo, Zumar M.A., and RomeelaMohee. (2018a). Ultrasound-assisted biological conversion of biomass and waste materials to biofuels: A review. Ultrasonics Sonochemistry, 40: 298–313.

Bundhoo, Zumar MA. "Microwave-assisted conversion of biomass and waste materials to biofuels." Renewable and Sustainable Energy Reviews 82 (2018b): 1149-1177.

Carrillo-Reyes, Julián, Martin Barragán-Trinidad, and Germán Buitrón. "Biological pretreatments of microalgal biomass for gaseous biofuel production and the potential use of rumen microorganisms: A review." Algal research 18 (2016): 341-351.

Cebreiros, Florencia, Mario Daniel Ferrari, and Claudia Lareo. "Cellulose hydrolysis and IBE fermentation of eucalyptus sawdust for enhanced biobutanol production by Clostridium beijerinckii DSM 6423." Industrial Crops and Products 134 (2019): 50-61.

Chi, Xue, Jianzheng Li, Shao-Yuan Leu, Xin Wang, Yafei Zhang, and Ying Wang. (2018). Features of a staged acidogenic/solventogenic fermentation process to improve butanol production from rice straw. Energy & Fuels, 33(2): 1123–1132.

Chundawat, S.P.S., Bals, B., Campbell, T., Sousa, L., Gao, D., and Jin, M. (2013). Primer on ammonia fiber expansion pretreatment. Aqueous Pretreatment of Plant Biomass for Biological and Chemical Conversion to Fuels and Chemicals, 169–200.

da Costa Sousa, Leonardo, Marcus Foston, Vijay Bokade, Ali Azarpira, Fachuang Lu, Arthur J. Ragauskas, John Ralph, Bruce Dale, and Venkatesh Balan. (2016a). Isolation and characterization of new lignin streams derived from extractive-ammonia (EA) pretreatment. Green Chemistry, 18(15): 4205–4215.

da Costa Sousa, Leonardo, MingjieJin, Shishir PS Chundawat et al. (2016b). Next-generation ammonia pretreatment enhances cellulosic biofuel production. Energy & Environmental Science, 9(4): 1215–1223.

Devi, Arti, Anita Singh, SomvirBajar, Deepak Pant, and Zaheer Ud Din. (2021). Ethanol from lignocellulosic biomass: An in-depth analysis of pre-treatment methods, fermentation approaches and detoxification processes. Journal of Environmental Chemical Engineering, 9(5): 105798.

Devi, Arti, SomvirBajar, HavleenKour, Richa Kothari, Deepak Pant, and Anita Singh. (2022). Lignocellulosic biomass valorization for bioethanol production: a circular bioeconomy approach. Bioenergy Research, 1–22.

Dheeran, Pratibha, and Lalini Reddy. (2018). Biorefining of lignocelluloses: an opportunity for sustainable biofuel production. In Biorefining of Biomass to Biofuels, pp. 1–23. Springer, Cham.

Dhyani, Vaibhav, and Thallada Bhaskar. (2018). A comprehensive review on the pyrolysis of lignocellulosic biomass. Renewable Energy, 129: 695–716.

Duque,Aleta,PalomaManzanares,andMercedesBallesteros.(2017).Extrusionasapretreatment for lignocellulosic biomass: Fundamentals and applications. Renewable Energy, 114: 1427–1441.

Dutta, Sambit, Shiladitya Ghosh, Dinabandhu Manna, and Ranjana Chowdhury. (2021). Energy and environmental performance of a near-zero-effluent rice straw to butanol production plant. Indian Chemical Engineer, 63(2): 139–151.

Ellis, Joshua T., Neal N. Hengge, Ronald C. Sims, and Charles D. Miller. (2012). Acetone, butanol, and ethanol production from wastewater algae. Bioresource Technology, 111: 491–495.

Ezeji, T.C., Nasib Qureshi, and H.P. Blaschek. (2005). Continuous butanol fermentation and feed starch retrogradation: butanol fermentation sustainability using *Clostridium beijerinckii* BA101. Journal of Biotechnology, 115(2): 179–187.

Farmanbordar, Sara, Hamid Amiri, and Keikhosro Karimi. (2018). Simultaneous organosolv pretreatment and detoxification of municipal solid waste for efficient biobutanol production. Bioresource Technology, 270: 236–244.
Fu, Hongxin, Jialei Hu, Xiaolong Guo, Jun Feng, Shang-Tian Yang, and Jufang Wang. (2021). Butanol production from Saccharina japonica hydrolysate by engineered *Clostridium tyrobutyricum*: The effects of pretreatment method and heat shock protein overexpression. Bioresource Technolog, 335: 125290.
Galbe, Mats, and Ola Wallberg. (2019). Pretreatment for biorefineries: A review of common methods for efficient utilisation of lignocellulosic materials. Biotechnology for Biofuels 12(1) : 1–26.
Gatt, Etienne, Luc Rigal, and Virginie Vandenbossche. (2018). Biomass pretreatment with reactive extrusion using enzymes: A review. Industrial Crops and Products, 122: 329–339.
Gottumukkala, Lalitha Devi, Anil K. Mathew, Amith Abraham, and Rajeev Kumar Sukumaran. (2019). Biobutanol production: microbes, feedstock, and strategies. In Biofuels: Alternative Feedstocks and Conversion Processes for the Production of Liquid and Gaseous Biofuels, pp. 355–377. Academic Press.
Haldar, Dibyajyoti, and Mihir Kumar Purkait. (2021). A review on the environment-friendly emerging techniques for pretreatment of lignocellulosic biomass: Mechanistic insight and advancements. Chemosphere, 264: 128523.
Jang, Myung-Oh, and Gyunghyun Choi. (2018). Techno-economic analysis of butanol production from lignocellulosic biomass by concentrated acid pretreatment and hydrolysis plus continuous fermentation. Biochemical Engineering Journal, 134: 30–43.
Jędrzejczyk, Marcin, Emilia Soszka, MartynaCzapnik, Agnieszka M. Ruppert, and Jacek Grams. (2019). Physical and chemical pretreatment of lignocellulosic biomass." In Second and Third Generation of Feedstocks, pp. 143–196. Elsevier.
Jiang, Yu, Jinle Liu, Weihong Jiang, Yunliu Yang, and Sheng Yang. (2015). Current status and prospects of industrial bio-production of n-butanol in China. Biotechnology Advances, 33(7): 1493–1501.
Jin, Qing, Nasib Qureshi, Hengjian Wang, and Haibo Huang. (2019). Acetone-butanol-ethanol (ABE) fermentation of soluble and hydrolyzed sugars in apple pomace by *Clostridium beijerinckii* P260. Fuel, 244: 536–544.
Jung, Young Hoon, and KyoungHeon Kim. (2015). Acidic pretreatment. In Pretreatment of biomass, pp. 27–50. Elsevier.
Kim, Jun Seok, Y.Y. Lee, and Tae Hyun Kim. (2016). A review on alkaline pretreatment technology for bioconversion of lignocellulosic biomass. Bioresource Technology, 199: 42–48.
Kumar, R., Strezov, V., Weldekidan, H., He, J., Singh, S., Kan, T., and Dastjerdi, B. (2020). Lignocellulose biomass pyrolysis for bio-oil production: A review of biomass pre-treatment methods for production of drop-in fuels. Renewable and Sustainable Energy Reviews, 123: 109763.
Li, Jing, Suan Shi, Maobing Tu, Brain Via, FubaoFuelbio Sun, and Sushil Adhikari. (2018). Detoxification of organosolv-pretreated pine prehydrolysates with anion resin and cysteine for butanol fermentation. Applied biochemistry and Biotechnology, 186(3): 662–680.
Li, Yuqiang, Wei Tang, Yong Chen, Jiangwei Liu, and F. Lee Chia-fon. (2019a). Potential of acetone-butanol-ethanol (ABE) as a biofuel. Fuel, 242: 673–686.
Li, Jing, Yinming Du, Teng Bao, Jie Dong, Meng Lin, Hojae Shim, and Shang-Tian Yang. (2019b). n-Butanol production from lignocellulosic biomass hydrolysates without detoxification by Clostridium tyrobutyricum Δack-adhE2 in a fibrous-bed bioreactor. Bioresource Technology, 289: 121749.

Liu, Qian, Tao Yuan, Qin-jin Fu, Yuan-yuan Bai, Feng Peng, and Chun-li Yao. (2019). Choline chloride-lactic acid deep eutectic solvent for delignification and nanocellulose production of moso bamboo. Cellulose 26(18): 9447–9462.

MacAskill, J.J., Suckling, I.D., Lloyd, J.A., and Manley-Harris, M. (2018). Unravelling the effect of pretreatment severity on the balance of cellulose accessibility and substrate composition on enzymatic digestibility of steam-pretreated softwood. Biomass and Bioenergy, 109: 284–290.

Mankar, Akshay R., Ashish Pandey, Arindam Modak, and K.K. Pant. (2021). Pre-treatment of Lignocellulosic biomass: A review on recent advances. Bioresource Technology, 125235.

Mirfakhar, Moein, Mohammad Ali Asadollahi, Hamid Amiri, and Keikhosro Karimi. (2020). Co-fermentation of hemicellulosic hydrolysates and starch from sweet sorghum by *Clostridium acetobutylicum*: A synergistic effect for butanol production. Industrial Crops and Products, 151: 112459.

Mondal, Subhadeep, Sourav Santra, SubhamRakshit, Suman Kumar Halder, Maidul Hossain, and Keshab Chandra Mondal. (2022). Saccharification of lignocellulosic biomass using an enzymatic cocktail of fungal origin and successive production of butanol by *Clostridium acetobutylicum*. Bioresource Technology, 343: 126093.

Nakasu, P.Y.S., Pin, T.C., Hallett, J.P., Rabelo, S.C., and Costa, A.C. (2021). In-depth process parameter investigation into a protic ionic liquid pretreatment for 2G ethanol production. Renewable Energy, 172: 816–828.

Nanda, Sonil, Ajay K. Dalai, and Janusz A. Kozinski. (2014). Butanol and ethanol production from lignocellulosic feedstock: biomass pretreatment and bioconversion. Energy Science & Engineering, 2(3): 138–148.

Oconnor, James J. (1972). Ammonia explosion pulping-new fiber separation process. Tappi, 55(3): 353–358.

Patil, Ravichandra C., Pravin G. Suryawanshi, RupamKataki, and Vaibhav V. Goud. (2019), Current challenges and advances in butanol production. Sustainable Bioenergy, 225–256.

Pratto, Bruna, Vijaya Chandgude, Ruy de Sousa Junior, Antonio José Gonçalves Cruz, and Sandip Bankar. (2020). Biobutanol production from sugarcane straw: defining optimal biomass loading for improved ABE fermentation. Industrial Crops and Products, 148: 112265.

Putrino, Fernando Marques, Marcela Tedesco, Renata Barbosa Bodini, and Alessandra Lopes de Oliveira. (2020). Study of supercritical carbon dioxide pretreatment processes on green coconut fiber to enhance enzymatic hydrolysis of cellulose. Bioresource Technology, 309: 123387.

Qi, Gaoxiang, Dongmei Huang, Jianhui Wang, Yu Shen, and Xu Gao. (2019). Enhanced butanol production from ammonium sulfite pretreated wheat straw by separate hydrolysis and fermentation and simultaneous saccharification and fermentation. Sustainable Energy Technologies and Assessments, 36: 100549.

Raquez, Jean-Marie, Ramani Narayan, and Philippe Dubois. (2008). Recent advances in reactive extrusion processing of biodegradable polymer-based compositions. Macromolecular Materials and Engineering, 293(6): 447–470.

Saadatinavaz, Fateme, Keikhosro Karimi, and Joeri FM Denayer. (2021). Hydrothermal pretreatment: An efficient process for improvement of biobutanol, biohydrogen, and biogas production from orange waste via a biorefinery approach. Bioresource Technology, 341: 125834.

Seifollahi, Mehran, and Hamid Amiri. (2020). Enhanced production of cellulosic butanol by simultaneous co-saccharification and fermentation of water-soluble cellulose oligomers obtained by chemical hydrolysis. Fuel, 263: 116759.

Serna, L.V. Daza, OrregoAlzate, C.E., and Cardona Alzate, C.A. (2016). Supercritical fluids as a green technology for the pretreatment of lignocellulosic biomass. Bioresource Technology, 199: 113–120.

Shallom, Dalia, and Yuval Shoham. (2003). Microbial hemicellulases. Current Opinion in Microbiology, 6(3): 219–228.

Shamjuddin, Amnani, Nurul Suhada Ab Rasid, MakamMba Michele Raissa, Muhammad Anif Abu Zarin, Wan Nor Nadyaini Wan Omar, ArdiyansyahSyahrom, Mohd Al FatihhiMohdSzaliJanuddi, and Nor Aishah Saidina Amin. (2021). Kinetic and dynamic analysis of ozonolysis pre-treatment of empty fruit bunch in a well-mixed reactor for sugar production. Energy Conversion and Management, 244: 114526.

Sharma, Hem Kanta, Chunbao Xu, and Wensheng Qin. (2019). Biological pretreatment of lignocellulosic biomass for biofuels and bioproducts: an overview. Waste and Biomass Valorization, 10(2): 235–251.

Singh, Anita, SomvirBajar, Arti Devi, and Deepak Pant. (2021). An overview on the recent developments in fungal cellulase production and their industrial applications. Bioresource Technology Reports, 14: 100652.

Tan, Yee Tong, Gek Cheng Ngoh, and Adeline Seak May Chua. (2019). Effect of functional groups in acid constituent of deep eutectic solvent for extraction of reactive lignin. Bioresource Technology, 281: 359–366.

Teh, Soek Sin, Soh KheangLoh, and Siau Hui Mah. (2019). Development of choline-based deep eutectic solvents for efficient concentrating of hemicelluloses in oil palm empty fruit bunches. Korean Journal of Chemical Engineering, 36(10): 1619–1625.

Valles, A., Capilla, M., Álvarez-Hornos, F.J., García-Puchol, M., San-Valero, P. and Gabaldón, C. (2021). Optimization of alkali pretreatment to enhance rice straw conversion to butanol. Biomass and Bioenergy, 150: 106131.

Valles, Alejo, F. Javier Álvarez-Hornos, Vicente Martínez-Soria, Paula Marzal, and Carmen Gabaldón. (2020). Comparison of simultaneous saccharification and fermentation and separate hydrolysis and fermentation processes for butanol production from rice straw. Fuel, 282: 118831.

Vivek, Narisetty, Lakshmi M. Nair, Binoop Mohan, Salini Chandrasekharan Nair, Raveendran Sindhu, Ashok Pandey, NarasinhaShurpali, and Parameswaran Binod. (2019). Bio-butanol production from rice straw–recent trends, possibilities, and challenges. Bioresource Technology Reports 7: 100224.

Wang, Yue, Wanqian Guo, Chieh-Lun Cheng, Shih-Hsin Ho, Jo-Shu Chang, and Nanqi Ren. (2016). Enhancing bio-butanol production from biomass of Chlorella vulgaris JSC-6 with sequential alkali pretreatment and acid hydrolysis. Bioresource Technology, 200: 557–564.

Xia, Menglei, Mingmeng Peng, Danni Xue, Yang Cheng, Caixia Li, Di Wang, Kai Lu, Yu Zheng, Ting Xia, and Min Wang. (2020). Development of optimal steam explosion pretreatment and highly effective cell factory for bioconversion of grain vinegar residue to butanol. Biotechnology for Biofuels, 13(1): 1–17.

Xiao, Min, Lan Wang, Youduo Wu, Chi Cheng, Lijie Chen, Hongzhang Chen, and Chuang Xue. (2019). Hybrid dilute sulfuric acid and aqueous ammonia pretreatment for improving butanol production from corn stover with reduced wastewater generation. Bioresource Technology, 278: 460–463.

Xu, Huanfei, Jianjun Peng, Yi Kong, Yaoze Liu, ZhenningSu, Bin Li, Xiaoming Song, Shiwei Liu, and Wende Tian. (2020). Key process parameters for deep eutectic solvents pretreatment of lignocellulosic biomass materials: A review. Bioresource Technology, 310: 123416.

Yan, Xu, Dongna Li, Xiaojun Ma, and Jianing Li. (2021). Bioconversion of renewable lignocellulosic biomass into multicomponent substrate via pressurized hot water pretreatment for bioplastic polyhydroxyalkanoate accumulation. Bioresource Technology, 339: 125667.

Yang, Haiping, Rong Yan, Hanping Chen, Dong Ho Lee, and Chuguang Zheng. (2007). Characteristics of hemicellulose, cellulose and lignin pyrolysis. Fuel 86(12-13): 1781–1788.

Yang, Ming, Minyuan Xu, Yufei Nan, SuviKuittinen, Md. Kamrul Hassan, JoukoVepsäläinen, Donglin Xin, Junhua Zhang, and Ari Pappinen. (2018). Influence of size reduction treatments on sugar recovery from Norway spruce for butanol production. Bioresource Technology, 257: 113–120.

Zabed, Hossain M., SuelyAkter, Junhua Yun, Guoyan Zhang, Faisal N. Awad, Xianghui Qi, and J.N. Sahu. (2019). Recent advances in biological pretreatment of microalgae and lignocellulosic biomass for biofuel production. Renewable and Sustainable Energy Reviews, 105: 105–128.

Zadeh, Zahra Echresh, Ali Abdulkhani, Omar Aboelazayem, and BasudebSaha. (2020). Recent insights into lignocellulosic biomass pyrolysis: A critical review on pretreatment, characterization, and products upgrading. Processes 8(7): 799.

Zhao, Chao, Qianjun Shao, and Shishir P.S. Chundawat. (2020). Recent advances on ammonia-based pretreatments of lignocellulosic biomass. Bioresource Technology, 298: 122446.

Zheng, Jun, and Lars Rehmann. (2014). Extrusion pretreatment of lignocellulosic biomass: a review. International Journal of Molecular Sciences, 15(10): 18967–18984.

Zhou, Zhi-you, Shang-tian Yang, Curtis D. Moore, Qing-hua Zhang, Shuai-ying Peng, and Han-guang Li. (2020). Acetone, butanol, and ethanol production from puerariae slag hydrolysate through ultrasound-assisted dilute acid by *Clostridium beijerinckii* YBS3. Bioresource Technology, 316: 123899.

11

Genetic Engineering in Butanol Production:

Recent Trends

Japleen Kaur,[1] *Zaheer Ud Din Sheikh,*[1] *Anita Singh,*[1,*] *Somvir Bajar*[2] and *Meenakshi Suhag*[3]

1. Introduction

Microbial engineering enables biologically emanated alternates concerning diesel, gasoline, and aviation fuel (Fischer et al. 2008). Most of the analysis has concentrated on ethanol as biogasoline. Several other biofuels increase energy density, lower the freezing point, and enhance compatibility with the regular fuel storage and distribution infrastructure (Ramos et al. 2002). Next-generation biofuels, such as long-chain alcohols, can be synthesized by microorganisms. These biofuels evolved as supplements or drop-in replacements for existing petroleum fuels. Next-generation biofuels are very advantageous, but they are primarily toxic to microbial species. The inherently low tolerance of bacteria limits their overall production potential. Microorganisms that persist in a hydrocarbon-rich environment have been isolated, but these are rarely suitable for biofuel production (Dunlop et al. 2011). Now, it is possible to transfer the tolerance

[1] Department of Environmental Sciences, Central University of Jammu, Jammu-180011, Jammu and Kashmir, India.
[2] Department of Environmental Science and Engineering, J.C. Bose University of Science and Technology, YMCA, Faridabad-121006, Haryana, India.
[3] Institute of Environmental Studies, Kurukshetra University, Kurukshetra.
* Corresponding author: anitasaharan@gmail.com

mechanism to a suitable biofuel-producing strain. The host for biofuel production should be a well-studied organism with the availability of genetic tools which could be efficiently engineered for butanol tolerance and biofuel production (Dunlop et al. 2011).

Among all the next-generation biofuels, butanol gets more attention because of its dominance over other biofuels. Biobutanol is a four carbon alcohol having a straight-chain structure with the presence of –OH (hydroxyl) group at the terminal carbon. 1-Butanol contains a broad range of commercial applications. It is often used for latex surface coatings. The other by-products of butanol are butyl acetate, butyl glycol ether, and plasticizers. It is proves an excellent solvent for the production of industrial antibiotics, hormones, and vitamins. Due to lower vapor pressure and properties similar to gasoline, butanol is utilized as pure fuel in vehicles without engine modifications. Butanol is less corrosive than ethanol and contains 25% more energy (Dahman et al. 2019). All these advantages permit butanol to overcome ethanol as an alternative fuel. Butanol production is carried out either by chemical synthesis via fossil fuel or from microbial fermentation. Due to rapid fossil fuel depletion, the current focus will move towards biobutanol production through microbial fermentation.

Louis Pasteur first reported the production of butanol through the microbial fermentation process in 1861, and now it has been extensively studied for the past few years (Ndaba et al. 2015). Historically n-butanol is produced by species of *Clostridium* through Acetone-Butanol-Ethanol (ABE) fermentation which results in the formation of the acidogenesis and solventogenesis phase. It is the most common biological route for the production of butanol, in which the main products are acetone, butanol, and ethanol with their generation ratio 3:6:1 in the fermentation (Xue et al. 2019). During the first phase of ABE fermentation, the growing native strain synthesizes butyric acid and acetic acid. When the strain reaches the end exponential phase, it reassimilates primary metabolites into acetone, ethanol, and n-butanol.

To improve the production of n-butanol, various approaches have been implemented nowadays in which there is inactivation of the acid-synthesis pathways, supplementation of butyric acid in the solventogenesis phase, increase in the production efficiency of the acidogenesis phase, and further reconstructions in strain with the help of metabolic pathway engineering (Birgen et al. 2019). Although the generation of 1-butanol through the ABE fermentation process is generally restricted by the lack of physiological information and genetic techniques for the *Clostridium* species, various synthetic biology approaches has been applied in non-native strains *in vitro* where the microbial species are genetically modified to produce n-butanol with a high yield with increased tolerance level (Liang et al. 2020).

Due to the vast knowledge on it and the availability of computational biology tools for genetic modifications, *E. coli* is considered as the model organism for butanol production (Koppolu et al. 2016). All the genomic data of *E. coli* is available for genetic modifications. *E. coli* serve to be an inherent host for genetic and metabolic engineering by expressing and cloning biofuel forming genes to produce alcohols via biosynthetic pathways for amino acids (e.g., isoleucine, leucine, phenylalanine, threonine, and valine (Adrio et al. 2010)).

The new trends of metabolic engineering and genetic engineering of butanol production have overcome many limitations and increased the overall productivity and yield. Metabolic engineering targets both native butanol-producing bacteria and other potential microbial strains capable of fabricating butanol synthesis pathways. Metabolic and genetic studies usually focus on high cell density, aerobic conditions tolerance, tolerance to alcohols, and primary utilization of lignocellulosic biomass (Nanda et al. 2017). Furthermore, a highly advanced technique such as recombinant DNA technology, proteomics, and transcriptomics have been used to reshape the targeted metabolic pathways in butanol-producing microbes (Gong et al. 2012).

Over the last few years, many new approaches have been proposed for better production such as phenotype simulation of microorganisms in various distinct environmental and genetic conditions. The *in silico*-aided metabolic engineering shows promising results due to its accuracy in predicting the effect of manipulation of target gene (Bro et al. 2006). These tools have been applied to genetically modified organisms such as *S. cerevisiae, B.* subtilis, and *E. coli.* Figure 1 shows the use of genetic advancement technology for butanol production.

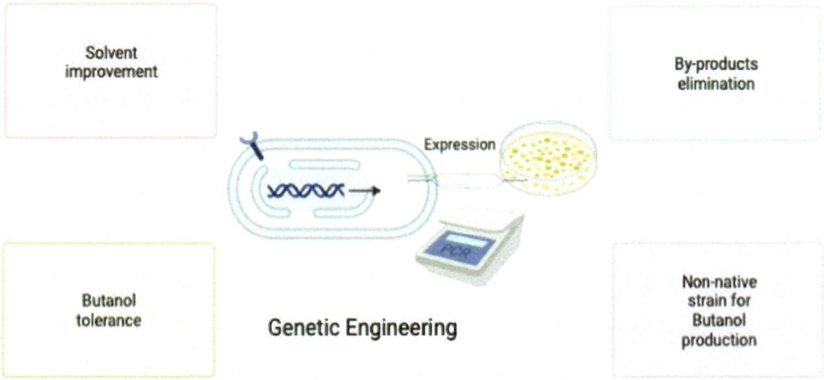

Figure 1. The role of genetic and metabolic engineering at the cellular level. Alterations in the cellular level to improve butanol tolerance and restoration of metabolic pathways in non-native strains.

2. Mechanism of butanol production through *Clostridium*

Butanol-producing strains are broadly classified into two different categories: the wild type and the genetically modified butanol-producing strain. The synthesis of butanol is exclusively done by the genus *Clostridium* (Xin et al. 2018). They are known as strict anaerobe, rod-shaped, spore-forming bacteria. It was studied that very few species of *Clostridia* under suitable conditions produce a significant amount of butanol. To overcome this limitation, genetically engineered non-*Clostridium* strains are being used for butanol production. Biphasic fermentation is considered the typical characteristic for the synthesis of butanol (Birgen et al. 2019). It is comprises of two different phases: acidogenic phase and solventogenic phase. By following different pathways, pentoses and hexoses sugars metabolize into pyruvate. By using the Embden-Meyerhof-pathway (EMP), hexose sugar metabolizes into 2 mol of pyruvate with the total production of 2 mol of NADH and ATP, and pentose sugar is degraded into pyruvate through glycolic pentose phosphate pathways (Kolesinska et al. 2019). When pyruvate is formed in the metabolic synthesis pathways, strain enters into an acidogenic phase in which through an exponential growth process, butyrate is formed (Gheshlaghi et al. 2009). The important switch in the whole process of butanol synthesis is the conversion of acetyl-coenzyme, which differentiates solvent- and acid-producing pathways.

The conversion of acetate from acetyl-CoA by acetate kinase and phosphotransacetylase leads to the formation of butyrate. The whole process takes place in a fermentation broth with favorable conditions. The external pH of the broth is decreased due to the accumulation of acid. The organic solvents such as acetone, butanol, and ethanol form in the ratio 3:6:1 in the solventogenic phase (Nanda et al. 2017). In the whole process of the butanol synthetic pathway, several necessary enzymes play an essential role (Ezeji et al. 2007). The conversion of acetyl-CoA to biobutanol involves six essential genes: thiolase *(thl)*, 3-hydroxybutyryl-CoA dehydrogenase *(hbd)*,crotonase *(crt)*, butyryl-CoA dehydrogenase *(bcd)*, butanol dehydrogenase *(bdh)*, and butyraldehyde dehydrogenase (Bao et al. 2021). Figure 2 shows the butanol synthesis pathways in *C. acetobutylicum*. For both acidogenic and solventogenic production, the thiolase enzyme plays key role (Inui et al. 2008). Electron flow also plays a vital role in the EMP metabolic pathway, which straight way regulating the formation of acetate, butanol, and ethanol formation (Gheshlaghi et al. 2009). The amount of biobutanol production depends on the NADH molecules. The generation of ATP is seen during acidogenic phase whereas the NADPH molecule induction is noticed in the solventogenic phase. In *Clostridium*, substrate utilization is seen in diverse forms as it can utilize a wide variety of carbon substrates such as polysaccharides, monosaccharides, and oligosaccharides. Conventional substrates for

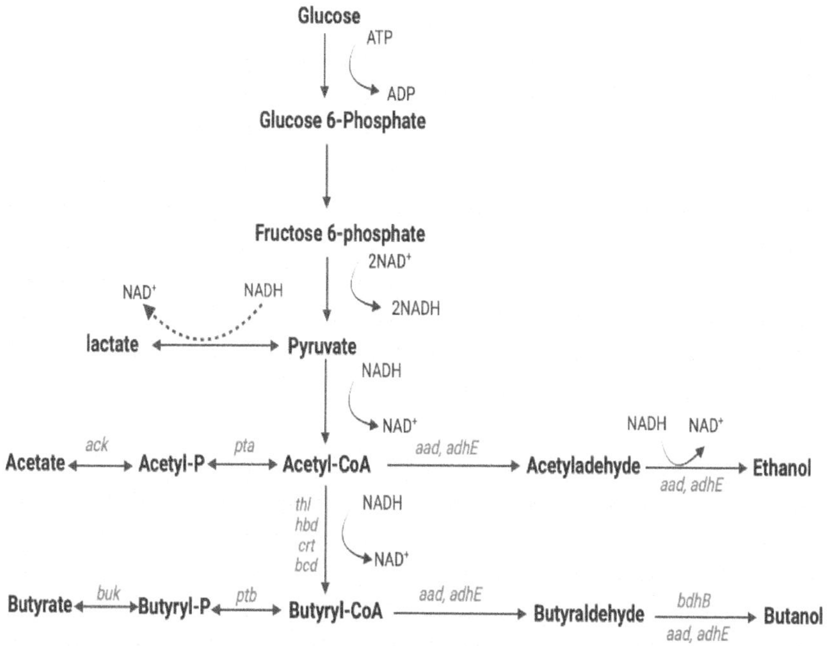

Figure 2. Butanol synthesis pathways in *C. acetobutylicum* with essential genes. The butanol forming genes in *C. acetobutylicum* involve thiolase *(thl)*, 3-hydroxybutyryl-CoA dehydrogenase *(hbd)*, crotonase *(crt)*, butyryl-CoA dehydrogenase *(bcd)*, butanol dehydrogenase *(bdh)*, and butyraldehyde dehydrogenase.

butanol fermentation include corn, potato, molasses but these substrates are not economically fit for butanol synthesis (Wen et al. 2014). To overcome this, renewable substrates, specifically lignocellulosic feedstock has been used, which is cost-effective and sustainable for the fermentation of butanol. For the utilization of lignocellulosic feedstock, hydrolysis and pretreatment plays immense role. The pretreatment process has drawback as it forms some inhibitory components such as alkaline peroxides, furfural, formic acid, and acetic acid (Gunasekaran et al. 2021).

3. Genetic and metabolic engineering in the butanol-producing strain

Various genetic and metabolic engineering pathways have been integrated in the non-native butanol-producing strains. Genetic engineering tools are the most efficient and reliable tools. Overexpressing or inserting a gene of interest correlated with the solvent formation and gene deletion with acidogenesis, that is, acid formation, are the most recurrent reported methods for butanol synthesis (Nawab et al. 2020). These days, various novel strategies have been implemented for reconstructing strains in which the most frequently practised approach is the integral plasmid approach, which disrupts metabolic pathways. For butanol production, two types of vectors (replicative and non-replicative plasmids) have been evolved for gene inactivation in *C. acetobutylicum*. For a more effective tool for targeting gene inactivation, the group II intron-based TargeTron technique has been developed in *C. acetobutylicum*, which replace low frequencies of recombination and time bond for double-crossing over-integration. Furthermore, other genetic engineering tools such as antisense RNA are advantageous (Joseph et al. 2018). The antisense RNA is a flexible and efficient tool for shaping microorganism metabolism and sustains efficiency in gene inactivation in metabolic engineering pathways. It is highly advantageous as it can alter protein expression rapidly without changing its gene regulation.

4. Solvent improvement through genetic engineering

Genetic engineering helps in the improvement of solvent in the bacterium species. In *C. acetobutylicum*, the production of solvent is increased by disturbing metabolic pathways (Schalck et al. 2021). It can also be enhanced by overexpressing butanol-producing genes. Harris et al. (2002) reported, through genetic engineering, the *C. acetobutylicum* strains were genetically modified by cloning the spo0A gene, and two recombinant strains were generated: the SKO1 inactivated strain and Spo0A overexpressed strain. The higher concentration of butanol is obtained from the Spo0A overexpressed strain while other SKO1 inactivated strain is inadequate in solvent formation. The *spoOA* enhanced and increases spore formation by overexpressing and inducing solvent production. This proves that *spoOA* plays a crucial role in solvent production and sporulation in *C. acetobutylicum*. Further studies identify that overexpression of a molecular chaperone *groESL* increases 30–40% of the final concentration of solvent in plasmid control and wild type strains (Tomas et al. 2003). Furthermore, by utilizing antisense RNA, the acetone group is improved 1.6-fold whereas the production of butanol declined 75.6% as compared to the control strain (Nakayama et al. 2008).

5. Genetic engineering to improve butanol tolerance

As we know, biobutanol has higher energy and a large amount of drop-in fuel capacity, drawing more attention in the past two decades. The natural process of biobutanol production through *Clostridia* which ferments starch and other sugars has been discussed earlier.

Over the past few years, significant progress has been made to improve butanol tolerance. Wu et al. (2021) used a genetically engineered thermotolerant *Clostridium acetobutylicum* to convert pretreated corn stover into butanol. Similarly, Valles et al. (2020) used *Clostridium beijerinckii* to convert pretreated rice straw to butanol yielding 51 g/kg raw rice straw. These experimental approaches make the butanol producing strains more cost-effective and also commercially available. In one study, the two most convenient genetically modified strains that model for solventogenic *Clostridia* were *Clostridium acetobutylicum* and *Clostridium beijerinckii.* The total ABE (acetone-butanol-ethanol) titer in this experiment ranged from 21 g/L in *Clostridium beijerinckii* to *Clostridium acetobutylicum*, which produced 13 g/L of ABE (Lutke et al. 2011). To scale-up the overall production, genetic and metabolic engineering techniques has been applied.

The unusual butanol production inside the cell inhibits the growth of the microorganism. The high butanol concentration leads to the change in the structure of membrane which disturbs the normal function. For commercial production of butanol through genetic engineering, it is necessary to lessen the sensitivity of butanol-producing strain to butanol. In *C. acetobutylicum,* the overexpression of the *spoOA* gene through metabolic engineering indicates enhanced tolerance and prolonged metabolism to butanol stress. Furthermore, overexpression of a molecular chaperone *groESL* also improves butanol tolerance; it helps prevent the aggregation of proteins. In several studies, strategies like genomic library enrichment are used to identify the butanol-tolerant gene in *Escherichia coli.* Overexpression of *E. coli* genes (*entC* and *feoA*) increased the tolerance of butanol by 32.8% and 49.1%, respectively (Zingaro et al. 2013). The deletion of the *astE* protein-encoding gene shows an overall 48.7% increased butanol tolerance (Reyes et al. 2011). Notable other non-native butanol synthesizing strains are genetically constructed to increase butanol tolerance. Strains like *E. coli* and *Lactobacillus* sp. tolerate 2–4% of butanol (Liu et al. 2017). The highest butanol tolerance is seen in *Pseudomonas putida*; it can tolerate 6% butanol (Rühl et al. 2009). The native strain *Bacillus subtilis* proves as an efficient host for butanol production tolerate up to 2% of butanol. The discussed strains shows a low yield of butanol production. Genetically modified microbes with their butanol tolerance are illustrated in Table 1. Classification of the butanol-tolerant gene and its overexpression in the host, which are heterologous or butanol-tolerant microbes through genetic engineering, and manipulating the metabolic pathway through metabolic engineering synergistically integrated with system biology and synthetic biology provide a great scope of vastly increasing the production of butanol (Liu et al. 2017).

Table 1. The table illustrates butanol tolerance and butanol production in various genetically modified butanol strains.

Microorganism	Butanol tolerance	Butanol production	References
Bacillus subtilis	2%	2.62 g/L	Li et al. 2011
Saccharomyces cerevisiae	2%	835 mg/L	Azambuja et al. 2020
Zymomonasmobilis	N/A	4.0 g/L	Qiu et al. 2020
Clostridium acetobutylicum	4%	15.3 g/L	Liu et al. 2013
Escherichia coli	1.5%	N/A	Lee et al. 2011
Pseudomonas putida	6%	N/A	Rühl et al. 2009
Lactobacillus brevis	3%	300 mg/L	Liu et al. 2012
Lactobacillus buchneri	3%	66 mg/L	Liu et al. 2010

6. Genetic engineering to increase the ratio of butanol with by-products elimination

The synthetic biology approach and metabolic engineering improve the overall productivity and make butanol production feasible at a commercial level by eliminating by-products' formation. The primary by-product of butanol fermentation is acetone. The ratio of butanol is increased by slowing down the acetone formation. To eliminate acetone as a by-product, TargeTron technology is used to disrupt *adc* (acetoacetate decarboxylase gene) in *C. acetobutylicum*, which increases the butanol ratio up to 80.85%; the reconstructed strain produces significantly less amounts of acetone (0.21 g/l) (Jiang et al. 2009). With the antisense RNA techniques, a strain of C. *acetobutylicum* was developed which reduced acetone formation (Tummala et al. 2003). By simultaneous disruption of *pta* and *buk* genes coding for phosphotransacetylase and butyrate kinase, while overexpressing adhE1^{D485G} gene, encoding a mutated aldehyde/alcohol dehydrogenase, butanol production increased by 160% compared to wild strain (Jang et al. 2012). Many genetically engineered strains eliminate by-product formation by targeting specific genes.

7. Engineering non-native strain for butanol production

Many of the non-native hosts for butanol production are very advantageous as they require simple nutrients, grow rapidly, and are easy to manipulate because of their known genetic background, but all these conditions differ from strain to strain. They are suitable for the large-scale industrial production of butanol. The synthetic and metabolic engineering approaches discussed earlier create an easy way to produce butanol commercially. In this process, the strain is engineered by successfully transferring the butanol pathways in non-native hosts such as *Saccharomyces cerevisiae* and *Escherichia coli*. Other non-native hosts such as *L. brevis*, *B. subtilis*, and *P. putida* are significantly engineered for higher butanol tolerance (Nawab et al. 2020).

7.1 Genetic and metabolic engineering in different strains of E. coli to produce butanol

Microorganisms that are metabolically engineered to switch renewable carbon sources to desired fuel products are considered as the best choice to obtain high volumetric productivity and yield. Considering the availability and the vast knowledge about genomics, metabolic fronts, and data availability in bioinformatics, *Escherichia coli* (*E. coli*) is considered the primary choice for the production of biofuels. It is a gram-negative, rod-shaped, facultative anaerobic bacterium. Most of the *E. coli* strains harmlessly colonize the gastrointestinal tract of humans and animals as a normal type of flora. *E. coli* has the unique advantage of being the best-studied model organism in terms of gene-regulation and expression and also with largest molecular tools available for genetically engineering the organism. *E. coli* strains have ability to utilize a variety of carbon sources (also sugars and sugar alcohols), is suitable in both aerobic and anaerobic conditions, and also best suited to a variety of industrial products. In biotechnology, *E. coli* is recognized as a workhorse because it can be easily modified genetically and cultured (Idalia et al. 2017). Furthermore, all the computational data, fermentation methods, and physiological information are easily available. For the improvement of production efficiency, the commonly adopted approach is metabolic engineering (Cronan et al. 2014). Strains are modified by the addition of desirable pathways and deletion of undesired pathways and removal of bottleneck nodes. To reconstruct the strain, *adh2* and *kivD* genes were interjected from *Lactobacillus lacti* and *S. cerevisiae*, sequentially (Ye at al. 2016). Further, these genes are converted into butanol, and *Adh2* converts into isobutanol and *kivD* converts into methyl butanol. The production of butanol in the strain of *E. coli* depends on the level of ketoacids. By diverting the amino acid biosynthetic pathway to 2-keto acid intermediates, the *E. coli* strain produced 22 g/L of isobutanol (Atsumi et al. 2008b). In other studies, the biobutanol synthesis pathway of the native *Clostridium* strain entered into a non-native strain of *E.coli* by inserting butanol-producing genes; *thl, crt, hbd, bcd, etfAB, atoB, and adhE2* (Shen et al. 2008). Figure 3 illustrates the butanol synthesis pathways in *E.coli*. The carbon source use in this synthetic butanol synthesis pathway is glucose under anaerobic conditions. The final strain of *E. coli* with induced *Clostridium* synthetic butanol pathway produced 6.1 g/l butanol titer (Koppolu et al. 2016).

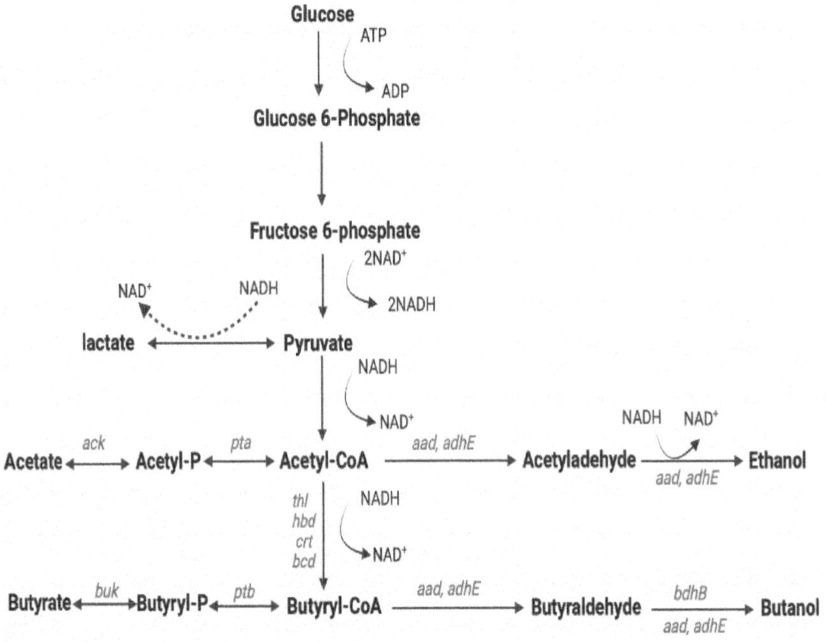

Figure 3. The butanol synthesis pathway in *E. coli* is reconstructed by the expression of the ethanologenic pathways of *Clostridium*.

In the research of Atsumi et al. (2008), the *E. coli* MG1655 strain was engineered in which the BW25113 strain was taken as wild strain. In this study, the *E. coli* genes *adhE, ldhA, frdB, pflB* were knocked out and further, the expression of the *Clostridium acetobutylicum* pathway in *E. coli* led to an increase in the butanol production. This research involves the transfer of the biosynthetic pathway from the native producer to the non-native producer. Overexpression of non-native pathways disturb the metabolism of the host and may affect the growth. But the vast knowledge of physiology and genetics of *E. coli* make it an easily manipulated organism for butanol production (Atsumi et al. 2008).

In a study by Jawed et al. (2020), the butanol production has been improved by FASII pathways in *E. coli* MG165. The butanol was produced via the FASII pathway with the help of thioesterase TesBT to produce the butyric acid, carboxylic acid reductase (CAR), and phosphopantetheinyl transferase for the formation of butyraldehyde and further conversion into the butanol. Three different gene configurations had been ptimized here: monocistronic, operon, and pseudo-operon with the expression of *Adh2* gene from *S. cerevisiae* (Saini et al. 2015). For improving the butanol production, two strains (one as the upstream strain which produces butyric acid and other a downstream strain which produces butanol

Table 2. Representations of diverse genetically modified *E. coli* strains and their butanol-producing ability.

S. No.	*E. coli* strain	Genetic Modifications	Butanol production	References
1.	*E. coli* (MG1655)	The *E. coli* strain has been engineered by expressing the native FASII pathway.	2.0 g/L	Jawed et al. 2020
2.	*E. coli*(BW25113)	Tn5 transposon mutations has been performed and genes *atoB, hbd, crt, ter, adhE2* and *fdh* inserted into the chromosomes of *E. coli*.	20 g/L	Dong et al. 2017
3.	*E. coli* K12	The butanol pathway in *E. coli* k12 was generated by a hypergraph algorithm and by expressing butanol genes in *E. coli* K12 and cultivation,	85 ± 1 mg/L	Ferreira et al. 2019
4.	*E. coli* (BL-A4)	The strain is reconstructed by knocking out unnecessary genes from *E. coli* (BL-A4) and the addition of butanol producing gene *atoDA* and *adhE2* from *Clostridium*.	5.5 g/L	Saini et al. 2015

form butyrate) had been optimised. Further, the co-cultivation of both the strains under the fed batch reactor produced 2 g/L of butanol. In this way, with the help of the synthetic and molecular biology approach, *E. coli* strains have been engineered genetically to produce butanol in desired amount (Jawed et al. 2020). Table 2 illustrates different butanol producing strains of *E. coli* with the genetic modifications brought into them.

7.2 Genetic and Metabolic Engineering in Bacillus subtilis for butanol production

Bacillus subtilis is a rod-shaped gram-positive bacterium with dormant spores. It comes under the category of non-pathogenic bacteria. The genome of *B. subtilis* is easily manipulated and serves as a model organism for study. It is engineered as the cell factory for isobutanol production due to its sensitivity to the n-butanol. For efficient production of butanol in strain the heterologous Ehrlich pathway with P_{43} promoter is introduced into the *Bacillus subtilis* and the overexpression of acetolactate synthase give rises to 2.8-fold of butanol production. The modifications in strain were performed by overexpressing the de novo biosynthetic pathway with the KIV precursor (Li et al. 2011). Furthermore, other vital factors such as metabolic profile, genetic stability, along with fermentation techniques

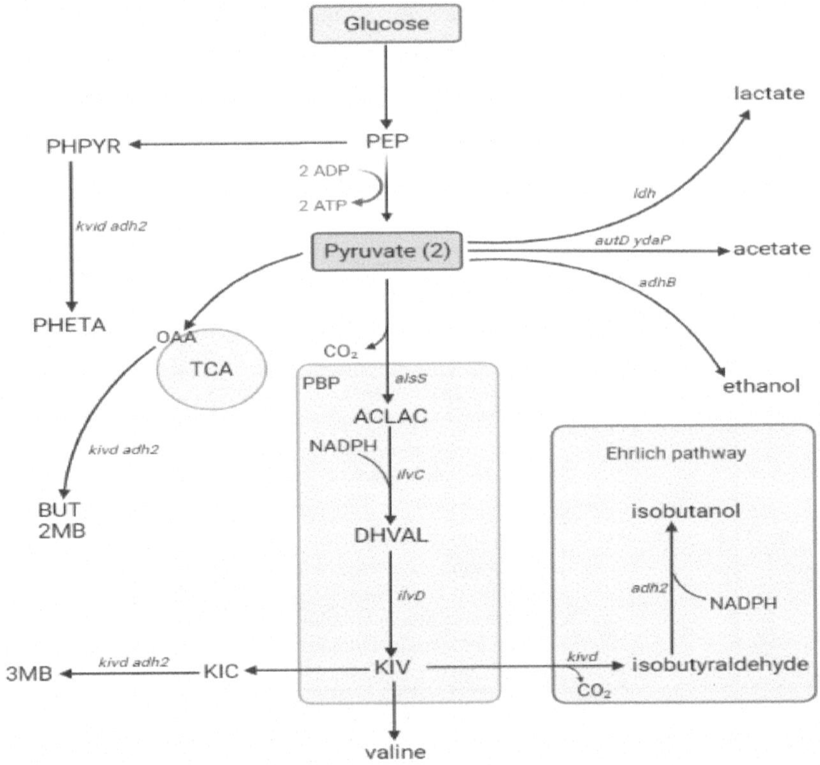

Figure 4. Butanol synthesis pathway in *B. subtilis*. Further reconstruction of pathways improve the butanol synthesis.

are also being investigated. The results show that, other than *E. coli*, *B. subtilis* has a higher solvent tolerant capacity and is considered the best cell factory for isobutanol production (Atsumi et al. 2008b). The Ehrlich pathway was constructed by adding *kivd* and *adh2* in sequence with well-known efficient promoter P_{43} which help to intensify the gene expression. The fermentation technology has also been taken into consideration. The engineered strain produced 2.62 g/L butanol in fed-batch fermentation, which was 21.3% higher than batch fermentation (Li et al. 2011). Figure 4 shows the butanol synthesis pathways in *Bacillus subtilis*.

Furthermore, the butanol-producing strain is also synthesized by Liu and team in which the *B. subtilis* 168 trpC2 strain is used as a wild type strain for recombinant strain (Li et al. 2004). The *Bacillus subtilis* was chosen as an easy host here because it has good heterologous expression capability and the whole genomic data has been sequenced so far. In this particular study, the isobutanol biosynthesis pathway is reconstructed in *Bacillus subtilis*, which is commonly known as a solvent-tolerant host. Isobutanol synthesis pathway was constructed in a strain of *Bacillus subtilis*.

After 35-h cultivation under sterile conditions, the maximum butanol production reaches 0.607 g/L with a pH of 7.0. It is clear from the result, the recombinant *B. subtilis* strain produces a low amount of isobutanol (Jia et al. 2012).

7.3 Genetic and metabolic engineering in *Zymomonas mobilis* for butanol production

Zymomonas mobilis is a facultative anaerobic, gram-negative bacteria with many beneficial industrial aspects such as high glucose consumption, low amount of biomass production, low aeration cost, and most importantly, high tolerance to ethanol. *Z. mobilis* is genetically engineered to ferment pentose sugars such as xylose and arabinose. The bioinformatical data such as genome sequences, proteins structural data, and metabolic models are now available for *Z. mobilis*.

7.4 Genetic framework of *Z. mobilis* and specific gene knock out for Butanol production

From three decades, studies on *Zymomonas mobilis* have been going on for the synthesis of biofuels. Extensively genetically modified *Z. mobilis* strains have been designed so far for the biobutanol and ethanol production. Highly extensive studies on various synthetic and genetic engineering techniques include expression system, cloning, gene knock-out, gene transformation and gene functions improve *Z. mobilis* butanol producing efficiency for biotechnological industries. The era of genomics and transcriptomics began in 2005 and many new strains have been identified. The three subspecies (subsp.) of *Z. mobilis* have been discovered till date such as *Zymomonas mobilislis* subsp. *pomaceae* and *Zymomonas mobilis* subsp. *Francensi* (He et al. 2014). From all these strains, strain ATCC 31821 (ZM4), ATCC 29192(ZM6), ATCC 10988 (ZM1) from *Z. mobilis* sups. *mobilis* were reconstructed genetically for butanol production (Lee et al. 2010). These species of *Z. mobilis* are considered as model organism in biotechnological and microbiological industries. In this way *Z. mobilis* plays a crucial role in various research and development sectors for the replacement of petrochemical products. The genetic engineering for the desired product is performed by addition and deletion of certain genes. This technology targeted the metabolic pathways in *Z. mobilis*. The different mutations techniques employed for inactivating genes of *Z. mobilis* include: site-specific FLP recombinase, fusion-PCR-based construction, transposons mutations, point deletion, and many other single or double mutations (Keasling et al. 2010). Till now the identified target genes for genetic engineering in *Z. mobilis* include Pyruuvate decarboxylase (*Pdc*-ZMO1360), Pyruuvatedecarbox *adhB*-ZMO1596 (alcohol dehydrogenases), lactate

Table 3. The table illustrates distinct strains of *Z. mobilis* and their origin species.

S. No	*Z. mobilis* strain	Source	References
1.	CP4	*Z. mobilis* subsp. *mobilis*	He et al. 2014
2.	ATCC 31821 (ZM4)	*Z. mobilis* subsp. *mobilis*	He et al. 2014
3.	ATCC 10988 (ZM1)	*Z. mobilis* subsp. *mobilis*	He et al. 2014
4.	ATCC 29192	*Z. mobilis* subsp. *pomaceae*	He et al. 2014
5.	NCIMB 11163	*Z. mobilis* subsp. *mobilis*	He et al. 2014

dehydrogenase *ldhA*-ZMO1237 (lactate dehydrogenase), *hfq*-ZMO0347 (RNA-binding protein *hfq*), *ndhA*-ZMO0117 (hydroxylamine reductase), gfo-ZMO0689 (glucosefructose oxidoreductase), *himA*-ZMO0976 (aldo-keto reductase. These are the specific phenotypes (Table 3) which are mainly edited for biofuel production in *Z. mobilis* (He et al. 2014).

Also a new era of technology such as CRISPR-Cas tool kits for identification and characterization facilitates heterologous metabolic pathway engineering for improving imbalances in reactions and robustness. The butanol synthesizing pathway of native *Z. mobilis* involves *Als* acetoacetate synthase, *IlvC* ketol-acid Reduce to isomerase, *IlvD* dihydroxy acid, *kdcA* 2-ketoacid decarboxylase, *Pdc* pyruvate dehydrogenase, *Adh* alcohol dehydrogenase, *Adhs* alcohol dehydrogenases (Qiu et al. 2020). The enzyme *Kdc,* for butanol synthesis is not found in most of the strains of *Z. mobilis,* which is the most essential known enzyme in the biobutanol pathway. To increase productivity the *KdcA* enzyme from *L. lactis* is introduced into a native strain of *Z. mobilis,* and a recombinant strain is generated. The recombinant strain produced butanol up to 4.01 g/L in yield (Qiu et al. 2020).

8. Computational approaches for *in silico* metabolic engineering for biofuel production

It has been 20 years since the evolution of metabolic engineering. Metabolic engineering is different from genetic engineering as it mainly focuses on biosynthetic and metabolic pathways (Chan et al. 2013). It involves the adaptation of regulatory and metabolic processes to upgrade the targeted cellular behaviours such as the synthesis of proteins and chemicals. It centers on enhancing the design of metabolic engineering pathways in microorganisms. For biofuel production, both native and non-native strains have been engineered as we discuss earlier. As we know, for large-scale production, microorganisms such as *E. coli* and *S. cerevisiae* are commonly used due to the presence of well-established genetic tools (Alper et al. 2009). To study the metabolic behaviour of microorganisms, a large amount of omics data has been created. This kind of research is called computational and *in silico* metabolic engineering. It involves modelling simulation and optimization of microbe with whole metabolic engineering extracted data. This kind of simulation minimize time, research expenditure, and labour. The complete genome sequencing of various microorganisms encourages microbiology and biotechnology researchers to study *in silico* metabolic behaviour of microorganisms (Chan et al. 2013). The metabolic model offers virtual screening of microbes for optimization and improvement of strain (Kanehisa et al. 2007). *In silico* metabolic engineering optimize strains by two different ways: predictive and determinative (Pfau et al. 2011). The predictive category is further divided into an optimization-based approach and a pathway-based approach. Nowadays research mainly focuses on optimization-based approach. This specific topic elaborates and illustrates the computational-based approaches in metabolic engineering pathways (Fong et al. 2014).

9. Approaches for recognizing Intracellular metabolic states

Various type of experimental approaches have been performed to compute the position of regulatory and metabolic networks, which mainly include analysis of flux, expression of genes, expression of proteins, concentration of metabolites, activities of enzymes, and assays of DNA bindings. These metabolic computational models are very helpful for analyzing and integrating datasets to quantify fluxes in metabolic pathways and also uncover their regulatory properties (Reed et al. 2010).

10. Analysis of metabolic flux

The capacity to quantitatively outline intracellular fluxes by employing metabolic flux analysis (MFA) is crucial for classifying pathway bottlenecks and elucidating system management in biological systems, particularly in engineered microorganisms with non-native metabolic capabilities (Sauer et al. 2006).^{13}C-MFA tests require supporting isotopically identified substrates to tissues, cells, and entire organisms and perform mapping of models of isotopes in intracellular metabolites or secreted products (Golubeva et al. 2017). Spectrometry techniques such as mass spectrometry (MS) and nuclear magnetic resonance (NMR) are mainly required for the quantification of relative abundance of various isotopomers associated with micro-molecules (Reed et al. 2010). The conventional method for calculating a flux map requires a nonlinear least-squares regression to minimize the lack-of-fit between (i) computationally simulated data (ii) experimentally measured data. In this way, various *in silico* approaches are applied to molecular simulations. For reconstructing desired microbial strain, *in silico* engineering plays a vital role. Many *E. coli* strains and *Clostridium* strains have been identified, which are ethanologenic. The bioinformatical data of all these strains have been characterized for development of genetically modified strains for butanol production.

11. Conclusion

Butanol is confirmed to be an essential biofuel. Nowadays, it is produced by genetically modified organisms. The synthetic biology and metabolic engineering approaches permit the production of the desired amount as well as improve productivity. The different native strains have been identified metabolically, and pathways have been reconstructed to overcome several limitations, such as tolerance and toxicity. In this manner, genetically modified strains are produced commercially for biofuel production. It has been extrapolated from various analyses that the presence of bioinformatical data, omics data, and computational biology approaches make it easier to reconstruct non-native strains of microorganisms that produce the desired amount of butanol.

Acknowledgement

The authors (Anita Singh and Japleen Kaur) would like to acknowledge SERB grant (ECR/2018/000672) for providing financial support to carry out this work.

References

Adrio, Jose-Luis, and Arnold L. Demain. (2010). Recombinant organisms for production of industrial products. Bioengineered Bugs, 1(2): 116–131.

Alper, Hal, and Gregory Stephanopoulos. (2009). Engineering for biofuels: exploiting innate microbial capacity or importing biosynthetic potential? Nature Reviews Microbiology, 7(10): 715–723.

Atsumi, Shota, Anthony F. Cann, Michael R. Connor, Claire R. Shen, Kevin M. Smith, Mark P. Brynildsen, Katherine J.Y. Chou, TaizoHanai, and James C. Liao. (2008). Metabolic engineering of *Escherichia coli* for 1-butanol production. Metabolic Engineering, 10(6): 305–311.

Atsumi, Shota, TaizoHanai, and James C. Liao. (2008b). Non-fermentative pathways for synthesis of branched-chain higher alcohols as biofuels. Nature, 451(7174): 86–89.

Azambuja, Suéllen P.H., and Rosana Goldbeck. (2020). Butanol production by Saccharomyces cerevisiae: perspectives, strategies and challenges. World Journal of Microbiology and Biotechnology, 36(3): 1–9.

Bao, Teng, Wenjie Hou, Xuefeng Wu, Li Lu, Xian Zhang, and Shang-Tian Yang. (2021). Engineering *Clostridium cellulovorans* for highly selective n-butanol production from cellulose in consolidated bioprocessing. Biotechnology and Bioengineering.

Birgen, Cansu, Peter Dürre, Heinz A. Preisig, and Alexander Wentzel. (2019). Butanol production from lignocellulosic biomass: revisiting fermentation performance indicators with exploratory data analysis. Biotechnology for Biofuels, 12(1): 1–15.

Bro, Christoffer, BirgitteRegenberg, Jochen Förster, and Jens Nielsen. (2006). *In silico* aided metabolic engineering of *Saccharomyces cerevisiae* for improved bioethanol production. Metabolic Engineering, 8(2): 102–111.

Chan, H., Weng, Mohd S. Mohamad, SafaaiDeris, and Rosli M. Illias. (2013). A review of computational approaches for *In silico* metabolic engineering for microbial fuel production. Current Bioinformatics, 8(2): 253–258.

Chen, Chang-Ting, and James C. Liao. (2016). Frontiers in microbial 1-butanol and isobutanol production. FEMS Microbiology Letters, 363(5): fnw020.

Cronan, John E. (2014). *Escherichia coli* as an Experimental Organism. eLS.

Dahman, Yaser, Kashif Syed, Sarkar Begum, Pallavi Roy, and BanafshehMohtasebi. (2019). Biofuels: Their characteristics and analysis. In Biomass, Biopolymer-Based Materials, and Bioenergy, pp. 277–325. Woodhead Publishing.

Dong, Hongjun, Chunhua Zhao, Tianrui Zhang, Huawei Zhu, Zhao Lin, Wenwen Tao, Yanping Zhang, and Yin Li. (2017). A systematically chromosomally engineered *Escherichia coli* efficiently produces butanol. Metabolic Engineering, 44: 284–292.

Dunlop, Mary J. (2011). Engineering microbes for tolerance to next-generation biofuels. Biotechnology for Biofuels, 4(1): 1–9.

Ezeji, Thaddeus Chukwuemeka, Nasib Qureshi, and Hans Peter Blaschek. (2007). Bioproduction of butanol from biomass: from genes to bioreactors. Current opinion in Biotechnology, 18(3): 220–227.

Ferreira, Sofia, Rui Pereira, Filipe Liu, Paulo Vilaça, and Isabel Rocha. (2019). Discovery and implementation of a novel pathway for n-butanol production via 2-oxoglutarate. Biotechnology for Biofuels, 12(1): 1–14.

Fischer, Curt R., Daniel Klein-Marcuschamer, and Gregory Stephanopoulos. (2008). Selection and optimization of microbial hosts for biofuels production. Metabolic Engineering, 10(6): 295–304.

Fong, Stephen S. (2014). Computational approaches to metabolic engineering utilizing systems biology and synthetic biology. Computational and Structural Biotechnology Journal, 11(18): 28–34.

Gheshlaghi, R.E.Z.A., Scharer, J.M., Murray Moo-Young, and Chou, C.P. (2009). Metabolic pathways of clostridia for producing butanol. Biotechnology Advances, 27(6): 764–781.

Golubeva, L.I., Shupletsov, M.S. and Mashko, S.V. (2017). Metabolic flux analysis using 13 C isotopes (13 C-MFA). 1. Experimental basis of the method and the present state of investigations. Applied Biochemistry and Microbiology, 53(7): 733–753.

Gong, Fuyu, Yanping Zhang, Yin Li, Hongjun Dong, Wenwen Tao, Zongjie Dai, and Liejian Yang. (2012). Biotechnology in China III: Biofuels and Bioenergy 128: 85.

Green, Edward M., Zhuang L. Boynton, Latonia M. Harris, Frederick B. Rudolph, Eleftherios T. Papoutsakis, and George N. Bennett. (1996). Genetic manipulation of acid formation pathways by gene inactivation in *Clostridium acetobutylicum* ATCC 824. Microbiology, 142(8): 2079–2086.

Gunasekaran, M., Gopalakrishnan Kumar, ObulisamyParthiba Karthikeyan, and Sunita Varjani. (2021). Lignocellulosic biomass as an optimistic feedstock for the production of biofuels as valuable energy source: Techno-economic analysis, Environmental Impact Analysis, Breakthrough and Perspectives. Environmental Technology & Innovation, 24: 102080.

Harris, L.M., Welker, N.E., and Papoutsakis, E.T. (2002). Northern, morphological, and fermentation analysis of spo0A inactivation and overexpression in *Clostridium acetobutylicum* ATCC 824. Journal of Bacteriology, 184(13): 3586–3597.

He, Ming Xiong, Bo Wu, Han Qin et al. (2014). *Zymomonas mobilis*: a novel platform for future biorefineries. Biotechnology for Biofuels, 7(1): 1–15.

Idalia, Vargas-Maya Naurú, and Franco Bernardo. (2017). *Escherichia coli* as a model organism and its application in biotechnology. Recent Adv. Physiol. Pathog. Biotechnol. Appl. Tech Open Rij. Croat, 253–274.

Inui, Masayuki, Masako Suda, Sakurako Kimura, Kaori Yasuda, Hiroaki Suzuki, Hiroshi Toda, Shogo Yamamoto, Shohei Okino, Nobuaki Suzuki, and Hideaki Yukawa. (2008). Expression of *Clostridium acetobutylicum* butanol synthetic genes in *Escherichia coli*. Applied Microbiology and Biotechnology, 77(6): 1305–1316.

Jawed, Kamran, Ali SamyAbdelaal, Mattheos A.G. Koffas, and Syed Shams Yazdani. (2020). Improved butanol production using FASII pathway in *E. coli*. ACS Synthetic Biology, 9(9): 2390–2398.

Jia, Xiaoqiang, Shanshan Li, Sha Xie, and Jianping Wen. (2012). Engineering a metabolic pathway for isobutanol biosynthesis in *Bacillus subtilis*. Applied Biochemistry and Biotechnology, 168(1): 1–9.

Jiang, Yu, Chongmao Xu, Feng Dong, Yunliu Yang, Weihong Jiang, and Sheng Yang. (2009). Disruption of the acetoacetate decarboxylase gene in solvent-producing *Clostridium acetobutylicum* increases the butanol ratio. Metabolic Engineering, 11(4-5): 284–291.

Jang, Y.S., Lee, J.Y., Lee, J., Park, J.H., Im, J. A., Eom, M.H., ... and Lee, S.Y. (2012). Enhanced butanol production obtained by reinforcing the direct butanol-forming route in *Clostridium acetobutylicum*. MBio, 3(5): e00314-12.

Joseph, Rochelle C., Nancy M. Kim, and Nicholas R. Sandoval. (2018). Recent developments of the synthetic biology toolkit for *Clostridium*. Frontiers in Microbiology, 9: 154.

Kanehisa, Minoru, Michihiro Araki, Susumu Goto et al. (2007). KEGG for linking genomes to life and the environment. Nucleic Acids Research, 36(suppl_1): D480–D484.

Keasling, Jay D. (2010). Manufacturing molecules through metabolic engineering. Science, 330(6009): 1355–1358.

Kolesinska, Beata, Justyna Fraczyk, Michal Binczarski, Magdalena Modelska, Joanna Berlowska, Piotr Dziugan, Hubert Antolak, Zbigniew J. Kaminski, Izabela A. Witonska, and Dorota Kregiel. (2019). Butanol synthesis routes for biofuel production: trends and perspectives. Materials, 12(3): 350.

Koppolu, Veerendra, and Veneela K.R. Vasigala. (2016). Role of *Escherichia coli* in biofuel production. Microbiology Insights, 9: MBI-S10878.

Lee, Ju Young, Kyung Seok Yang, Su A. Jang, Bong Hyun Sung, and Sun Chang Kim. (2011). Engineering butanol-tolerance in *Escherichia coli* with artificial transcription factor libraries. Biotechnology and Bioengineering, 108(4): 742–749.

Lee, Kyung Yun, Jong Myoung Park, Tae Yong Kim, Hongseok Yun, and Sang Yup Lee. (2010). The genome-scale metabolic network analysis of *Zymomonas mobilis* ZM4 explains physiological features and suggests ethanol and succinic acid production strategies. Microbial Cell Factories, 9: 1–12.

Li, Shanshan, Jianping Wen, and Xiaoqiang Jia. (2011). Engineering *Bacillus subtilis* for isobutanol production by heterologous Ehrlich pathway construction and the biosynthetic 2-ketoisovalerate precursor pathway overexpression. Applied Microbiology and Biotechnology, 91(3): 577–589.

Li, Weifen, Xuxia Zhou, and Ping Lu. (2004). Bottlenecks in the expression and secretion of heterologous proteins in *Bacillus subtilis*. Research in Microbiology, 155(8): 605–610.

Liang, Liya, Rongming Liu, Emily F. Freed, and Carrie A. Eckert. (2020). Synthetic biology and metabolic engineering employing *Escherichia coli* for C2–C6 Bioalcohol Production. Frontiers in Bioengineering and Biotechnology, 8: 710.

Linville, Jessica L., Miguel Rodriguez, Steven D. Brown, Jonathan R. Mielenz, and Chris D. Cox. (2014). Transcriptomic analysis of *Clostridium thermocellum* Populus hydrolysate-tolerant mutant strain shows increased cellular efficiency in response to Populus hydrolysate compared to the wild type strain. BMC Microbiology, 14(1): 1–17.

Liu, Siqing, Kenneth M. Bischoff, Nasib Qureshi, Steven R. Hughes, and Joseph O. Rich. (2010). Functional expression of the thiolase gene thl from *Clostridium beijerinckii* P260 in *Lactococcus lactis* and *Lactobacillus buchneri*. New Biotechnology, 27(4): 283–288.

Liu, Siqing, Kenneth M. Bischoff, Timothy D. Leathers, Nasib Qureshi, Joseph O. Rich, and Stephen R. Hughes. (2012). Adaptation of lactic acid bacteria to butanol. Biocatalysis and Agricultural Biotechnology, 1(1): 57–61.

Liu, Siqing, Nasib Qureshi, and Stephen R. Hughes. (2017). Progress and perspectives on improving butanol tolerance. World Journal of Microbiology and Biotechnology, 33(3): 51.

Liu, Xiao-Bo, Qiu-Ya Gu, and Xiao-Bin Yu. (2013). Repetitive domestication to enhance butanol tolerance and production in *Clostridium acetobutylicum* through artificial simulation of bio-evolution. Bioresource Technology, 130: 638–643.

Lutke-Eversloh, T. (2011). Metabolic engineering of *Clostridium acetobutylicum*: recent advances to improve butanol production. Curr. Opin. Biotechnol., 22: 1–14.

Nakayama, Shun-ichi, Tomoyuki Kosaka, Hanako Hirakawa, Kentaro Matsuura, Sadazo Yoshino, and Kensuke Furukawa. (2008). Metabolic engineering for solvent productivity by downregulation of the hydrogenase gene cluster hupCBA in *Clostridium saccharoperbutylacetonicum* strain N1-4. Applied Microbiology and Biotechnology, 78(3): 483–493.

Nanda, Sonil, DasantilaGolemi-Kotra, John C. McDermott, Ajay K. Dalai, Iskender Gökalp, and Janusz A. Kozinski. (2017). Fermentative production of butanol: perspectives on synthetic biology. New Biotechnology, 37: 210–221.

Nawab, Said, Ning Wang, Xiaoyan Ma, and Yi-Xin Huo. (2020). Genetic engineering of non-native hosts for 1-butanol production and its challenges: a review. Microbial Cell Factories, 19(1): 1–16.

Ndaba, Busiswa, IdanChiyanzu, and Sanette Marx. (2015). n-Butanol derived from biochemical and chemical routes: A review. Biotechnology Reports, 8: 1–9.

Pfau, Thomas, Nils Christian, and Oliver Ebenhöh. (2011). Systems approaches to modelling pathways and networks. Briefings in Functional Genomics, 10(5): 266–279.

Qiu, Mengyue, Wei Shen, Xiongyin Yan et al. (2020). Metabolic engineering of *Zymomonas mobilis* for anaerobic isobutanol production. Biotechnology for Biofuels, 13(1): 1–14.

Ramos, Juan L., Estrella Duque, Maria-Trinidad Gallegos, Patricia Godoy, Maria Isabel Ramos-Gonzalez, Antonia Rojas, Wilson Terán, and Ana Segura. (2002). Mechanisms of solvent tolerance in gram-negative bacteria. Annual Reviews in Microbiology, 56(1): 743–768.

Reed, Jennifer L., Ryan S. Senger, Maciek R. Antoniewicz, and Jamey D. Young. (2010). Computational approaches in metabolic engineering. Journal of Biomedicine and Biotechnology 2010.

Reyes, Luis H., Maria P. Almario, and Katy C. Kao. (2011). Genomic library screens for genes involved in n-butanol tolerance in *Escherichia coli*. PloS one 6(30: e17678.

Rosano, Germán L., and Eduardo A. Ceccarelli. (2014). Recombinant protein expression in *Escherichia coli*: advances and challenges. Frontiers in Microbiology, 5: 172.

Rühl, Jana, Andreas Schmid, and Lars Mathias Blank. (2009). Selected *Pseudomonas putida* strains able to grow in the presence of high butanol concentrations. Applied and Environmental Microbiology, 75(13): 4653–4656.

Saini, Mukesh, Min Hong Chen, Chung-Jen Chiang, and Yun-Peng Chao. (2015). Potential production platform of n-butanol in *Escherichia coli*. Metabolic Engineering, 27: 76–82.

Sauer, Uwe. (2006). Metabolic networks in motion: 13C-based flux analysis. Molecular Systems Biology, 2(1): 62.

Schalck, Thomas, Bram Van den Bergh, and Jan Michiels. (2021). Increasing solvent tolerance to improve microbial production of alcohols, terpenoids and aromatics. Microorganisms, 9(2): 249.

Shen, Claire R., and James C. Liao. (2008). Metabolic engineering of *Escherichia coli* for 1-butanol and 1-propanol production via the keto-acid pathways. Metabolic Engineering, 10(6): 312–320.

Tomas, Christopher A., Neil E. Welker, and Eleftherios T. Papoutsakis. (2003). Overexpression of groESL in *Clostridium acetobutylicum* results in increased solvent production and tolerance, prolonged metabolism, and changes in the cell's transcriptional program. Applied and Environmental Microbiology, 69(8): 4951–4965.

Tummala, Seshu B., Neil E. Welker, and Eleftherios T. Papoutsakis. (2003). Design of antisense RNA constructs for downregulation of the acetone formation pathway of *Clostridium acetobutylicum*. Journal of Bacteriology, 185(6): 1923–1934.

Valles, A., Álvarez-Hornos, F.J., Martínez-Soria, V., Marzal, P., and Gabaldón, C. (2020). Comparison of simultaneous saccharification and fermentation and separate hydrolysis and fermentation processes for butanol production from rice straw. Fuel, 282: 118831.

Wen, Zhiqiang, Mianbin Wu, Yijun Lin, Lirong Yang, Jianping Lin, and Peilin Cen. (2014). Artificial symbiosis for acetone-butanol-ethanol (ABE) fermentation from alkali extracted deshelled corn cobs by co-culture of *Clostridium beijerinckii* and *Clostridium cellulovorans*. Microbial Cell Factories, 13(1): 1–11.

Wu, Y., Wang, Z., Ma, X., and Xue, C. (2021). High temperature simultaneous saccharification and fermentation of corn stover for efficient butanol production by a thermotolerant *Clostridium acetobutylicum*. Process Biochemistry, 100: 20–25.

Xin, Fengxue, Wei Yan, Jie Zhou, Hao Wu, Weiliang Dong, Jiangfeng Ma, Wenming Zhang, and Min Jiang. (2018). Exploitation of novel wild type solventogenic strains for butanol production. Biotechnology for Biofuels, 11(1): 1–8.

Xue, Chuang, Youduo Wu, Y. Gu, W. Jiang, H. Dong, Y. Zhang, C. Zhao, and Y. Li. (2019). 3.07-Biofuels and Bioenergy: Acetone and Butanol. Comprehensive Biotechnology, 79–100.

Ye, Wei, Weimin Zhang, Taomei Liu, Guohui Tan, Haohua Li, and Zilei Huang. (2016). Improvement of ethanol production in *Saccharomyces cerevisiae* by high-efficient disruption of the ADH2 gene using a novel recombinant TALEN vector. Frontiers in Microbiology, 7: 1067.

Zingaro, Kyle A., and Eleftherios Terry Papoutsakis. (2013). GroESL overexpression imparts *Escherichia coli* tolerance to i-, n-, and 2-butanol, 1, 2, 4-butanetriol and ethanol with complex and unpredictable patterns. Metabolic Engineering, 15: 196–205.

12

Biobutanol Production Using Nanotechnology: A Way Forward

Renu Singh,[1,*] *Sibananda Darjee,*[1] *Bharti Rohtagi,*[3] *Ashish Khandelwal,*[1] *Sapna Langyan,*[2] *Amit Kumar Singh,*[4] *Manoj Shrivastava,*[1] *Anu Bharti,*[5] *Har Mohan Singh*[6] and *Sujata Kundan*[7]

1. Introduction

Over the past decade, interest in preparing chemicals as well as fuels out of renewable materials has grown. The major causes for this increased interest involve ever-increasing concerns regarding astral warming, climate change, changes in oil abundance, rising crude oil prices, and

[1] Division of Environment Science, ICAR-Indian Agricultural Research Institute, New Delhi 110012.
[2] Division of Germplasm Evaluation, ICAR-National Bureau of Plant Genetic Resources, New Delhi 110012, India.
[3] Division of Microbiology, ICAR-Indian Agricultural Research Institute, New Delhi 110012.
[4] Department of Cellular and Bioengineering, Sam Higginbottom University of Agriculture, Technology and Sciences, Allahabad.
[5] Department of Environmental Sciences, Central University of Jammu, Rahya Suchani, (Bagla) Samba, (J & K), 181143, India.
[6] School of Energy Management, Shri Mata Vaishno Devi University, Katra, (J & K), 182320, India.
[7] Department of Chemistry and Chemical Sciences, Central University of Jammu, Rahya Suchani, (Bagla) Samba, (J & K), 181143, India.
* Corresponding author: renu_icar@yahoo.com

legislation keeping track of the excessive utilization of nonrenewable sources of energy. The technological advances and demands of the twenty-first century have created alarming conditions of global warming and the reduction of fossil fuels. Thus, the potential substitutes of energy sources to reduce dependency on petro-based fuels have basically influenced the interest in biofuels such as bioethanol and biobutanol.

Biobutanol is considered a highly utilized fuel additive in reference to its ecofriendly as well as renewable nature. Biobutanol is getting more attention due to its universal acceptance as fuel, unlike bioethanol. Biobutanol is an alcohol that offers many advantages as a transport fuel, basically due to its high energy content (up to 29.2 MJ/L), lower vapor pressure, and higher hydrophobicity. It gives better fuel economy, blending ability, has a close resemblance to modern gasoline, and offers usability in terms of safety and distribution along with other crucial characteristics (Figure 1). In addition, it has low deficit problems as the thermal conductivity of butanol is less than half of that of ethanol, which is why a butanol-powered engine is easier to start in colder climates as compared to one powered by ethanol or methanol. Thus, the past two decades have seen increasing commercial interest in fermentative biobutanol production.

The production of biobutanol utilizes fermentation technology for transforming rich carbohydrates into biofuels, although the expensive nature of production along with other technical problems are seen as barriers to its wide-scale adoption. The conventional large-scale process

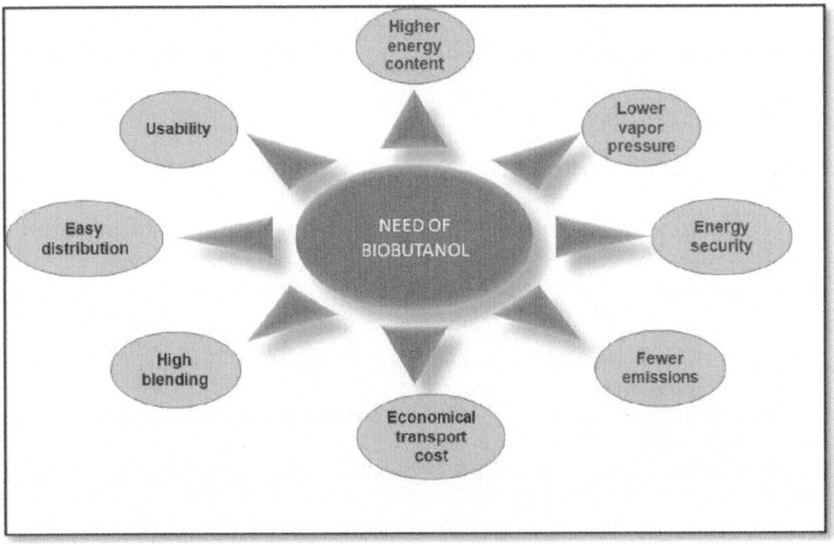

Figure 1. Benefits of biobutanol.

of acetone-butanol-ethanol fermentation using waste biomass still has many limitations to overcome. These limitations are associated with a low reaction rate, inhibitors of fermentation, the requirement of pretreatment, and complexity of downstream processing resulting in high cost of processing and low yield (Meramo-Hurtado et al. 2020). Biobutanol has a tendency of mixing easily with gas and petrol, at any desired ratio, which can be further utilized as fuel. However, biofuels like this can only become competitive at a global level when they are delinked from food crops. Generally, most of the feeds related to the non-food domain for the generation of bioalcohols (such as bioethanol and biobutanol) face the problem of transformation from cellulose to simpler sugars, which can further be subjected to the process of fermentation to obtain the desired end products. The widely utilized methods that are presently in practice for converting the cellulosic feedstocks having complexity in their structures into sugars is an expensive ordeal, therefore, making the biofuels produced via these methods not economical.

Considering the present environmental and economic issues, serious efforts are being made at the global level to obtain bioethanol and biobutanol from second- and third-generation feedstocks. Several technologies, including the utilization of nanotechnology, offer a valuable approach that have been developed to minimize the complex nature as well as the cost factor involved in various stages for obtaining bioalcohol. For example, utilizing suitable nanoparticles can help in minimizing the abovestated issues. There is a need for a three-way approach for comprehending cost-effective generation of bioalcohol that includes technology meant for a good crop yield, better processing of the feedstock, and developing several new biofuels like biobutanol. Several nanoparticles including nickel cobaltite, and zinc oxide along with various other nanocomposites are being utilized for the generation of new biofuels. Utilizing such nanomaterials during several phases in bioconversion operations offers a sustainable path by decreasing the costs involved during the processing of raw biomass and biofuel production and thus, reducing the severe environmental effects. Recently, the field of "nanotechnology" has grown to be an influential research area that has impacted almost every nook and corner of scientific research and has many crucial roles in everyday life also.

Nanoparticles are materials that have a diameter less than 100 nm and their fundamental characteristic and physical and chemical profile depends on the material of origin. Nanoparticles have variable applications in material science, medicine, and research and product development related to the welfare of humankind. Nanotechnology can help in overcoming challenges along with helping enable biofuel production via a sustainable

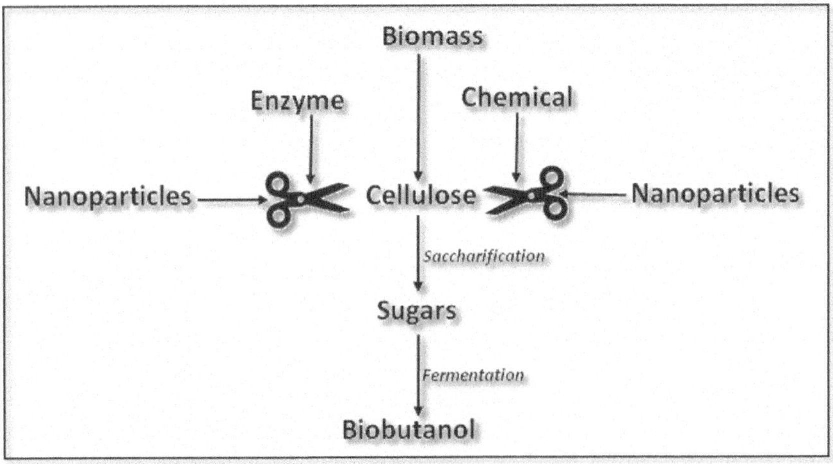

Figure 2. Role of nanoparticles in biobutanol production.

route. Several nanoparticles as well as nanomaterials are being reported to contribute to the production of biobutanol.

Nanotechnology can influence biobutanol production in the following ways (Figure 2):

1. Manipulate cell wall structure in pretreatment
2. Immobilize the catalyst/enzyme for hydrolysis, enhancing its stability and activity
3. Nanobiological system creates biobutanol

This chapter will give an overview of nanomaterials with their application in biobutanol production.

2. Nanomaterials and their characteristics

Nanomaterials differ from conventional materials with respect to size and provide a very large surface-area-to-volume ratio resulting in a larger active site. Nanotechnology works at the microscopic level and can alter the electrical, chemical, magnetic, mechanical, or biological characteristics without changing the chemical composition of a substance, thus influencing the flexibility, strength, conductivity, surface tension, as well as the color of a material. Nanoparticles enhance the heat and mass transfer rate of any system due to their Brownian motion and diffusiophoresis (Kumar et al. 2021).

3. Types and synthesis of nanomaterials

Nanoparticles are basically of two types; organic and inorganic. Organic nanoparticles include micelles, liposomes, compact polymeric, and hybrid compounds. Silica, quantum dots, fullerenes, and metal nanoparticles are the inorganic type. Nanoparticles can also be categorized into different groups (Figure 3). Nanoparticles that have potential application in biobutanol production include magnetic nanoparticles, carbon nanotubes, and inorganic nanoparticles. Magnetic nanoparticles have offered an approach that enables easy separation and recycling and are widely used for enzyme immobilization (Rai et al. 2016). Oxides of metals and non-metals (TiO_2, Fe_3O_4, ZnO, CaO, SiO_2, Al_2O_3) have high-temperature stability.

Carbon tubes (CNT, graphene, MWCNT, fullerene) are inert with high stability and thermal conductivity. Besides, hybrid (Ag/SiO_2, Au/SiO_2, Ni/SiO_2, Fe_3O_4/SiO_2, Au/TiO_2, Fe/C, $FeNi/SiO_2$, ZnO/SiO_2) and core-shell (FeMo, CuMo) nanoparticles can also be used in biofuels production (Kumar et al. 2021). These materials can be synthesized by the two most basic pathways, that is, top-down as well as bottom-up approaches (Table 1). These two approaches are different in aspects such as the method, energy demand, toxicity to the environment, and time involved. Nanomaterials are either prepared by breaking down larger particles or

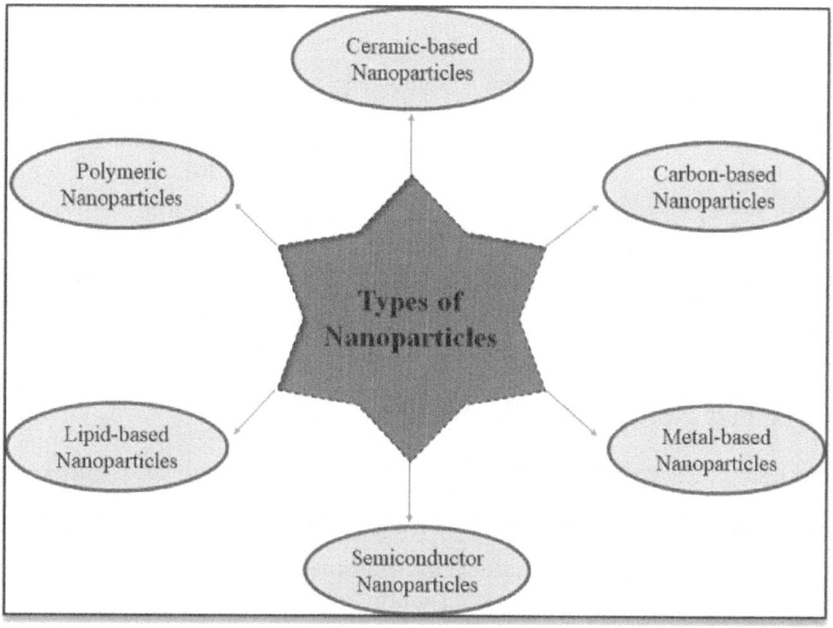

Figure 3. Types of nanoparticles.

Table 1. Synthesis approaches for nanoparticles

Characteristic	Top-down Approach	Bottom-up Approach	
		Chemical methods	Green technologies
Method of preparation	Physical methods such as grinding, milling, laser ablation, sputtering, etc.	Either physical, chemical, or biochemical methods. For example, laser evaporation, chemical vapor deposition (CVD), sol-gel, hydrothermal, supercritical fluid synthesis, and magnetron sputtering or molecular condensation	Various prokaryotic bacteria and eukaryotic organisms. For example, S-layer bacteria (gypsum and calcium carbonate layers) magnetotactic bacteria (MNPs) diatoms (siliceous materials)
Energy demand	High	Low to high	Low
Toxicity to environment	No toxic product	Capping agents are generally required that are toxic for the environment. For example, mercapto acetate, thiourea, mercaptoethanol, N,N-dimethylformamide cetyl trimethyl ammonium bromide, thioglycerol, polyvinylpyrrolidone, etc.	Cofactors act as capping agents that are safe for the environment
Level of operation	Easy and fast	Fast and easy to scale up	Difficult, slow, and highly dependent on operational variables. Difficult to scale up

assembling smaller particles. The method of nanoparticle preparation target the specific size, properties, and composition of a material as per the application. Magnetic nanoparticles (MNPs) are the most frequently used nanoparticles for cellulase immobilization owing to their easy separation and reusability. However, they are easily oxidized and thus it is important to functionalize them (Singh et al. 2020). Different methods of functionalization (Table 1) include acid-functionalization, alkyl-sulfonic acid functionalization, and amino-, silica-, chitosan-functionalization.

4. Methods for biobutanol production

Biobutanol is mostly chemically synthesized through the Reppe or Crotonaldehyde hydrogenation method and oxo method. However, the cost-benefit ratio discourages these methods for the production of biobutanol as a source of alternative biofuel. The most beneficial and economical method for the production of biobutanol is based on the fermentation process that can rely on a wide range of biomass and the application of *Clostridium* spp. This method is commonly known as acetone–butanol–ethanol fermentation (ABE). ABE needs a lot of R&D in the domain of pretreatment to achieve low-cost fermentation substrates, minimization of biobutanol inhibitory effects, high biobutanol yield, and lowering of downstream processing inputs.

4.1 Biobutanol production through chemical processes

Biobutanol production though the chemical route is achieved by adopting three basic steps as follows (Figure 4):

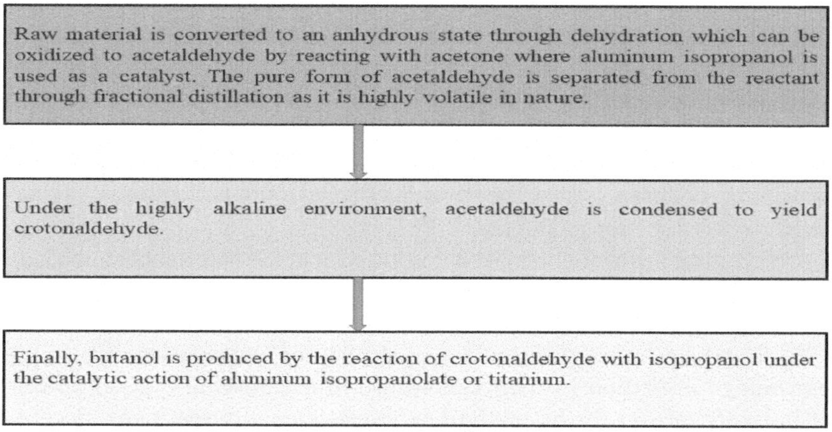

Figure 4. Production of biobutanol through the chemical route.

This whole production procedure is known as the Gerber reaction, which is very efficient in the conversion of a range of alcohols due to such catalyzed reaction series. In fact, direct catalytic dimerization can be applied for the production of 1-butanol from ethanol using zeolites, solid bases, carbon sources, and metals like Ni, Co, Ru, along with a modified catalyst (KOH, CeO_2). One step butanol production from bioethanol is also achieved in the sub-/supercritical state, which gives better diffusion, solubility, and lower viscosity (Lu et al. 2002).

4.2 Biobutanol production through biotechnological processes

Generally, butanol on an industrial scale is produced from anaerobic spore-forming bacteria of the *Clostridium* genera. *Clostridium* spp. of bacteria can produce acids, solvents, organic compounds, and alcohols through fermentation, using a wide range of carbohydrate-based substrates, Specifically, butyrate, which can be produced by saccharolytic and a few mesophilic species that can achieve higher fermentation efficiency such as *Clostridium beijerinckii, C. tetanomorphum, C. acetobutylicum,* and

C. aurantibutylicum (Kushwaha et al. 2018). Biobutanol can be produced through fermentation in two steps: (i) sugar conversion into organic acids such butyasrate and acetate in the acidogenesis step where pH decreases to 4.5. This decrease in pH also slows the acidogenesis process. (ii) Solventogenesis, that is, where the solvent is produced from butyrate and acetate consumption induced by solventogenic enzymatic action. The second step produces butanol as the major product where ethanol and acetone are also present in the admixture. This whole process stops due to the inhibitory effect of butanol concentration on bacterial metabolism. The Embden-Meyerhof-Parnas pathway governs the degradation of sugars to pyruvate in the acidogenesis and solventogenesis steps. Complex sugars like pentoses and hexoses are utilized by *Clostridium* spp. bacteria by the action of extracellular enzymes (glucoamylase, pullulanase, amylase, saccharase, glucosidase, and amylopullulanase), which convert them into simple sugars (xylose, glucose, arabinose). The enzymes produced by *Clostridium* facilitates butanol generation from varied carbon sources including potatoes, sugar beet, millet, corn, millets, tapioca, cassava, artichokes, whey permeate, etc.

Biobutanol production from polysaccharides namely, starch and xylan is very common. Algal biomass as the source of biobutanol has the advantage of being autotrophic, that is, be capable of consuming CO_2, thus helping in greenhouse gas mitigation. Algae contain a higher percentage of carbon in their cell walls, which provide a good amount of polysaccharides for biobutanol synthesis after pretreatment. Recently *Nannochloropsis* sp. *Dunaliella*, *Arthrospira platensis*, and wastewater algae were found to be economical substrates for biobutanol generation. The agricultural, agro-industrial, and forest waste which is higher in lignin, cellulose, and hemicellulose can also be used to obtain fermentable sugars after proper chemical, physical, biological, or combined pretreatment. Waste from vegetables and fruits and grain markets that otherwise create environmental pollution can be used in this system after identifying the suitable microbial strain for biobutanol synthesis.

The strategies for biobutanol production (Figure 5) through clostridial fermentation (Van Den Berg et al. 2013) can be improved by focusing research in:

a. Better bacterial strains having tolerance to a higher concentration of biobutanol
b. Utilization of homogenous and low-cost carbonaceous substrates
c. Application of metabolic engineering for better understanding of the process so that interventions can be applied
d. Improvements in biobutanol recovery
e. Upscaling of the production system
f. Recyclizng of biomass

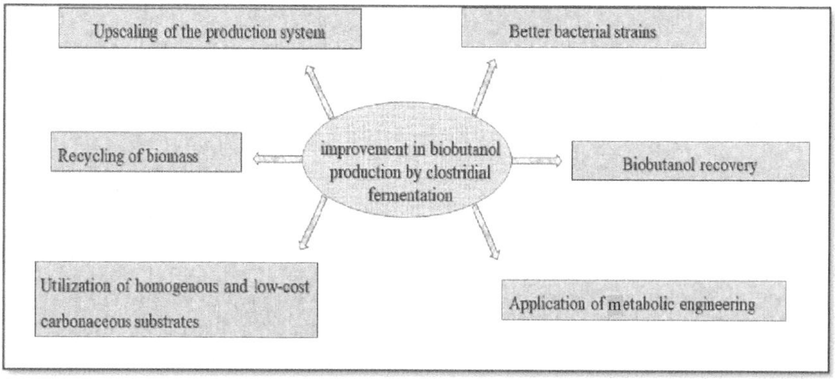

Figure 5. Strategies for biobutanol production.

4.3 Biobutanol production by nanotechnological approaches

Extensive research initiatives have been taking place in areas of biological and chemical sciences. Researchers are considering it as a boon for efficient modifications in biocatalysts that are target specific for assisting bioconversion operations for obtaining biofuels in a sustainable way (Finnigan et al. 2021). The utilization of nanoparticles in the bioenergy domain for obtaining a sustainable energy supply along with the long-term sustainability of the environment has actually gained interest among researchers. Nanoparticle application during the production of bioalcohol has a role in enhancing the procedural effectivity by improving the pretreatment, enzymatic hydrolysis, as well as the reaction rate throughout the fermentation process. Reaction time, size, surface area, morphology, nature and type of biomass are the important deciding factor for the production of end product (Singh et al. 2017). The nanotechnology utilization is helpful because there are several shortcomings associated with conventional techniques for obtaining bioethanol or biobutanol such as a lower rate of reaction, high biomass processing cost of biomass, and less yield of product (Bharathiraja et al. 2015). Therefore, for overcoming such problems, nanoparticles are making their way to elevate productivity by successfully being utilized as a sustainable alternative for bioethanol production.

Although the step involving the biomass pretreatment is crucial, it is an expensive step. Therefore, an economical alternative for biomass preprocessing is the priority step for obtaining bioethanol or biobutanol. The utilization of nanoparticles in this regard along with other sustainable approaches for the purpose of biomass pretreatment results in increasing the efficiency of the process. Nanoparticles are effective in eliminating the pollution that is usually caused when subjected to chemical pretreatment (Yan et al. 2021). They also enhance the chemistry present at the molecular level. Metal nanoparticles have especially shown more effectiveness when it comes to the penetration inside the cell wall of raw biomass (Giannakis 2019). This is because they possess a small structure that has the capability of interacting effortlessly with the biomolecules for releasing carbohydrates that can further be utilized for biobutanol generation. Razack et al. 2016 obtained a total carbohydrate yield of approximately 15.26% from *Chlorella vulgaris* biomass by 150 lg/g of silver nanoparticles that were synthesized via the biological mode. The yield was achieved in 40 min of incubation at 100 rpm (Razack et al. 2016). The high number of nanoparticles reduce the incubation period to break more parts of the cell wall and release some intracellular components such as carbohydrates and lipids. Since there is very limited literature available to show research work being done to enhance the biobutanol yield by utilizing nanotechnology. So, there exists a potent space for conducting research work in the said field. However, there are certain studies (Table 2) that have reported enhancement in the yield of different types of bioalcohol.

The formation of liquid fuels via the process of fermentation involves the impact of nanoparticles on biochemical conversions because they are capable of influencing the enzymatic activity or even the gas-liquid mass transfer rate.

Several nanoparticles having a metallic origin have been shown to impact the catalytic constituents in the domain of renewable energy generation. Kim et al. 2014 obtained improved generation of bioethanol (166.1%) with nanoparticles having methyl-functionalized silica as the main constituent in the fermentation of syngas. They utilized six varieties of nanoparticles in their study, which included palladium on carbon,

Table 2. Nanomaterials utilized to enhance the bioalcohol yield.

S. No	Nanomaterials utilized	Bioalcohol production increased	References
1.	Chitosan-coated Fe_3O_4 nanoparticles	51.5%	Javid et al. 2022
2.	Nickel oxide (NiO) nanoparticles (NPs) as a biocatalyst	19%	Sanusi et al. 2020
3.	Methyl-functionalized cobalt-ferrite-silica (CoFe2O4@SiO2-CH3) nanoparticles.	213.5%	Kim and Lee 2016
4.	Methyl-functionalized silica nanoparticles	166.1%	Kim et al. 2014
5.	Nanodisperse ZIF-8/PDMS hybrid membranes	-	Fan et al. 2014
6.	Rh particles inside CNTs	-	Pan et al. 2007

palladium on alumina, silica, hydroxyl-functionalized single-walled carbon nanotubes, alumina, and iron (III) oxide. Silica nanoparticles out of all varieties showed their effectiveness in enhancing the gas-liquid mass transfer that was again doped with hydrophobic functional groups such as methyl and isopropyl for enhancing the operation. Kim and Lee (2016) utilized methyl-functionalized magnetic nanoparticles to achieve increased production of bioethanol in the course of syngas fermentation. Approximately 213.5% more generation was achieved on utilizing methyl-functionalized cobalt-ferrite-silica nanoparticles. Moreover, the impactful recovery along with the capacity to reuse these nanoparticles resulted in making the process cost-effective. Ivanova et al. 2011 achieved an enhanced fermentation of bioethanol with the application of alginate-magnetic nanoparticles that were trapped with yeast cells and subjected to immobilization on chitosan-magnetite microparticles as well as cellulose-coated magnetic nanoparticles via a covalent bond. The researchers found in their study that the effectiveness of entrapped yeast cells magnetic nanoparticles was good as compared to the situation when two of these were utilized in a column reactor. Nearly 91% ethanol production was achieved in this instance. Another main advantage of utilizing nanoparticles for obtaining bioethanol or biobutanol involves compound detection via immobilized metal nanoparticles applied over the structures made of nanosheet.

4.3.1 Fermentation nanotechnology

The toxicity of alcoholic products for the microorganism itself is the primary issue with large-scale biobutanol production. Nanoparticles not only influence certain biological processes but also microorganisms and their environment. They either participate in the enzymatic reactions or act as catalysts themselves. Skrotskyi et al. 2019 studied the effect of nanoparticles of gold, silver, cerium, gadolinium, and iron oxide on butanol biosynthesis by two strains of acetonebutyl bacteria, *Clostridium beijerinckii* IMV B-7806 and *C. acetobutylicum* IMV B-7807. They have reported an increase in butanol yield with NPs of iron oxide and cerium by the later strain. Biohydrogen production has been achieved with the *Clostridium beijerinckii* NCIMB8052 strain immobilized on magnetite nanoparticles functionalized with chitosan and alginic acid (Seelert et al. 2015).

4.3.2 Downstream processing

In acetone–butanol–ethanol (ABE) fermentation, butanol, the main product becomes toxic to microorganisms at 1 wt % concentration and therefore it is necessary to remove butanol *in situ* during fermentation. This will not only limit product inhibition but also enhance butanol productivity. For volatile compounds in alcoholic fermentation, pervaporation (a combination of permeation and vaporization) is preferred over distillation for selective butanol removal from the fermentation broth. It offers two advantages: firstly, it poses the least harm to microbes, and secondly, it needs a lesser amount of energy. Nanoparticles can improve the separation performance of pervaporation membranes by providing a better polymer-particle interfacial area and a thinner separation layer (Liu et al. 2014). Nanotechnology can also make this pervaporation process economically viable for industries as it involves large energy demands (Figure 6).

Russell (2012) tried to increase the energy efficiency of pervaporation by plasmonic heating of gold nanoparticle-functionalized, polymer nanocomposite membranes (AuNCMs) with little success. Azimi et al. (2018) used activated carbon nanoparticles in the matrix of commonly used PDMS (poly-dimethylsiloxane) membranes to enhance the butanol separation. Butanol production through the fermentation process is gaining an increasing amount of interest because of the rising issues related to the environment, such as pollution as well as fossil fuel shortage. However, the main hindrances to the path of industrialization of ABE fermentation involve the expensive price of raw materials along with food scarcity. Therefore, economic and sustainable routes should be kept in mind before applying this technology.

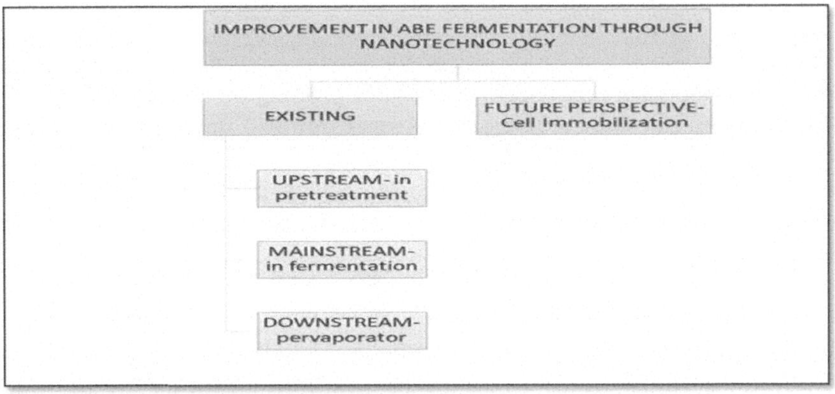

Figure 6. Improvements in acetone–butanol–ethanol (ABE) fermentation by using nanoparticles.

5. Environmental impacts of biobutanol

Biobutanol is an efficient renewable energy source having high energy density, making it a potential candidate that has the capability of replacing petro-based fossil fuels for long-distance transport (Shanmugan et al. 2021). Besides being an efficient alternative for meeting the demands of an ever-growing population, there still can be some issues associated with this type of biofuel depending upon the method of its generation. For example, if it is subjected to chemical pretreatment, there is a fair chance of the equipment getting corroded, which can ultimately harm the environment when it gets disposed of (Sharma et al. 2020). Generation of biobutanol by utilizing nanotechnology no doubt is being looked upon as an efficient, economic, and sustainable method. However, it can still have minor impacts on the environment, depending upon the type of nanoparticle utilized such as metal oxides nanoparticles, CNTs, etc. During the processing of raw material, there is a possibility of release of some gases into the atmosphere, metals, and chemicals into the surrounding environment (Figure 7). Therefore, problems such as acidification, eutrophication, ecotoxicological effects, as well as other associated environmental effects can arise to some extent. Hence, certain things such as proper management of waste-containing toxic metal nanoparticles should be considered before commercializing it on a large scale. There is also a requirement of feedstock or substrates that are economical along with being a sustainable feedstock causing minimum or no harm to the environment, as there shall be no purpose of generating some other alternatives of fossil fuels if the generated ones are also causing harm to our environment.

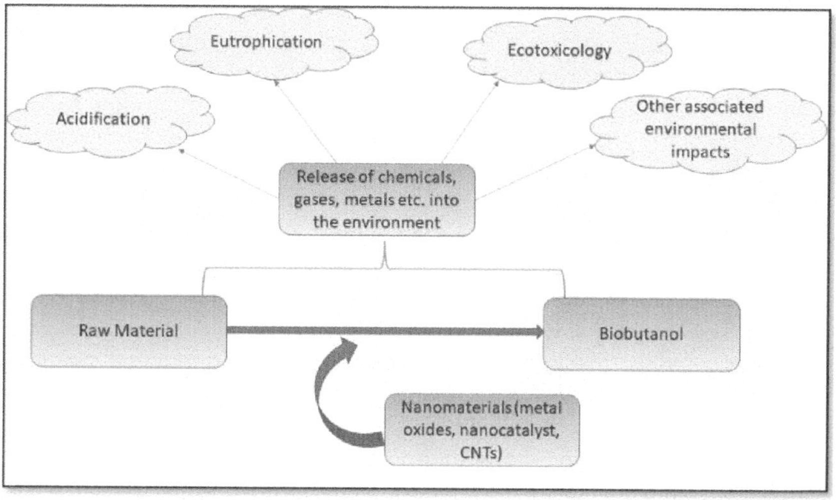

Figure 7. Impacts of nano-based biobutanol on the environment.

6. Conclusion and future perspectives

It is quite evident that a wide range of nanomaterials is utilized for biomass pretreatment and further its conversion into liquid fuels. Their effectiveness is widely known now, especially at the laboratory scale. The nanomaterials based technologies have potential due to its suitability, efficiency and re-usability. Nanomaterials application for generating biofuels out of renewable feedstocks in a sustainable way has basically gained tremendous interest presently and moreover, lab-scale work has proved that nanotechnology possesses the capacity to provide new paths for maintaining the sustainability of the energy sector through addressing the issues related to the biomass preprocessing leading to good biofuel yield. Nanomaterials can be crucial in obtaining bioethanol or biobutanol at an industrial scale in an economical way. They have the capability to increase the production of bioalcohols via the process of fermentation. The application of nanomaterials in the production of biobutanol from feedstocks having sustainable impacts has generated interest. Presently, there is a body of work on NP-assisted biomass processing, fermentation of pretreated feedstock, along with detecting the fermentation products for other biofuels like biodiesel, bio-ethanol, and bio-hydrogen. The utilization of nanoparticles to immobilize enzymes can offer a new path to production technology by enhancing the availability as well as stability of immobilized enzymes, decreasing the cost included in the production via reusing the nanomaterials, and offering larger surface area for reaction, etc.

The utilization of nanomaterials for producing biobutanol via fermentation still requires a lot of work to be implement successfully. More research is required in this regard for exploring new ways that are economical to prepare, monitor, as well as control the fermentation process and optimize the cost factor. Nanomaterials are also utilized for detecting several process intermediates as well as end products.

Hence, it can be concisely summarized that noteworthy research has been made in the domain of nanotechnology in some past years and NPs have proved to have enormous potential for producing biofuels via green methods for biomass processing in order to achieve better product selectivity as well as yield in an economical way. However, several challenges are still present that need to be overcome. Some of the major steps (Figure 8) that are required include synthesizing more nanoparticles that are versatile in nature, which can further be utilized for biomass processing, developing nanocatalysts that are selective in nature and can transform bio-derived sugars into biofuels, more greenways of nanoparticles synthesis to be utilized as biocatalysts carriers such that they are easy in utilization for the fermentation of bioalcohols, etc.

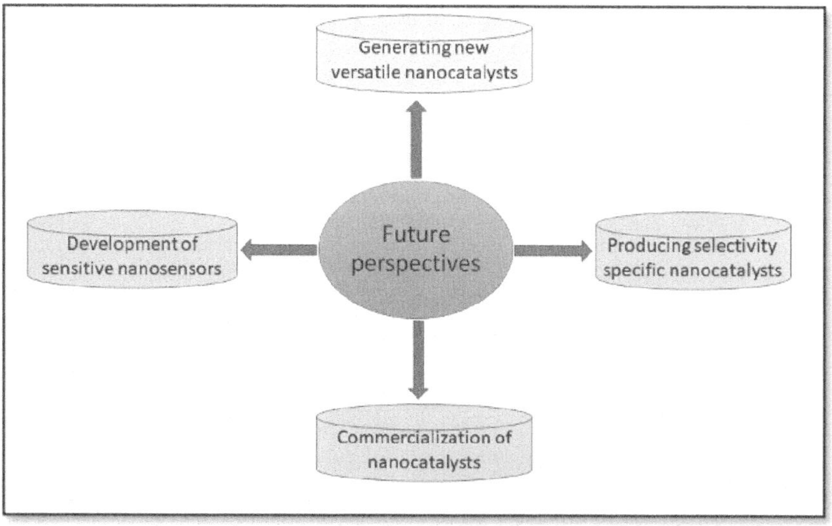

Figure 8. Future perspectives associated with biobutanol generation through nanotechnology.

References

Azimi, H., Handan, T.F., and Jules, T. (2017). Effect of embedded activated carbon nanoparticles on the performance of Polydimethylsiloxane (PDMS) membrane for pervaporation separation of butanol: Performance of composite membrane (PDMS-AC) for separation of butanol. Journal of Chemical Technology & Biotechnology.

Azimi, H., Ebneyamini, A., Tezel, F.H., and Thibault, J. (2018). Separation of organic compounds from ABE model solutions via pervaporation using activated carbon/PDMS mixed matrix membranes. Membranes, 8: 1–15.

Bharathiraja, B., Chakravarthy, M., Kumar, R.R., Yogendran, D., Yuvaraj, D., Jayamuthunagai, J., Kumar, R.P., and Palani, S. (2015). Aquatic biomass (algae) as a future feed stock for bio-refineries: A review on cultivation, processing and products. Renewable and Sustainable Energy Reviews, 47: 634–653.

Fan, H., Wang, N., Ji, S., Yan, H., and Zhang, G. (2014). Nanodisperse ZIF-8/PDMS hybrid membranes for biobutanol permselective pervaporation. J. Mater. Chem. A, 2(48): 20947–20957.

Finnigan, W., Hepworth, L.J., Flitsch, S.L., and Turner, N.J. (2021). RetroBioCat as a computer-aided synthesis planning tool for biocatalytic reactions and cascades. Nature Catalysis, 4(2): 98–104.

Giannakis, S. (2019). A review of the concepts, recent advances and niche applications of the (photo) Fenton process, beyond water/wastewater treatment: surface functionalization, biomass treatment, combatting cancer and other medical uses. Applied Catalysis B: Environmental, 248: 309–319.

Ivanova, V., Petrova, P., and Hristov, J. (2011). Application in the ethanol fermentation of immobilized yeast cells in matrix of alginate/magnetic nanoparticles, on chitosan-magnetite microparticles and cellulose-coated magnetic nanoparticles. arXiv preprint arXiv:1105.0619.

Javid, A., Amiri, H., Kafrani, A.T., and Rismani-Yazdi, H. (2022). Post-hydrolysis of cellulose oligomers by cellulase immobilized on chitosan-grafted magnetic nanoparticles: A key stage of butanol production from waste textile. Int. J. Biol. Macromol. 207: 324–332.

Kim, Y.K., Park, S.E., Lee, H., and Yun, J.Y. (2014). Enhancement of bioethanol production in syngas fermentation with *Clostridium ljungdahlii* using nanoparticles. Bioresource Technology, 159: 446–450.

Kim, Y.K., and Lee, H. (2016). Use of magnetic nanoparticles to enhance bioethanol production in syngas fermentation. Bioresource Technology, 204: 139–144.

Kumar, A., Singh, S., Tiwari, R., Goel, R., and Nain, L. (2017). Immobilization of indigenous holocellulase on iron oxide (Fe2O3) nanoparticles enhanced hydrolysis of alkali pretreated paddy straw, Int. J. Biol. Macromol., 96 538–549.

Kumar, Y., Yogeshwar, P., Bajpai, S., Jaiswal, P., Yadav, S., Pathak, D.P., Sonker, M., and Tiwarya, S.K. (2021). Nanomaterials: stimulants for biofuels and renewables, yield and energy optimization. Mater Adv., 2: 5318–5343.

Kushwaha, D., Srivastava, D., Mishra, I., Upadhyay, S.N., and Mishra, P.K. (2018). Recent trends in biobutanol production. Rev. Chem. Eng.

Kushwaha, D., Upadhyay, S.N., and Mishra, P.K. (2018). Nanotechnology in bioethanol/biobutanol production springer international publishing AG. Srivastava et al. (eds.), Green Nanotechnology for Biofuel Production, Biofuel and Biorefinery Technologies, 5: 115–127.

Ladole, M.R., Mevada, J.S., and Pandit, A.B. (2017). Ultrasonic hyperactivation of cellulose immobilized on magnetic nanoparticles. Bioresour. Technol., 239: 117–126.

Liu, S., Liu, G., Shen, J., and Jin, W. (2014). Fabrication of MOFs/PEBA mixed matrix membranes and their application in bio-butanol production. Separation and Purification Technology, 133: 40–47.

Lu, J., Boughner, C.E., Liotta, C.L., and Eckert, C.A. (2002). Nearcritical and supercritical ethanol as a benign solvent:Polarity and hydrogen-bonding. Fluid Phase Equilib., 198: 37–49.

Meramo-Hurtado, S.I., González-Delgado, A.D., Rehmann, L., Quiñones-Bolaños, E., and Mehrvar, M. (2020). Comparison of biobutanol production pathways via acetone–butanol–ethanol fermentation using a sustainability exergy-based metric. ACS Omega, 5: 18710–18730

Pan, X., Fan, Z., Chen, W., Ding, Y., Luo, H., and Bao, X. (2007). Enhanced ethanol production inside carbon-nanotube reactors containing catalytic particles. Nat. Mater. 6(7): 507–511.

Poorakbar, E., Shafiee, A., Saboury, A.A., Rad, B.L., Khoshnevisan, K., Ma'mani, L., Derakhshankhah, H., Ganjali, M.R., and Hosseini, M. (2018). Synthesis of magnetic gold mesoporous silica nanoparticles core shell for cellulase enzyme immobilization: Improvement of enzymatic activity and thermal stability. Process Biochem., 71: 92–100.

Rai, M., Santos, J.C., Soler, M.F., Marcelino, P.R.F., Brumano, L.P., Ingle, A.P., Gaikwad, S., Gade, A., and Silva, S.S. (2016) Strategic role of nanotechnology for production of bioethanol and biodiesel. Nanotechnol. Rev., 5(2): 231–250.

Razack, S.A., Duraiarasan, S., and Mani, V. (2016). Biosynthesis of silver nanoparticle and its application in cell wall disruption to release carbohydrate and lipid from *C. vulgaris* for biofuel production. Biotechnology Reports, 11: 70–76.

Russell, A. (2012). Plasmonic Pervaporation via Gold Nanoparticle-Functionalized Nanocomposite Membranes. Theses and Dissertations. 476.

Sánchez-Ramírez, J., Martínez-Hernández, J.L., Segura-Ceniceros, P., López, G., Saade, H., Medina-Morales, M.A., Ramos-González, R., Aguilar, C.N., and Ilyina, A. (2017). Cellulases immobilization on chitosan-coated magnetic nanoparticles: application for Agave *Atrovirens lignocellulosic* biomass hydrolysis. Bioprocess Biosyst. Eng., 40(1): 9–22.

Sanusi, I.A., Suinyuy, T.N., Lateef, A., and Kana, G.E. (2020). Effect of nickel oxide nanoparticles on bioethanol production: process optimization, kinetic and metabolic studies. Process Biochem. 92: 386–400.

Seelert, T., Ghosh, D., and Yargeau, V. (2015). Improving biohydrogen production using *Clostridium beijerinckii* immobilized with magnetite nanoparticles. Appl. Microbiol. Biotechnol., 99: 4107–4116.

Shanmugam, S., Hari, A., Kumar, D., Rajendran, K., Mathimani, T., Atabani, A.E., Brindhadevi, K., and Pugazhendhi, A. (2021). Recent developments and strategies in genome engineering and integrated fermentation approaches for biobutanol production from microalgae. Fuel, 285: 119052.

Sharma, A., Singh, G., and Arya, S.K. (2020). Biofuel from rice straw. Journal of Cleaner Production, 277: 124101.

Singh, G., Kim, I.Y., Lakhi, K.S., Srivastava, P., Naidu, R., and Vinu, A. (2017). Single step synthesis of activated bio-carbons with a high surface area and their excellent CO2 adsorption capacity. Carbon, 116: 448–455.

Singh, N., Dhanya, B.S., and Verma, M.L. (2020). Nano-immobilized biocatalysts and their potential biotechnological applications in bioenergy production. Materials Science for Energy Technologies, 3: 808–824.

Skrotskyi, S., Voychuk, S., Khomenko, L., Vasyliuk, O., and Pidgorskyi, V. (2019). Influence of nanoparticles on the solventogenesis of bacteria *Clostridiumbeijerinckii* IMV B-7806, *Clostridium acetobutylicum* IMV B-7807. Ukrainian Food Journal. 110–118.

Van Den Berg, C., Heeres, A.S., van der Wielen, L.A., and Straathof, A.J. (2013). Simultaneous clostridial fermentation, lipase-catalyzed esterification, and ester extraction to enrich diesel with butyl butyrate. Biotechnology and Bioengineering, 110(1): 137–142.

Yan, H., Lai, C., Wang, D., Liu, S., Li, X., Zhou, X., Yi, H., Li, B., Zhang, M., Li, L., and Liu, X. (2021). *In situ* chemical oxidation: peroxide or persulfate coupled with membrane technology for wastewater treatment. Journal of Materials Chemistry A, 9(20): 11944–11960.

Index

S

T

About the Editors

Dr. Anita Singh is working as an Assistant Professor at Department of Environmental Sciences, Central University of Jammu (J&K). She is having more than 9 years of teaching and research experience. Dr. Singh did her Ph.D. in bioethanol production from lignocellulosic biomass from GJUST, Hisar, Haryana, India. She did her post-doctoral research at the University of Georgia, USA, on the selective removal of inhibitors from lignocellulosic biomass under the UGC Raman Post-doctoral fellowship program. Dr. Singh has published 50 research papers in international and national journals as well as book chapters in edited books. Dr. Singh completed two sponsored research projects from UGC and DST-SERB.

Dr. Richa Kothari is working as an Associate Professor at the Department of Environmental Sciences, Central University of Jammu (J&K), India. She was a WARI Fellow (IUSSTF) at the University of Nebraska-Lincoln, Lincoln, USA. She was selected in the list of World's top 2% scientists by Stanford's University, USA for significant research work, 2020 and 2021. She has been actively involved in research in the areas of clean and green energy production and wastewater treatment. She has published more than 100 research papers/book chapters in SCI/Scopus journals. She has delivered lectures at national and international levels.

Dr. Somvir Bajar is currently working as an Assistant Professor in the Department of Environmental Science and Engineering, J.C. Bose University of Science and Technology, YMCA, Faridabad, Haryana (India). His work is focused on the exploration of energy recovery possibilities from waste sectors and sustainable solutions for the abatement of environmental pollutants. He has more than 10 years of teaching and research experience and has published 35 research papers in peer-reviewed journals of international repute and edited books. He has been also involved in providing environmental consultancy to more than 600 industries and played a considerable role in the development and implementation of a self-sustaining business model with consultancy activities in his career.

Dr. Vineet Veer Tyagi is currently working as Head at the School of Energy Management, Shri Mata Vaishno Devi University, Katra, Jammu (J&K), India. He was Post Doctoral Research Fellow at UMPEDAC, University of Malaya, Kuala Lumpur, Malaysia. He was the recipient of prestigious research fellowships from MNRE and CSIR at IIT-Delhi, New Delhi, India. He has published more than 150 publications in reputed international peer-reviewed (SCI/Scopus) journals with more than 15000 citations. He was selected in the list of the World's top 2% scientists by Stanford University, USA for significant academic and research work in 2020 and 2021. His work is focused on renewable energy technologies and energy storage.